INDUSTRIAL WATER PURIFICATION

INDUSTRIAL WATER PURIFICATION

Louis F. Martin

NOYES DATA CORPORATION
Park Ridge, New Jersey London, England
1974

Library of Congress Catalog Card Number: 74-84302
ISBN: 0-8155-0554-X
Printed in the United States

Published in the United States of America by
Noyes Data Corporation
Noyes Building, Park Ridge, New Jersey 07656

FOREWORD

The detailed, descriptive information in this book is based on recent U.S. patents relating to industrial water purification.

This book serves a double purpose in that it supplies detailed technical information and can be used as a guide to the U.S. patent literature in this field. By indicating all the information that is significant, and eliminating legal jargon and juristic phraseology, this book presents an advanced, technically oriented review of modern methods of purifying polluted industrial water.

The U.S. patent literature is the largest and most comprehensive collection of technical information in the world. There is more practical, commercial, timely process information assembled here than is available from any other source. The technical information obtained from a patent is extremely reliable and comprehensive; sufficient information must be included to avoid rejection for "insufficient disclosure." These patents include practically all of those issued on the subject in the United States during the period under review; there has been no bias in the selection of patents for inclusion.

The patent literature covers a substantial amount of information not available in the journal literature. The patent literature is a prime source of basic commercially useful information. This information is overlooked by those who rely primarily on the periodical journal literature. It is realized that there is a lag between a patent application on a new process development and the granting of a patent, but it is felt that this may roughly parallel or even anticipate the lag in putting that development into commercial practice.

Many of these patents are being utilized commercially. Whether used or not, they offer opportunities for technological transfer. Also, a major purpose of this book is to describe the number of technical possibilities available, which may open up profitable areas of research and development. The information contained in this book will allow you to establish a sound background before launching into research in this field.

Advanced composition and production methods developed by Noyes Data are employed to bring our new durably bound books to you in a minimum of time. Special techniques are used to close the gap between "manuscript" and "completed book." Industrial technology is progressing so rapidly that time-honored, conventional typesetting, binding and shipping methods are no longer suitable. We have bypassed the delays in the conventional book publishing cycle and provide the user with an effective and convenient means of reviewing up-to-date information in depth.

The Table of Contents is organized in such a way as to serve as a subject index. Other indexes by company, inventor and patent number help in providing easy access to the information contained in this book.

15 Reasons Why the U.S. Patent Office Literature Is Important to You –

1. The U.S. patent literature is the largest and most comprehensive collection of technical information in the world. There is more practical commercial process information assembled here than is available from any other source.

2. The technical information obtained from the patent literature is extremely comprehensive; sufficient information must be included to avoid rejection for "insufficient disclosure."

3. The patent literature is a prime source of basic commercially utilizable information. This information is overlooked by those who rely primarily on the periodical journal literature.

4. An important feature of the patent literature is that it can serve to avoid duplication of research and development.

5. Patents, unlike periodical literature, are bound by definition to contain new information, data and ideas.

6. It can serve as a source of new ideas in a different but related field, and may be outside the patent protection offered the original invention.

7. Since claims are narrowly defined, much valuable information is included that may be outside the legal protection afforded by the claims.

8. Patents discuss the difficulties associated with previous research, development or production techniques, and offer a specific method of overcoming problems. This gives clues to current process information that has not been published in periodicals or books.

9. Can aid in process design by providing a selection of alternate techniques. A powerful research and engineering tool.

10. Obtain licenses — many U.S. chemical patents have not been developed commercially.

11. Patents provide an excellent starting point for the next investigator.

12. Frequently, innovations derived from research are first disclosed in the patent literature, prior to coverage in the periodical literature.

13. Patents offer a most valuable method of keeping abreast of latest technologies, serving an individual's own "current awareness" program.

14. Copies of U.S. patents are easily obtained from the U.S. Patent Office at 50¢ a copy.

15. It is a creative source of ideas for those with imagination.

CONTENTS AND SUBJECT INDEX

INTRODUCTION

The Federal Water Pollution Control Act (PL 92-500) passed on October 18, 1972 will have a far-reaching effect on all industry. The pretreatment standards now being published as a result of this act will have an increasingly heavy impact on process design and expansion plans in the years ahead. Only within the last year or two, many local municipal agencies have adopted new industrial waste ordinances that reflect the stringent requirements and far-reaching ramifications of this new law. These ordinances enable the local authorities to initiate source control programs and enforce the Environmental Protection Agency (EPA) pretreatment requirements on toxic substances.

The *Guidelines for the Pretreatment of Discharges of Publicly-Owned Works* published by the EPA calls for the prevention of the introduction of pollutants which would pass through a municipal facility inadequately treated. Industry must meet these fundamental standards if the municipal treating plant standards imposed by EPA under the National Pollutant Discharge Elimination System are to be reached.

Thus, all existing facilities and new plant design must include specific downstream water purification processes before the effluent is sewered or discharged into public water systems. Since 1970, many agencies, as a condition for receiving federal grant funds for the construction of water pollution control facilities, have instigated sewage service charges based on both the volume and nature of the industry's discharge.

The industry's concern and the need to move rapidly in these water purification processes is evident in the patent literature where literally hundreds of pollution control related patents are granted yearly. The patent office has initiated a priority system in this area which should significantly reduce the time between filing and the awarding of such a patent.

SOLID-LIQUID SEPARATION PROCESSES

FLOCCULATION

Methylamine-Epichlorohydrin Polymers

According to a process described by *H.P. Panzer and R. Rabinowitz; U.S. Patent 3,741,891; June 26, 1973; assigned to American Cyanamid Company* the cationic flocculant obtained by reacting a monoalkylamine with an epihalohydrin is quaternized with quaternizing agents, such as dimethyl sulfate, methyl chloride and ethylene oxide, producing a polyquaternary polymer. The polymer shows improved flocculating power in flocculating aqueous dispersions, such as raw river water, digestion liquors from ilmenite and sulfuric acid, enzyme mashes, sewage flocculation and sewage sludge dewatering and are also useful in systems which contain chlorine as the quaternary groups are nonreactive with chlorine.

In the examples the viscosity is determined in aqueous solution at 25°C and containing an amount of the polymer corresponding to 20% by weight of the polymer minus the anion. This is referred to as "real organic polymer". When the product of Example 1 is tested under the conditions of 20% real organic polymer, its viscosity is over 200 cs. This product is referred to as "Example 1".

Example 1: To 100 grams of methylamine (3.25 mols) dissolved in 400 grams of water in a reaction flask provided with stirrer, thermometer and reflux condenser is added 260 grams of epichlorohydrin (2.8 mols, equivalent to 0.87 mol/mol of amine) over 60 minutes, cooling (by ice bath) being applied as necessary to keep the temperature of the reaction mixture between 25° to 40°C at the first half of the reaction and at 50° to 80°C during the second half of the reaction. 160 grams of 36% aqueous sodium hydroxide solution (1.44 mols) at 85°C is then added to the reaction mixture.

The reaction mixture is tnen heated to 95°C and epichlorohydrin is added in 1 ml ($\frac{1}{214}$ mol) portions and the viscosity of the reaction mixture is followed by filling a 6 mm (i.d.) vertical glass tube with the hot solution and noting the number of seconds required for the level of the solution to fall 13 inches when the bottom of the tube is opened. Results are as follows:

Time (minutes)[1]	Ml. epi. added	Temp. (° C.)	Viscosity (seconds)[2]
110	1	95	4
115	1	95	5.5
123	1	95	10.0

[1] From start of reaction.
[2] Of reaction mixture, by glass tube method.

After 132 minutes, the reaction mixture becomes very viscous and the viscosity continues to rise. There is then added 274 grams of cold water containing 0.8 ml of methylamine as reaction terminator, and the reaction mixture is heated to 94°C. The following viscosity changes occur:

Time (minutes)[1]	Temp. (° C.)	Viscosity (seconds)[2]
157	94	48
220	94	90
250	94	120
310	94	136
345	94	125
605	94	60

[1] From start of reaction.
[2] Of reaction mixture, by glass tube method.

The decrease in viscosity after 310 minutes is caused by scission of bonds within the polymer molecule. The reaction mixture, which has a pH of 8.7, is cooled, acidified to pH 6.3, and diluted with water. The product contains 19.3% polymer and has a viscosity of 900 cp at 20°C. The solution is stable indefinitely at 70°F and 135°F both at pH 8.7 and 4.5.

Example 2: An aqueous dispersion of Example 1 20% real organic polymer was introduced into a reaction vessel provided with good cooling and means for constantly monitoring pH. Dimethyl sulfate and 50% sodium hydroxide were slowly and simultaneously added, maintaining the temperature between 52° and 68°C and the pH between 7.0 and 7.8. The amount of the materials are typically 242 grams of Example 1, 83.6 grams dimethyl sulfate and 23 grams sodium hydroxide in a 50% aqueous solution.

At the start of the addition of the dimethyl sulfate and sodium hydroxide, Example 1 was warmed up to 50°C. Addition took about 45 minutes on the scale described. At the end of the addition, when the pH became constant, after 15 minutes, the reaction mixture was cooled and brought to a pH of 5.5 to 6.5 by the addition of a small amount of concentrated sulfuric acid. The product had a viscosity in excess of 300 cs.

Example 3: To a one liter flask equipped with Dry Ice condenser, thermometer, gas inlet, mechanical stirrer, addition funnel, and pH electrode was added 470 grams Example 1 (20.8% solids). Ethylene oxide, 80 grams, was added over 5 hours while maintaining the temperature at 50° to 55°C. The alkaline solution was then acidified with 50 grams 60% nitric acid. After 2 hours the pH was again adjusted to neutral with 6 grams nitric acid and the clear yellow product was cooled. The product had a viscosity of 225 cs at 25°C and 25% real organic polymer solids.

Example 4: Flocculation of ilmenite digestion liquor was tested as follows: Hot liquor at 50°C was poured into a 1,000 ml graduated cylinder and the particular flocculant added in the two dosages of 100 and 200 ppm real organic polymer. Mixture of the flocculant was effected with a perforated metal plunger, and after 15 minutes the level of the interface between lower and supernatant phases noted. After 4 hours of settling, a 50 ml sample of the supernatant was taken at the 750 ml level in the cylinder and its turbidity measured on a photoelectric colorimeter. The sample was then diluted with deionized water at a ratio of 1:4 and its turbidity again measured. The following table shows the results:

Polymer	Dosage p.p.m. real organic polymer	Turb. at 4 hrs.[1]
Example 1	60	970
Dimethyl sulfate quat—Example 2	50	600
Ethylene oxide quat—Example 3	77	635
Example 1	60	1,000+
Methyl chloride quat	64	585
Dimethyl sulfate quat—Example 2	51	670

[1] Klett-Summerson Colorimeter readings (2 cm. light path).

It will be noted that there are slight variations from test to test. This is normal, but it will be seen that, particularly the quaternized product with dimethyl sulfate of Example 2, gives much better clarification than Example 1 even though the dosage of the flocculant was actually lower, about 50 parts of real polymer instead of 60. The table also shows very extensive flocculation even at 4 hours, the results being, for the products of the process, satisfactory. Speed of flocculation is a very important factor for economic flocculation because the shorter time permits much more economical utilization of equipment, an important advantage of the products of the process.

According to a process described by *D.E. Nagy; U.S. Patent 3,755,159; August 28, 1973; assigned to American Cyanamid Company* a cationic, water-soluble, storage-stable, methylamine-epichlorohydrin polymer is prepared by reacting a quantity of a methylamine-epichlorohydrin polymer which is well below its gel point with successively added small amounts of epichlorohydrin until the polymer nears the stage at which it is an irreversible gel, and then reacting the polymer with sufficient methylamine to inactivate substantially all amine-reactive epichlorohydrin residues present. At least the last step of the reaction is performed at a temperature at which a partial depolymerization of the polymer occurs.

Water-soluble polymers of higher molecular weight and larger dimensions are produced by graft polymerizing one or more water-soluble vinyl monomers upon a hydrophilic water-dispersible cationic poly(hydroxyalkylene) polyamine, which is not necessarily the foregoing polyamine. The polymers are flocculants for suspended solids in sewage and mine effluent water and dry strength agents for paper.

Poly(Quaternary Ammonium) Polyether Polyelectrolytes

E. Witt; U.S. Patent 3,755,160; August 28, 1973; assigned to Polysar International, SA, Switzerland describes water-soluble poly(quaternary ammonium) polyether polyelectrolyte salts containing quaternary nitrogen atoms in the polymer backbone and chain-extended by ether groups which are prepared by treating the polymeric reaction product from an N,N,N',N'-tetraalkyl-hydroxy substituted diamine and an organic dihalide such as a dihaloalkane or a dihalo ether with an epoxyhaloalkane. These polyelectrolytes are used in processes for flocculating particles dispersed in aqueous media, for example, white water clarification, clay flocculation, sewage treatment, and wet-end addition, by adding the polyelectrolyte to the aqueous media in amounts sufficient to flocculate the dispersed particles.

Example 1: This example illustrates preparation of one of the polyquaternary ammonium salts used to prepare the chain-extended polyelectrolytes of this process. In a reaction vessel, 3.5 mols of dimethylamine (aqueous solution) were added to 1.0 mols of epichlorohydrin, with the temperature being maintained at 28° to 30°C during the addition of the epichlorohydrin. Then the contents of the vessel were heated to 65°C and held at this temperature for 4 hours.

After the reaction, 1.0 mol of sodium hydroxide was added to the reaction vessel. The reaction mixture was then vacuum stripped to remove water and excess dimethylamine. The residue, a salt slurry, was extracted with 1.35 mols of ethanol to remove the ditertiary amine product. The extract was flash distilled under vacuum to remove the ethanol, leaving a residue which was vacuum distilled at a pressure of 5 mm of mercury and at a temperature of 55° to 58°C to provide the product N,N,N',N'-tetramethyl-2-hydroxy-1,3-propane diamine in a yield of 53.6%. Upon analysis, the product showed a purity of 99+%.

To a reaction vessel were added 145 grams (0.993 mols) of the previously prepared N,N,N',N'-tetramethyl-2-hydroxy-1,3-propane diamine (referred to as THPDA); 150 grams (1.048 mols) of beta, beta'-dichlorodiethylether (referred to as DCEE); and 50 grams of water, with agitation. This mixture was heated to a temperature of 75° to 80°C and held at this temperature for a period of 5½ hours. Then an additional 150 grams of water were added and the temperature of the reaction mixture maintained at 75° to 80°C for an additional one-half hour. A polymeric polyquaternary ammonium salt product, containing repeating units of the formula shown on the next page which had a LVT Brookfield viscosity of 318 cp (No. 2 spindle at 30 rpm) and a solids content of 60.9%.

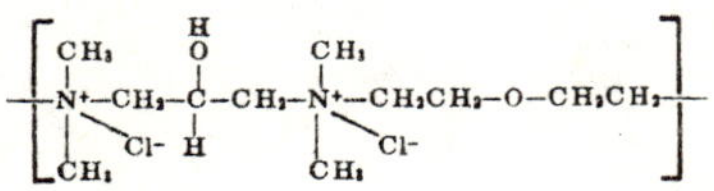

Example 2: Another salt having repeating units as shown in Example 1 was prepared in a 500-gallon reactor. 49 parts by weight of THPDA; 51 parts by weight of DCEE and 16.25 parts by weight of water were charged to the reactor and stirred. This mixture was heated to 176°F and maintained at this temperature for 6 hours. A pressure of 5 psig was maintained with an inert gas mixture including nitrogen.

The resulting product was steam stripped to remove unreacted DCEE. Additional water (49.5 parts by weight) was added to adjust the viscosity. The product had an LVT Brookfield viscosity of 520 centipoises at 23°C (No. 2 spindle at 30 rpm) and a solids content of 55.6%. This polymer solution is designated Polyquat 55.6. At 33.0% solids Polyquat 55.6 had a Brookfield viscosity of 52 cp.

Example 3: Another polymeric polyquaternary salt was prepared by following the procedure and conditions of Example 1 and by employing 0.993 mols of THPDA and 0.984 mols of ethylene dibromide (EDB). A polyquaternary polyammonium salt having a viscosity of 13.0 cp at 23°C (Number 1 spindle at 60 rpm), a solids content of 63.0%, and containing repeating units of the following formula:

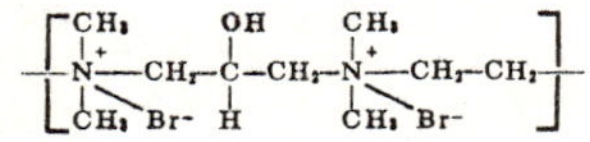

was obtained.

Example 4: A poly(quaternary ammonium) polyether polyelectrolyte was prepared in the following manner. Sodium hydroxide (8 grams) dissolved in 40 ml of water was added to 100 grams of Polyquat 55.6 stirred in a 500 ml reaction flask equipped with a mechanical stirrer, thermometer, reflux condenser, and a nitrogen-blanketing device. The mixture was heated to 70°C and 40 grams of epichlorohydrin (designated EPCHD) was added. A total of 125 ml of water was added to the stirred and heated mixture to control viscosity, as follows: 5 ml at 10 minutes after EPCHD addition, 15 ml at 15 minutes, 50 ml at 19 minutes, and 55 ml at 29 minutes. The resulting poly(quaternary ammonium) polyether polyelectrolyte solution contained 31.9% solids and had a Brookfield viscosity of 500 cp at 23°C using a Number 2 spindle at a spindle speed of 30 rpm at pH 8.3. This is a tenfold increase in viscosity over Polyquat 55.6.

The product has a structure containing repeating units of the following structure:

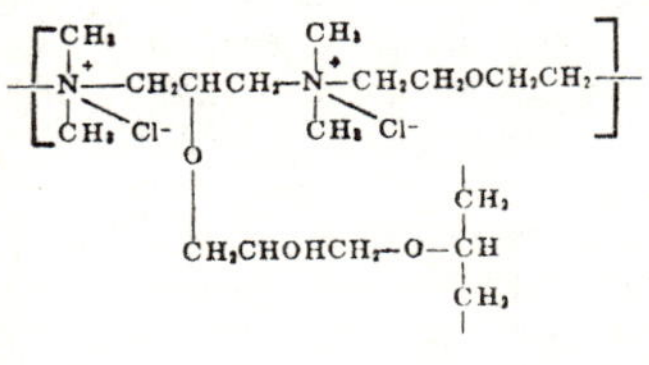

Example 5: The polyquaternary ammonium salt prepared in Example 3 is treated with 35 grams of EPCHD per 100 grams of polymer solution to yield a poly(quaternary ammonium) polyether polyelectrolyte of higher viscosity than the starting polymer. The structure of this polyquaternary polyether has repeating units of the following formula shown on the next page.

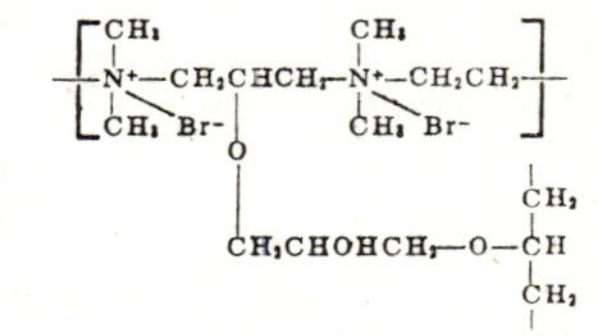

Example 6: This example illustrates the importance of the technique for adding water of dilution to the reaction mixture for preparing the poly(quaternary ammonium) polyether polyelectrolytes so that polymer solubility is maintained without deleterious effect on polymer properties. To 100 grams of Polyquat 55.6 in a stirred 500 ml reaction flask equipped with a mechanical stirrer, thermometer, reflux condenser, and nitrogen-blanketing means was added 8 grams of sodium hydroxide dissolved in the amounts of water shown in the following table. The mixture was heated to 70°C and EPCHD was added as indicated in the table.

Water in amounts varying from 125 to 225 ml was added to the reaction mixture continuously beginning immediately after the EPCHD was added with an initial amount of water, at a rate of 1.4 ml/min using a 1 rpm Emdeco intravenous injector pump. The stirred mixture was heated by a bath maintained at 70°C during the entire reaction period. The temperature of the mixture decreased by about 3°C when the EPCHD was added. After about 10 minutes, a maximum of about 77°C was attained; the temperature of the reaction mixture decreased to 71°C after about an hour. The temperature was kept at 70°C for at least 12 hours or until the pH of the mixture decreased below 8.5. In some cases more water of dilution was added after the continual metered amount.

Run No.	Charged [a]		Total additional water of dilution (ml.)	Percent solids	Brookfield viscosity at 23° C.		
	EPCHD (grams)	Water (ml.)			Cps.	Spindle No.	R.p.m.
1	20	36	[b] 175	30.6	460	2	30
2	20	37	[b] 125	31.9	7,850	4	30
3	20	37	[c] 387	17.0	1,850	2	6
4	20	39	[b] 125	33.4	75	1	30
5	20	39	[c] 425	26.3	52	1	30
6	15	34	[b] 200	30.2	12.5	1	60
7	15	34	[b] 125	30.8	7.5	1	60
8	15	34	[d] 175	36.6	11	1	60
9	15	38	[b] 225	32.2	11	1	60
10	15	38	[b] 125	28.7	8.5	1	60

[a] With NaOH.
[b] At 1.4 ml per minute.
[c] All above 125 milliliters at 1.4 ml. per minute as fast as possible.
[d] Incrementally 25 mils. at 37 minutes of reaction; 15 mils. at 65 minutes of reaction; 10 mils at 90 minutes of reaction.

It will be observed that at the higher EPCHD level, the viscosity was improved when the water was kept within certain levels. Thus, the data show that the amount of water which is added to a 100 gram charge of Polyquat 55.6 may vary within the range of about 100 to 400 ml of water. This corresponds to between 29 and 116 mols of water per mol of Polyquat 55.6 repeating unit. These experiments further show that adding large amounts of water early in the reaction is detrimental in that the solutions have lower viscosity than those obtained when water is added gradually during the course of a reaction run.

Example 7: The experiments conducted in this example show the effect on polymer solution viscosity of the amount of water initially charged to the reaction vessel in order to dissolve sodium hydroxide. In each case, the reaction was carried out at 70°C and a total of 125 ml of water was added at the rate of 1.4 ml/min. Polyquat 55.6 (100 grams), 8 grams of sodium hydroxide, and varying amounts of EPCHD and of water were charged to the reaction vessel, as indicated in the following table.

When less than 35 ml of water was charged with the indicated amounts of Polyquat 55.6 and sodium hydroxide at 40 grams of EPCHD charged, gelatinous masses were obtained. Thus, for each 100 grams of Polyquat 55.6 and 40 grams of EPCHD charged, the highest product viscosities are obtained when about 35 ml of water are charged; this corresponds to about 10 mols of water per mol of repeating Polyquat 55.6 units. At a charge of 100 grams of Polyquat 55.6 and 20 grams of EPCHD, the optimum ratio of water charged is about 10 to 11 mols of water per mol of repeating Polyquat 55.6 unit.

	Charged				Brookfield viscosity at 23° C.		
Run No.	EPCHD (grams)	Water (ml.)	Percent solids	pH (after 12 hours)	Cps.	Spindle No.	R.p.m
1	40	35	33.1	8.1	23,000	4	6
2	40	35	31.9	8.5	18,500	4	6
3	40	35	33.9	8.4	6,150	3	6
4	40	35	29.9	8.3	1,675[a]	2	6
5	40	37	32.4	8.3	320	1	6
6	40	40	35.0	7.9	215	1	6
7	20	35	29.8	9.2	1,200,000	4	3
8	20	37	31.9	10.2	7,850	4	30
9	20	38	34.0	9.8	340	1	6
10	20	39	33.4	9.5	75	1	30
11	20	40	31.0	7.1	335	2	30
12	15	34	30.8	10.4	7.5	1	60
13	15	37	29.6	10.6	8	1	60
14	15	38	28.7	10.7	8.5	1	60

[a] This sample was further diluted with 105 milliliters of water.

Wide variations in polymer solution viscosity, even under apparently identical conditions, show the extreme sensitivity of the reaction between Polyquat 55.6 and EPCHD to the amount of water initially present. These experiments further show that the amount of water initially charged may vary in the range between 10 and 12 mols per mol of Polyquat 55.6 units in order to obtain products which have higher viscosities than the Polyquat 55.6 solutions but are still fluid. It will be appreciated, however, when the rate of water addition is raised a lower viscosity product may be obtained even if the initial water content is lowered.

Example 8: This example shows the usefulness of polyether polyelectrolytes for clarifying white water from paper making. Evaluations were done in a laboratory simulation of a paper mill flotation unit in which fiber and pigments are lifted to the surface by air and the solids skimmed off. Ground wood fiber beaten to 50 ml Canadian freeness, clay (50% by weight of the pulp), rosin, aluminum sulfate at a level of 50 lb per ton of pulp, and sodium aluminate at a level of 1,000 ppm were made up into the synthetic wastewater mixture. The polymeric flocculant was added at the indicated level and then 500 ml water pressurized with air at 50 psi was added; the mixture was agitated and allowed to stand for one minute. A sample was taken at the midpoint and percent light transmission at 530 mμ was determined. The following data was obtained.

Source of Additive	Dosage, ppm	% Transmission at 530 mμ
None (raw)		10
Processed white water*		76
Polyquat 55.6	0.1	86
Polyquat 55.6	1.0	92
Example 7, run 3	0.1	89
Example 7, run 3	1.0	98
Tylite Number 9**	0.1	90

*Treated with air, no polymer present
**Tylite Number 9 is a cationic polymeric flocculant

These experiments show that both the poly(quaternary ammonium) polyelectrolytes used as starting materials in this process and the chain-extended polyether polyelectrolytes are efficient in white water clarification processes. The polyether polyelectrolyte has an efficiency comparable to another commercially useful material.

Aminohydroxyalkyl Esters of Carboxylic Acid Polymers

J.D. Buckman, F.W. Raths, P.P. Boggs, and R.L. Donerson; U.S. Patent 3,730,888; May 1, 1973; assigned to Buckman Laboratories, Inc. describe a method for preparing water-soluble acid salts of tertiary aminohydroxyalkyl esters of carboxylic acid polymers via the reaction between a glycidyl tertiary amine and the carboxyl groups of a homo- or copolymer containing at least 10% of the mers, i.e., the repetitive chemical structural units of a polymer, which are derived from an ethylenically unsaturated carboxylic acid corresponding to the formulas:

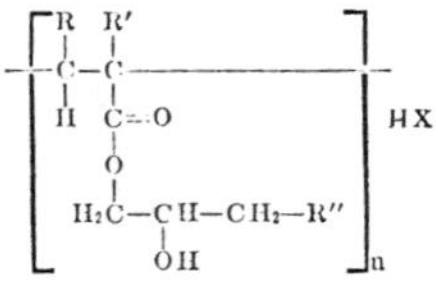

and

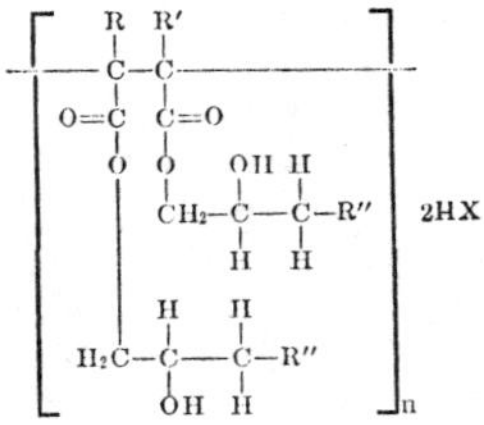

where n is an integer greater than 1, R and R' independently represent hydrogen, methyl, or ethyl; R" represents aziridyl, 2,6-dimethyl morpholyl, N-methyl piperazyl, N-2-hydroxyethyl piperazyl, piperidyl, pyrrolidyl, or

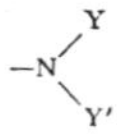

where Y and Y' independently represent alkyl and hydroxy substituted alkyl groups containing from 2 to 6 carbon atoms, and X represents an acidic anion. The tertiary aminohydroxyalkyl esters of carboxylic acid polymers possess the following advantageous properties:

(1) Effective as a flocculant:
   (a) At low concentrations
   (b) Over a pH range of 1 to 9
(2) Very soluble in water over a pH range of 1 to 9.
(3) Concentrated aqueous solutions have very low viscosities.
(4) Unaffected or enhanced in flocculating activity by the presence of metal ions such as Ca^{++}, Al^{+++}, and the like.

Example 1: A glass reaction flask was charged with 379 grams (0.5 mol) of a 12.3% sodium polyacrylate-water solution which was neutralized to pH 3 with 46 grams of concentrated hydrochloric acid. (The sodium polyacrylate solution had a viscosity of 239 cp at 25°C as a 22% aqueous solution.) To the stirred solution at room temperature was added 65 grams (0.5 mol) of 1-diethylamino-2,3-epoxypropane [prepared as described in the *Jour. of Amer. Chem. Soc.* 68: 1292 (1946)]; an exothermic reaction occurred and the temperature rose to 41°C. The reaction mixture was heated to 75° to 80°C for thirty minutes, cooled, and 0.5 mol of nitric acid was added. The resulting solution was effective as a flocculant.

Example 2: Into a one liter round bottomed flask equipped with a mechanical stirrer and a thermometer was placed 315 grams (3.0 mols) of diethanolamine. With stirring, 277.5 grams (3.0 mols) of epichlorohydrin was added over a period of three hours and at such a rate so as not to exceed 30°C. The reaction mixture was stirred an additional hour at room temperature to give 3-chloro-2-hydroxypropyl diethanolamine containing 0.96% ionic chloride.

Example 3: A glass reaction flask was charged with 43.7 grams (0.1 mol) of a 21.5% sodium polyacrylate-water solution and 19.9 grams (0.1 mol) of 3-chloro-2-hydroxypropyl diethanolamine. (The sodium polyacrylate solution had a viscosity of 294 cp at 25°C as a 22% aqueous solution.) The stirred reaction mixture was divided into two equal portions. One portion was tested as is and the other was neutralized with 4.5 grams (0.05 mol) of concentrated nitric acid prior to testing. Both portions were effective fluocculants.

Example 4: A glass reaction flask was charged with 42.6 grams (0.1 mol) of sodium polyacrylate-water solution, 10.5 grams (0.1 mol) diethanolamine, and 42.0 grams water. (The sodium polyacrylate solution had a viscosity of 226 cp at 25°C as a 22% aqueous solution.) To this stirred mixture was added dropwise 9.2 grams (0.1 mol) of epichlorohydrin. The reaction mixture was cooled with an ice bath in order to maintain 25° to 30°C during the 5 minute addition of epichlorohydrin. When the addition was complete, the reaction mixture was stirred at 25° to 30°C for 30 minutes and then refluxed for two hours. A sample was analyzed for ionic chloride and 3.4% chloride was found. Theory was 3.4% chloride. The reaction mixture was neutralized with concentrated nitric acid and the resulting mixture was an effective flocculant.

Polymers of Acrylamidosulfonic Acids

According to a process described by *D.I. Hoke; U.S. Patent 3,692,673; September 19, 1972; assigned to The Lubrizol Corporation* polymers of acrylamidosulfonic acids and their salts, especially 2-acrylamido-2-methylpropanesulfonic acid and its salts, are useful as flocculants for aqueous systems, especially in combination with inorganic coflocculants.

The process is illustrated by an experiment which shows the effects as flocculants of several polymers of sodium 2-acrylamido-2-methylpropanesulfonate. The polymers are used in combination with an inorganic coflocculant to effect precipitation of a suspension of 200 ppm of kaolin in water. The inorganic coflocculant is added as a 0.5% aqueous solution while stirring rapidly for three minutes; the suspension is then stirred slowly for 12 minutes, the polymer is added (as a 1% aqueous solution) during a second three minute period of fast stirring, the suspension is stirred slowly again for 12 minutes and allowed to stand for 15 minutes. The turbidity of the liquid is then measured using a Bryce-Phoenix light scattering photometer. The polymer solutions used are as follows.

(1) A homopolymer of sodium 2-acrylamido-2-methylpropanesulfonate, prepared by agitating an aqueous solution of the sulfonic acid monomer with an equivalent amount of sodium carbonate until polymerization is complete.

(2) A homopolymer of sodium 2-acrylamido-2-methylpropanesulfate prepared by dissolving the sulfonic acid monomer and an equivalent amount of sodium hydroxide in water and polymerizing in the presence of ammonium persulfate and sodium bisulfite.

(3) A copolymer prepared from 80 parts of sodium 2-acrylamido-2-methylpropanesulfonate and 20 parts of sodium acrylate, by polymerization in aqueous solution in the presence of ammonium persulfate and sodium bisulfite.

(4) A copolymer of 95 parts of sodium 2-acrylamido-2-methylpropanesulfonate and 5 parts of acrylamide, prepared in aqueous solution using an ammonium persulfate-sodium bisulfite initiator.

(5) A copolymer of 50 parts of 2-acrylamido-2-methylpropanesulfonate with 50 parts of N-(1,1-dimethyl-3-dimethylaminobutyl)acrylamide quaternized with dimethyl sulfate, the copolymer being prepared in aqueous solution using an ammonium persulfate initiator.

The results are given in the table below.

Polymer	Polymer, ppm	Salt	Salt, ppm	Turbidity
	0	-	0	7.96
	0	$Fe_2(SO_4)_3$	3	7.62
1	0.03	$Fe_2(SO_4)_3$	18	0.23
1	0.1	$Fe_2(SO_4)_3$	3	0.12
1	0.5	$Fe_2(SO_4)_3$	3	0.15
1	1.0	$Fe_2(SO_4)_3$	3	0.24
5	0.1	$Al_2(SO_4)_3$	10	0.33
5	0.8	$Al_2(SO_4)_3$	5	0.29

The table below gives similar results on flocculation of sewage obtained from the sewage treatment plant of Euclid, Ohio.

Polymer	Polymer, ppm	$Fe_2(SO_4)_3$, ppm	Turbidity
1	0.2	2	0.083
1	1.0	10	0.026
1	1.0	100	0.008
2	2.0	100	0.316
2	5.0	200	0.008
3	1.0	10	0.176
3	5.0	10	0.106
4	5.0	200	0.008

Flocculation and Electrolysis

L.B. Armstrong; U.S. Patent 3,664,951; May 23, 1972; assigned to Pollution Engineering International, Inc. describes a system designed to process large volumes of industrial wastewater that contain colloidal materials and other undesirable pollutants. Colloidal materials are finely disseminated materials in the water. They give color and odor to industrial wastewater. Colloidal materials remain suspended because of a small difference in specific gravity of the particles in the surrounding water. These particles are uniformly electrically charged, and many colloids, including proteins, starches, celluloses, polypeptides and other substances possess negative charges. Other colloids are positively charged.

These small particles can be removed by electrically disturbing this charge by an electrolysis process and causing the particles to coalesce. The particles so coalesced frequently do not have an adequate specific gravity for precipitation to the bottom of a tank in which the water is being treated, but they may be removed by foaming and mechanically removing the foam from the water surface.

Electrodes used in this electrolysis process can be of many materials, but since one electrode will be gradually disintegrated during the electrolysis process, the electrodes are put to work to perform other duties. As electric current passes between the electrodes, the colloidal particle sizes become larger and larger until they are sufficiently large to trap bubbles of oxygen and hydrogen that are being evolved at the electrodes. These bubbles float the particles to the surface of the water as foam and then may be removed by mechanical means above mentioned. The foam from most wastewater will be made up of organic material and can be burned or dried for recovery of valuable minerals or stored for disposal at approved disposal areas.

Demineralization and demetalization of the industrial wastewater can be accomplished at the same time that the colloidal organic material is being removed via the process of foaming. Electric current that causes the colloidal material to be removed will create ions of whatever material of which the electrodes are composed, and these ions will generate a floc which can be utilized to precipitate various minerals, dissolved metals and other inorganic materials present in the water. Since various metals and their ions will produce different types of floc for precipitating various types of minerals and inorganic matter, the electrodes can be selected to do specific jobs. Aluminum and carbon electrodes may be selected to remove temporary hardness, zinc electrodes may be used to reduce water toxicity, kill viruses and reduce floc settling time. Iron electrodes may be used if the water contains sulfides.

In addition to removing the colloidal material and precipitating the dissolved minerals and inorganic material, bacteria, algae and other oxygen-demanding matter may be destroyed by electrolysis. The oxygen generated and bubbling through the water makes intimate contact with organic material and will oxidize these organic materials as well as odorous gases of organic origin. Further, should the water have some chloride salts in it, these salts will break down and give off free chlorine gas which will also help kill bacteria and reduce odors.

The system utilizes flowmeters to measure the flow rate of the water passing through the apparatus for adjustment of the electrode apparatus and quantity of oxidant necessary to treat the water. The pH of the polluted wastewater is measured initially and is brought to a desired level, generally in the range of 5.0 to 5.5, by treatment in a preliminary treatment tank involving the addition of either necessary caustic or acid solutions to reach the desired pH level. Further, a certain quantity of seed floc is introduced into the preliminary treatment tank to initially precipitate a large quantity of organic and inorganic materials from the polluted water prior to treatment in the electrode tanks in order to lighten the load carried by the electrode tanks performing the electrolysis process, thereby drastically cutting the power requirements for the electrolysis process.

The resistivity of the water discharged from the preliminary treatment tank is measured and if the resistivity is too low, higher resistivity water from the system in an advanced state of treatment is recycled back to a mixing tank to raise the resistivity of the water prior to being applied to the first electrode tank assembly.

In the electrode tank assembly, organic materials are coalesced and bubbled to the surface as foam where they are removed by the action of mechanical paddles which sweep the foam from the surface of the electrode tank and discharge it into a suitable conveying means for transportation to a preferred form of disposal means, such as a burner, in which the organic materials may be dried and reclaimed later if desired or burned completely leaving an inorganic residue. Additional floc precipitation is induced in the first electrode tank, as hereinabove described, to reduce the quantity of inorganic materials and improve the color of the water. The water discharged from the electrode tank assembly is then delivered to a floc settling tank.

The water discharged from the first electrode tank assembly contains a quantity of floc and is used as a seeding floc in the first floc settling tank, allowing the water to circulate through the floc precipitate causing additional formation of floc and thereby removing additional organic and inorganic materials from the water. The water discharge from the first floc settling tank is then applied as an input to a second electrode tank assembly if one is necessary for further treatment of the water. In a second electrode tank assembly, a different combination of electrode materials may be utilized than was used in the first electrode tank assembly, in order to cause coalescing and foaming of additional organic materials and to cause additional floc precipitation of other organic and inorganic materials not removed in the first electrode tank assembly of the first floc settling tank.

For the first electrode tank assembly, the materials foamed to the surface of the second electrode tank are removed by a series of paddles that sweep the foam from the surface of the electrode tank and deposit it in a suitable conveying means for transfer to the desired disposal means. The output of the second electrode tank is discharged to a second floc settling tank along with a quantity of seeding floc from the electrode tank, as was done in discharging treated water to the first floc settling tank.

The second floc settling tank allows additional organic and inorganic materials to be precipitated from the water thereby improving its color and completing the demineralization and demetalizing process. However, there may be cases in which, because of the various and many combinations of minerals and dissolved metals and other materials in the polluted water, additional electrode tank assemblies and floc settling tanks may be necessary to accomplish the process. However, it has been found that the combination of two electrode tank assemblies and two floc settling tanks will handle substantially all industrial polluted wastewater presently known.

The water discharged from the second floc settling tank may be used directly in most instances for reuse as process water, cooling water, or as boiler feedwater in industrial applications. However, if the water is to be discharged to a stream or river, it would be necessary to put the water through a final pH and oxidation step in order to meet federal and/or state requirements. The Public Health Service requires a certain residual chlorine be present in water prior to disposing of waste fluids in public streams. This requirement is an attempt to kill any bacteria, reduce color and odors, and to reduce the oxygen demand of organic matter still present in the disposed water.

To provide for such final pH and oxidant adjustment, the process further includes the use of a pH and oxidant adjustment tank for receiving the water discharged from the final floc settling tank. The pH, resistivity and color are measured, and the results used to either add a caustic or acid solution from appropriate sources to adjust the pH to a final predetermined level. The required quantity of oxidant for oxidizing the water in the tank is also added. The water discharge from the pH and oxidant adjustment tank is discharged into a final polishing pond, merely a storage tank, in which the water is allowed to stand for a short period of time for allowing the oxidation material introduced in the oxidant adjustment tank to further act on the water in an attempt to kill additional bacteria, reduce odors and to reduce the oxygen demand of remaining organic material. The water discharged from the final polishing pond may then be reused in the industrial system, or discharged into public sewage systems, streams, lakes or rivers.

The system utilizes an oxidant generator assembly employing a chlorine generator for supplying chlorine as the oxidation material for use in the oxidant adjustment tank as well as a convenient means for producing the caustic and acid solutions for use in the pH adjustment of the water at various stages in the system.

The floc precipitate in the electrode tank assemblies and floc settling tanks will continue to build up in the bottom of the tanks and would eventually fill each of the tanks unless discharged. Accordingly, floc height sensors are utilized for monitoring the height of the floc precipitate in each tank, and when it has reached a predetermined level, a solenoid valve is actuated allowing a portion of the floc precipitate to be discharged from each of the tanks to a central floc accumulation tank for use as seeding floc in the preliminary treatment tank as hereinbefore described. The excess floc accumulation in the preliminary treatment tank and the floc accumulation tank, is discharged directly to the desired disposal means as waste material.

Apparatus for Dispersing Coagulant

According to a process described by *J.D. Walker; U.S. Patent 3,666,663; May 30, 1972; assigned to Chicago Bridge & Iron Company* minuscule quantities of additive solution are almost instantaneously distributed throughout every incremental portion of a huge water stream flowing through a large pipe or flume by dividing the additive into a plurality of carefully metered smaller streams, passing the individual added streams into the larger water stream at a plurality of points in the cross-section of the large stream, in a zone in which the entire water stream is subjected to turbulence as it passes the point of injection.

Referring to Figure 1.1, mixing station **15** is incorporated into raw water feedline **17**. Mixing station effluent proceeds to a clarifier **19**, diagrammatically shown, which includes relatively wide diameter basin **21**, overflow troughs **23** and gentle-agitation section **25**. From the clarifier overflow, the water proceeds to filitration stations in other conventional processing stations. Raw water line includes coupling section **27**, pipe **28** of the mixing station and clarifier input line **29**. The pipe of the mixing station is secured to sections **27** and **29** by conventional butting joint flanges **31**, secured by customary nuts and bolts.

However, interposed between the opposing flanges at both upstream joint **35** and downstream joint **37**, are baffle discs **40**, each having peripheral section **42**, in-line baffle section **44** and spaced openings **46** in the peripheral section for passage of bolts therethrough. In the examples illustrated in Figures 1.1b, 1.1c, 1.1d and 1.1g, vertical vane **50** and horizontal vanes **52** are secured to the baffle section by welds **54**. The horizontal vanes are also secured to the vertical vane by welds **56**. The vertical vanes include extensions or projections **60** extending from end **62** of vane **50** which is most distant from the disc.

Openings **65** are provided in the vertical vanes to permit passage therethrough of additive secondary feedlines which will be described in detail hereinafter. In the illustrated examples, two identical disc-vane assemblies **70** are utilized, one at the upstream joint with the vanes extending downstream, the other at the downstream joint with its respective vanes extending upstream.

FIGURE 1.1: APPARATUS FOR DISPERSING COAGULANT INTO WASH STREAM

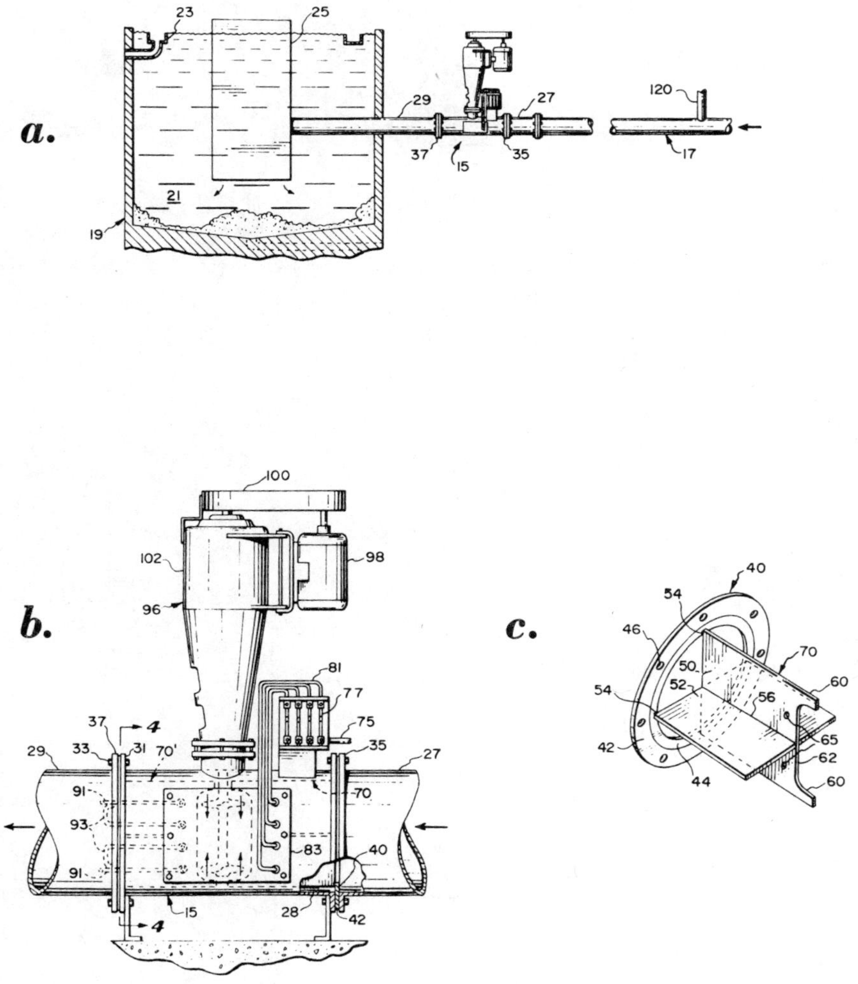

(a) Schematic View of Water Treatment System with Disperser
(b) Enlarged View of Dispersing Unit
(c) Disc-Vane Assembly

(continued)

FIGURE 1.1: (continued)

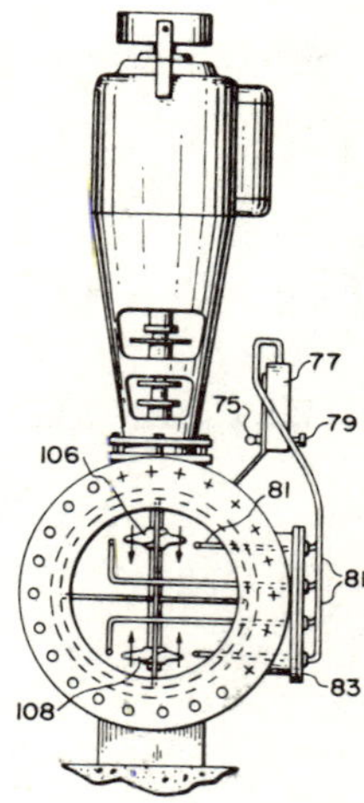

d.

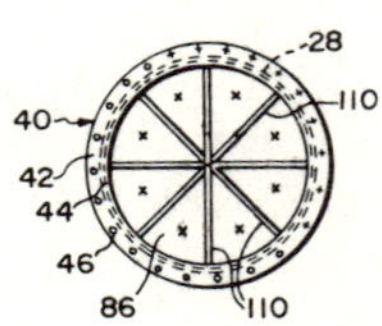

e.

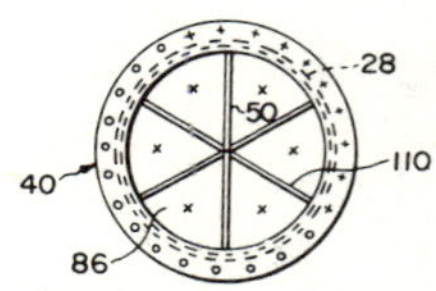

f.

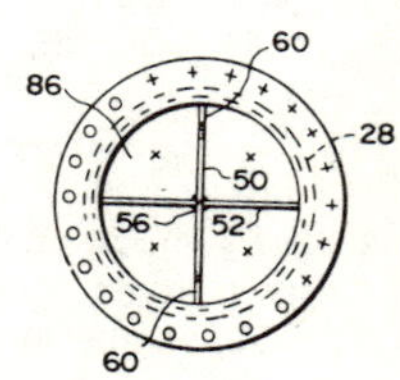

g.

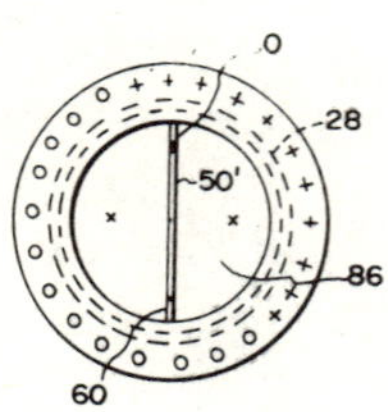

h.

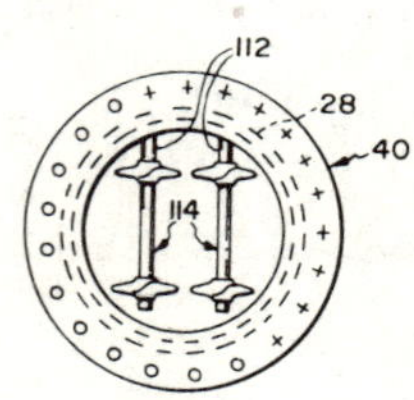

i.

(d) View Along Line **4—4** of Figure 1.1b
(e)-(i) Alternate Designs of Vane Systems

Source: J.D. Walker; U.S. Patent 3,666,663; May 30, 1972

Additive Injection System – Main additive feedline **75**, for example, a line connected to a source of coagulant-chemical solution in a water treatment plant, is connected to a plurality of valved flow meters **77**. The flow through each flow meter is accurately regulatable by its respective valve **79**. Branch additive lines **81** pass from the respective flow meters through access plate **83** to the interior of pipe **28**. The access plate is sealed around secondary feedlines **81**. Each of the branch feedlines ends in a respective section **86** of the interior of the pipe formed by the vanes, preferably discharging its stream of additive near the centroid of the cross section of its respective section, as perhaps best seen in Figure 1.1d. Each **x** shown in sections **86** in Figures 1.1e through 1.1h indicates the preferred approximate point of injection in each of the illustrated configurations.

Sections **86** are defined as that fractional portion of the cross-sectioned area of the stream pipe or flume, represented by the total cross-sectional area divided by the number of branch additive injection points. In the preferred case, each section is bounded by physical walls or dividers, e.g., the vanes, though the presence of the vanes is not absolutely essential. It is possible that the vanes can be omitted, provided that in some way, flow past, and in close proximity to, each additive injection point is made a consistent part of the whole, with sufficient equality between the flows to achieve the uniformity of distribution desired. It is important, however, that the chemical streams be injected at enough points distributed through the cross section of flow, with an approach to equalizing of dosing rates, and with sufficient agitation immediately upon injection, to achieve instantaneous mixing throughout the entire mainstream.

Flow meters **27** are preferably of the conventional type in which an indicator rises in a tapered tube, its height increasing as the flow increases. With the indicators all at the same level, the equality of flow usually desired is achieved. If any constructional features or other factors might induce less flow in some sections **86**, than others, an operator can experiment and see if the rate of injection to that section can be reduced without impairing the ultimate results. The equality of dosing rate desired is that which will amount to the same percentage of coagulant to the water flowing through each section.

Power Mixer – Power mixer, generally **96**, includes motor means **98**, transmission means such as pulley system **100**, gear box **102**, shaft **114**, upper propeller **106**, and lower propeller **108**. The propellers are pitched to propel water in opposite directions, preferably toward each other, as indicated in Figures 1.1b and 1.1d. It is noted that extensions **60** of vanes **50**, parallel to the general direction of flow liquid passing through the pipe, are located axially with respect to the axis of rotation of the impellers, and serve as rotation spoilers. Although they act directly on the water approaching the impellers axially, they reduce the rotation of the entire body of water surrounding each shaft.

Alternative Constructions – Disc-vane assembly **70** can be modified, depending on the size of the pipe and the quantity of water into which the additive is to be dispersed, to include a larger, or smaller, number of sections **86**. Each section would have a branch additive feedline **81**. Thus, as illustrated in Figures 1.1e and 1.1f vanes **110** may be used in a variety of angular spacings. In a relatively small pipe, one vane **50** will suffice, as illustrated in Figure 1.1h. If the shaft should be horizontal, it would also be horizontal. It is essential, in accordance with this process, that a branch additive feedline inject a proportional portion of additive in each of the sections **86**.

As illustrated in Figure 1.1i a plurality of mixers **112** on shafts **114** may be positioned along a plane perpendicular to the general direction of flow through the pipe, and this arrangement is particularly preferred in cases in which the pipe is of relatively large diameter. Rotation spoilers **60** should be provided for each shaft, as close to the plane as convenient. Separate pieces may be welded in place if no vane is reasonably near the plane.

Operation – Raw water, or other liquid through which a relatively small quantity of additive is to be dispersed, passes through feedline **17** in the direction indicated by the horizontal arrows of Figures 1.1a and 1.1b. Upstream baffle **40** causes considerable turbulence immediately downstream in the region between upstream and downstream end of the mixing station. It and the vanes following it also tend to ensure equal flow through the various sections **86**. A main stream of additive, passing through the main feedline is split into a plurality of branch streams, in the example of Figures 1.1b through 1.1d, into four streams, and the relative quantity of additive liquid passing in each of the branch streams is operator-controlled using the conventional flow meters such as rotameters to provide substantially equal secondary streams passing through respective branch additive lines. Thus, material passing through the lines and being discharged into the sections are discharged into a violently turbulent region within the mixing station.

When power mixer **96** is in operation, fluid is drawn from outer zones of the pipe and driven inward, axially of the shaft. This vigorous agitation complements the mixing action of the upstream baffle. At relatively low rates of flow, i.e., at rates of flow well below the design limit for the particular pipe, turbulence generated downstream of the upstream baffle, is insufficient for the desired quick and complete dispersal of additive throughout the mass of liquid passing through each of the sections. During these relatively low flows the mixer nevertheless provides virtually instantaneous homogenization, i.e., homogeneous dispersion of the additive solution throughout the water passing through the pipe. Thus, within a fraction of a second, perhaps even within a few microseconds, depending on the rate of flow through the feedline, the additive solution is completely dispersed throughout each increment of fluid passing through the mixing station.

Additional additive feedlines and stream sampling lines **93** may be added through access plate **83** as suggested at the phantom lines in Figure 1.1b. Lines **91**, **93**, however, terminate downstream of the downstream baffle, and hence can be used for sampling, or can be used to discharge a supplementary additive, e.g., alkaline agents, solid carbon in a slurry, polyelectrolytes, etc. into the main stream outside of the mixing station. Turbulence in the downstream region generated by the downstream baffle continues the homogenization or thorough mixing of the stream well past the mixing station. Also, upstream line **120** can be used to add pH modifiers, or other modifiers as required by the chemistry of the particular raw water stream.

Reports on actual tests in a water treatment plant indicate that using the preferred procedure illustrated in Figures 1.1a, 1.1b and 1.1d, the quality of water passing from a final filter of a water treatment plant was improved even though substantially 20% less coagulant chemicals than before were required. Also, in the same test, head loss of less than 1 foot was reportedly encountered through mixing station **15**.

The improved quality of the water with less coagulant appears to result from the quick uniform dispersal of the coagulant. Samples taken at the sample lines during operation, showed the complete equalization of the concentration in each section, of an additive added through the branch lines. As indicated above, operation of the mixing station, in accordance with this process, at relatively low flows through the pipe, i.e., at flows well below the maximum design capacity, does not provide sufficient turbulence or churning within the station due to the action of the upstream baffle alone. When the mixing station is operated under these relatively low flow conditions, therefore, use of the power mixer is essential.

Thus, in accordance with this process, virtually instantaneous mixing of additive, for example, chemical coagulants in water treatment streams is provided, with the result that decreased levels of additive have been found to be sufficient to provide improved water quality, and with the result that head losses encountered using the process are vastly and astoundingly less than the head loss encountered in using the prior art rapid mixers.

Decolorizing Liquors

P. Smith and H. Suzor; U.S. Patent 3,708,337; January 2, 1973; assigned to The Colonial Sugar Refining Company Limited, Australia describe a continuous column process for decolorizing an aqueous sugar liquor, in which the sugar liquor is fed downwardly through a floating bed of a flocculent polyamide. The polyamide is assisted initially to float in the liquor by being conditioned in an aqueous liquid, such as water. It has been found that a rate of liquor flow can be regulated such that (1) at least a portion of the bed is maintained generally in suspension, but (2) particles of spent polyamide are carried away in the color-reduced effluent (product). The particles of spent polyamide are separated from the effluent outside the column, and the polyamide is then regenerated and returned as required to the top of the floating bed. The polyamides used are water-insoluble, water-swellable and pulverulent (typically, nylon-66 and poly-N-vinyl polypyrrolidone), and cannot be used conventionally in a packed column fitted with a basal retaining plate. The process is particularly applicable to decolorizing the high density sugar liquors, such as raw washings. The following examples illustrate the process.

Example 1: A quantity of poly-N-vinyl polypyrrolidone (PVPP), 200 g, was first conditioned by immersing in excess raw washings of 68° to 70° Brix at 20°C for one day. By the end of this time, the PVPP was found to have swollen to a gelatinous caked mass of maximum volume. This was then filtered from the excess raw washings. A column (diameter 5.08 cm, height 91 cm) was set up, having an inlet at the top and an outlet at the bottom, and was held by a water jacket at 60°C. The gelatinous caked mass of PVPP was broken up and was suspended as a bed in the column in a further portion of the raw washings at 60°C. A feed supply of the same raw washings at 60°C was then pumped through the column at a rate of 10 ml/min.

At least a portion of the PVPP bed remained generally suspended in the column throughout the process and color-reduced effluent liquor was withdrawn from the outlet. Particles of PVPP were lost progressively by the bed as they became relatively spent and were entrained in the color-reduced effluent liquor. After withdrawal from the column, such particles of spent PVPP were readily separated from the liquor by a conventional filtration method (discharging from a goose-neck into a funnel fitted with a steel gauze).

With respect to the Attenuation Index at 420 mμ pH 9, the color of the feed raw washings was 41253. The color of the effluent liquor was determined at various stages in the throughput of the feed, and the results with respect to the cumulative throughput are given in the following table.

Color of Effluent	Cumulative Throughput, liters
21303	1
17387	2
18642	3
23266	4
31464	15.2

Example 2: Separate samples (each 25 grams) of PVPP and nylon-66 were first conditioned by immersing in excess water for 8 hours. The pH was then adjusted to 5 by the addition with stirring of a quantity of 0.1 N hydrochloric acid. The flocculent acidified polyamides were filtered to remove excess liquid, and were then treated by immersion in excess color-free sugar liquor of 66° Brix for 16 hours. The polyamides were again filtered to remove excess liquid. Two identical columns (diameter 2.54 cm, height 91 cm) were set up, having an inlet at the top and an outlet at the bottom, and were held by a water jacket at 80°C. The flocculent polyamide masses were broken up and suspended as beds in the columns in raw washings of 66° Brix at 80°C. A feed supply of the same raw washings at 80°C was then pumped through each column at a rate of 0.175 ml/min.

At least a portion of each polyamide bed remained generally suspended in its column throughout the process, and particles of relatively spent polyamide were entrained in the color-reduced effluent liquor withdrawn from the column. The spent material recovered by filtration was weighed at known cumulative throughputs. The color of the effluent liquor was determined for each column at the same cumulative throughputs and the results with respect to the cumulative throughputs of liquor and the cumulative mass of spent polyamide are given in the following table. The color of the effluent liquor is given as the reduced effluent color, i.e., the ratio of the color of the effluent liquor to the color of the influent raw washings. The cumulative mass of spent polyamide is expressed as a percentage of the total mass of polyamide initially suspended in the column.

Polyamide	Reduced effluent colour	Cumulative: Throughput (ml.)	Cumulative: Spent polyamide (percent)
PVPP	0.76	717	13.26
PVPP	0.80	20,34.4	98.44
Nylon-66	0.79	775.3	3.20
Do	0.89	1,828.6	68.16

When the example with PVPP was repeated in all respects except that the raw washings were replaced by a color-free sugar liquor of 66° Brix (solids provided by sucrose only), it was found that the maximum percentage of PVPP that could be entrained out of the column was 24%.

FILTRATION

Continuous Process

L.R. Bourdalé; U.S. Patent 3,741,892; June 26, 1973 describes a filtering method which is mainly characterized in that it consists in simultaneously creating two vacua, one on each side of the filtering surface, the first vacuum causing flow of the purified fluid, and the second withdrawing part of the fluid during its upstream travel for the purpose of carrying the particles towards an outlet for the particles, the upstream path of travel of the fluid being substantially parallel with the filtering surface, whereas the flow of fluid into this path of travel decreases constantly from the inlet for the laden fluid to the outlet for the particles so that the mean velocity is substantially constant and sufficient to entrain all the particles of which the path of travel is always in the same direction as the upstream path of travel of the fluid.

As the result of this arrangement, the particles are automatically progressively removed, and this method consequently enables filtering to be carried out on a continuous basis with a virtually constant loss of pressure.

The filter illustrated in Figure 1.2a and 1.2b is mainly constituted by a hollow body **1** to which are connected, at the top, an inlet port **2** for the laden fluid, at the side an outlet port **3** for the purified fluid and at the bottom, an outlet port **4** for the particles. This hollow body has two opposite lateral faces **5** of trapezoidal form, and therefore has a cross-section which diminishes constantly from the port **2** to the port **4**.

The interior of the body of the filter is partitioned by a certain number of filtering plates **6** which are arranged vertically and are separated from each other by gaps **7** alternating with smaller gaps **8**. These plates which are twelve in number in the example here described, are likewise of trapezoidal form and they are arranged parallel with the lateral faces of the body of the filter. It will also be seen that the end plates are separated from these side faces by a gap which is half the width of the gap **7**, for a reason which will be explained hereinafter.

The unoccupied spaces **7** between the filtering plates are blocked at the side presented to the outlet port **3** for the purified fluid by fluid-tight partitions **9**, whereas the unoccupied spaces **8** are blocked by fluid-tight partitions **10** at the side presented to the inlet port for the laden fluid. This arrangement defines an upstream zone, comprising the inlet port, the unoccupied spaces **7** and the outlet port **4**, and a downstream zone, comprising the unoccupied spaces **8** and the outlet port **3**. These upstream and downstream zones are separated by fluid-tight partitions **9** and **10** and by the filtering plates.

The filter functions in the following manner. The liquid or gaseous particle-laden fluid is admitted through the inlet port and then passes into the unoccupied spaces **7** in the body of the filter. The greater portion of the fluid passes through the filtering partitions, moves into the outlet port **3** which is subjected to a first reduction in pressure in relation to the inlet port, and purified fluid thus results. The remaining portion of the fluid continues its travel through the unoccupied spaces **7**, parallel with the filtering plates, and emerges through the port **4** which is subjected to a second reduction in pressure in relation to the inlet port.

FIGURE 1.2: CONTINUOUS FILTRATION PROCESS

a.

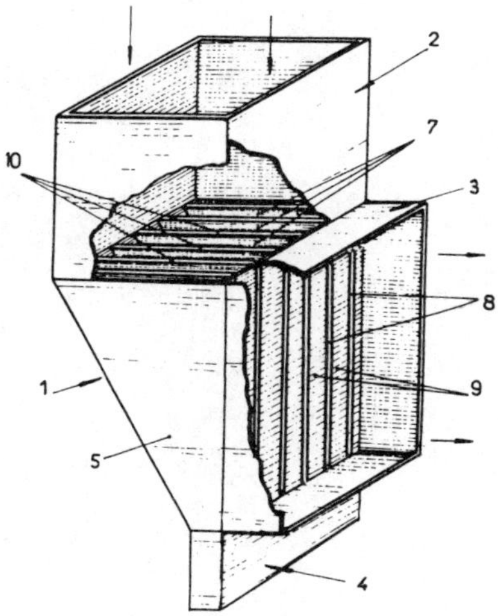

b.

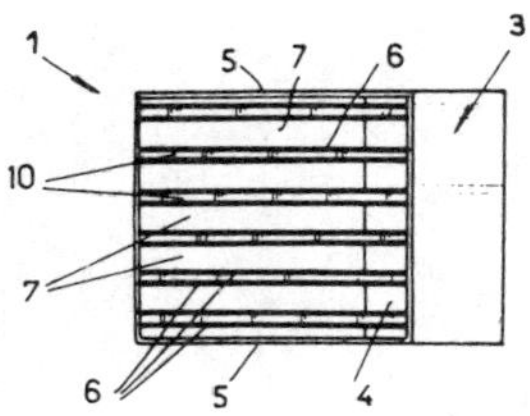

Source: L.R. Bourdalé; U.S. Patent 3,741,892; June 26, 1973

As the cross-section of the body of the filter decreases progressively and the gaps **7** are of equal width with the exception of the first and last, between the end plates and the side faces, which are half the width of the others, the flow of fluid in each of these unoccupied spaces **7** decreases progressively from the port **2** to the port **4**. The mean velocity of the fluid is thus substantially constant, and the system is so designed that this velocity is great enough to carry almost all if not all of the particles, suspended in the fluid admitted through the port **2**, towards the port **4**.

Thus, the static filter of the process enables a fluid and the particles suspended therein to be separated from each other on a continuous basis and with a virtually constant loss of pressure, practically all the particles, including the smallest normally falling within the particle-size range of the fluid treated, being removed efficiently. It will in fact be seen that the particles follow a path of travel which is always in the same direction as the upstream path of travel of the fluid, so that the smallest particles are also entrained, despite their low inertia.

Ultrasonic Screening for Microparticles

A process described by *H.W. Dietert; U.S. Patent 3,782,547; January 1, 1974; assigned to Harry W. Dietert Co.* relates to microparticle separation and refers more specifically to the filtering of particles of, for example, up to fifty microns in size through a filter screen with the particles in a liquid bath by vibrating the liquid bath and/or filter screen at an ultrasonic frequency to scrub the particles and maintain them and the filter screen clean and to therefore maintain the filter screen at maximum efficiency at all times.

The ultrasonic screening structure **10** for microparticle separation as shown in Figure 1.3a includes an outer container **12**, an inner filter container **14**, a source of ultrasonic alternating electrical energy of between fifteen and sixty thousand hertz **16** and an electrical frequency to mechanical vibration transducer **18**. The microparticle separation structure further includes the means **21** for providing a water bath **20** in the containers and the means **22** for placing additives in the containers.

FIGURE 1.3: ULTRASONIC MICROPARTICLE SCREENING PROCESS

a.

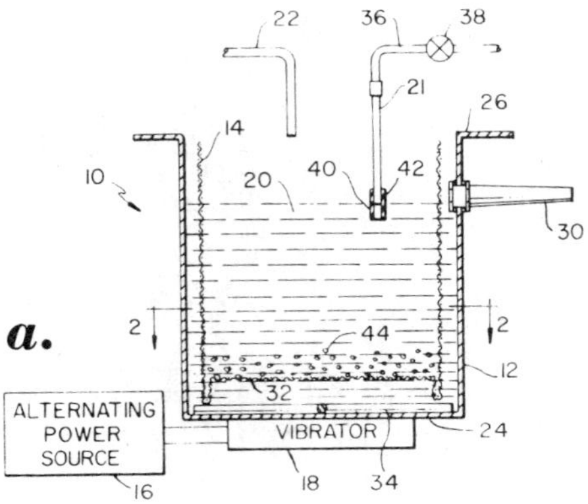

View of Ultrasonic Cleaning Structure Along Line **1–1** of Figure 1.3b

b.

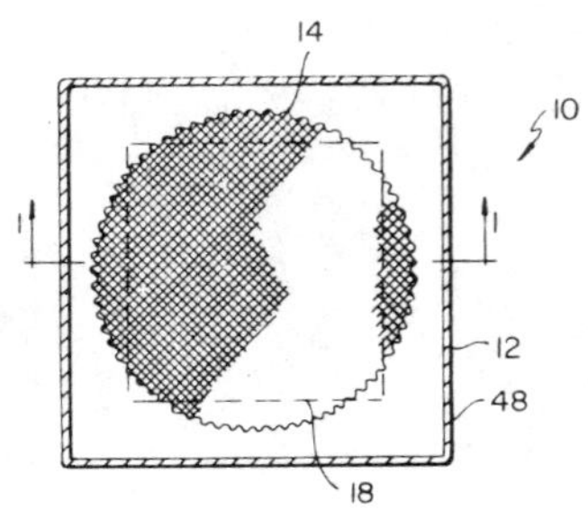

View Along Line **2–2** of Figure 1.3a

c.

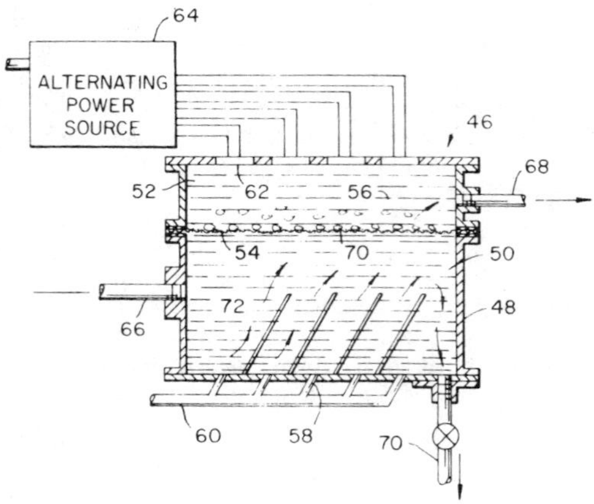

View of Structure Modified for Continuous Operation

Source: H.W. Dietert; U.S. Patent 3,782,547; January 1, 1974

The container **12** is substantially square and has a closed bottom end **24** and an open top end **26**. The bottom is relatively thin and is highly polished so as to transmit ultrasonic vibrations from the transducer to the water bath in the containers. An overflow outlet or drain **30** is provided in the container **12** adjacent the open end thereof. The filter container is constructed of a fine mesh screen such as twenty microns having suitable supporting structure. The filter container is separated from the bottom of the container **12** by suitable means such as the recessed bottom **32** of the filter container construction and/or the cross-shaped spacing member **34** on which the container **14** is positioned in the container **12**.

The structure **21** for maintaining a fluid bath in the containers, as shown in Figure 1.3a, may be a water conduit **36** having a valve **38** therein for regulating the water supply to the containers. A rubber tube **40** or the like is placed over the end **42** of the conduit to prevent injury to the fine mesh screen container **14** in use. The structure **22** for adding additives to the water bath **28** may be entirely similar to the structure **21**.

In operation of the microparticle separation structure illustrated in Figures 1.3a and 1.3b, a batch of material to be separated in accordance with the size of the particles therein is placed in the filter container and the container **14** is positioned within the outer container, as shown in Figure 1.3a. A water bath is provided in the containers through the conduit and the water bath is vibrated at an ultrasonic frequency by the transducer in response to the ultrasonic signal from the source of ultrasonic alternating electrical energy **16**.

Both the particles **44** to be separated in the water bath and the filter container are thus vibrated at an ultrasonic frequency whereby the particles scrub each other and the filter so that the particles are freed from air bubbles and extraneous materials such as clay or other fine material which may coat them and air bubbles and lodged particles too large to pass through the filter are removed from the filter. A pressure head may be maintained between the inside and outside of the filter container by regulating the valve in conjunction with the opening of the overflow from the outer container. Thus, the particles are urged to pass through the filter screen **32**.

In addition, due to the additives passed through the additive adding means into the water bath such as, for example, sodium pyrophosphate and sodium hydroxide, the particles may be wetted and/or dispersed in the water bath which also aids their filtering through the filter.

After a predetermined time of filtering the particles as indicated above, the filter container is removed from the outer container and the filter container is weighed. The difference in weight between the initial weight of the filter container having the particles to be separated therein and the weight after filtering will provide an accurate indication of the quantity of material filtered through the filter container, if such measurement is desirable. Alternatively or in addition, the overflow fluid from the overflow nozzle may be dried and weighed to determine the amount of particulate material passing through the filter screen. Thus, an accurate indication of the quantity of a material having less than a predetermined size may be determined by means of the structure illustrated in Figure 1.3a and the batch method described.

Should the continuous separation of fine material from coating and larger material be desired, the structure **46** illustrated best in Figure 1.3c is provided. The structure illustrated in Figure 1.3c includes a container **48** having a first chamber **50** and a second chamber **52** therein with a filter member **54** such as a twenty micron filter screen separating the chambers. The water bath **56** in the container **53** which is maintained by jets **58** opening into chamber **50** from supply conduit **60** along with the filter screen is again vibrated at ultrasonic frequencies of between fifteen and sixty thousand hertz by the transducers **62** excited by an alternating electric signal from the source **64**. Particulate material which may be carried by the water bath is introduced into chamber **50** through conduit **66**. Filtered particles and a portion of the water bath are withdrawn from the chamber **52** through condiut **68**, while the remainder of the water bath is withdrawn through the conduit **70** from the chamber **50**. In operation of the continuous ultrasonic screening structure **14** for microparticle separation in a water bath, the water bath in the chambers is vibrated at the ultrasonic frequency by the transducers. Again, the particles **70** are caused to scrub each other and pass through the filter screen which is maintained in a clean condition due to the vibration thereof with the water bath and a hydrostatic head is provided between the chambers due to regulation of the inlet pressure in the conduits **60** and **66** and the outlet pressures in the conduits **68** and **70**.

The particles as they pass into the chamber **50** through the conduit **66** are urged toward the filter screen by means of the water jets and the baffles **72** directing the water jets and thus the particles toward the screen. The particles which pass the screen and into chamber **56** are removed through conduit **68**. Those that do not pass through the screen pass out of chamber **50** through conduit **70**. Again, if it is desired to provide an absolute indication of the quantity of particles passing the screen or the material not passing the screen, or the relative percentage thereof, the material may be separated from the water bath withdrawn from the conduits, dried and weighed to provide a basis for any mathematical calculation.

Finely Divided Alumina

G. Hirs; U.S. Patent 3,669,882; June 13, 1972; assigned to Hydromation Engineering Company describes the use of granular, finely divided alumina (Al_2O_3) as an insoluble, chemically inert, high temperature filtration medium in systems requiring ultrafine filtration, such as in steam turbine power plants, chemical processing and pharmaceutical manufacturing.

A method and apparatus for utilizing this heavy and difficultly handled material is also proposed. The method and apparatus are characterized by high filtration rates, extreme clarity of effluent, rejuvenation of the medium, and the prevention of medium-contamination of the effluent.

Diatomaceous Earth Filter Aid

According to a process described by *D.R. MacPherson; U.S. Patent 3,680,699; August 1, 1972; assigned to Johns-Manville Corporation* water is clarified by passage through a sand filter having a thin precoat layer of coarse, porous, granular material such as flux-calcined diatomaceous earth. The material in the precoat layer should be coarser than the top layer of sand in the sand filter, and preferably comprises 8 to 100 mesh particles. A body feed of coarse, porous, granular materials may be added to the water to be clarified if a substantial amount of colloidal or particulate material must be removed.

Examples 1 through 4: Two pressure sand filtration columns (4¾ inches inside diameter, clear heavy-wall Lucite) are set up so that both can be run in parallel. Twenty-six inches of filter sand are placed in each column. This sand contains about 60% by weight of 20 to 30 mesh particles and about 40% by weight of 30 to 50 mesh particles. A high-pressure, six-stage Moyno pump system dispenses a body feed of flux calcined diatomaceous earth with a portion of the influent into the sand bed. The flow rates through the sand filters are measured with Hagan ring-balance meters. The pressure differentials across each of the sand filters are indicated with a conventional 24 hour recording Foxboro pressure meter placed across the influent and effluent lines for each filter.

In all tests, a filter aid is used in conjunction with one of the sand filters, and the other plain sand filter is run in parallel under identical conditions. Periodically, both sand filters are used for blank runs to determine that each filter is behaving in a similar manner. The filter aid utilized in Examples 1 through 4 is a flux-calcined diatomaceous earth (Celite) and having particles in the size ranges listed in the table below. The following table shows the gain in water throughput with granular precoat for parallel pressure sand filters using coarse, flux-calcined diatomaceous earth as precoat and body feed at constant filter rate of 12 gsfm.

	Raw Raritan River Water			Alum Flocculated Water
Example	1	2	3	4
Column designation				
A	Sand	Sand	Celite	Celite
B	Celite	Celite	Sand	Sand
Celite filter aid used	30/45	30/45	30/45	30/45
Influent				
Pressure, psi	40	40	80	80
Turbidity, ppm*	5-7	3-7	2-5	0.5-2
Length running, hours				
To breakthrough, sand**	6.5	6.5	6.0	3.6
Total	23.2	16.5	16.0	9.6
Diatomaceous earth used				
Precoat, psf	0.54	0.54	0.18	0.18
Body feed, psf	2.08	1,24	0.84	0.18
Total per 1,000 gal	0.157	0.149	0.091	0.052
Throughput, gsf				
Sand up to breakthrough	4650	4650	4310	2590
Diatomaceous earth	16,700	11,900	11,500	6,910***

*Measured with Hellige turbidimeter. Water quality of effluent as measured by turbidity through column including filter aid was 50 to 100% better than through sand.

**Pressure differentials at breakthrough varied from 0 to 3.0 psi.

***This test could have been continued much longer than 9.6 hours.

The results show the significant increase in throughput that can be achieved by the process while actually producing an increase in water quality.

Aluminum Sulfate

Y. Hayashi and A. Obara; U.S. Patent 3,719,592; March 6, 1973; assigned to Ebara Infilco Kabushiki Kaisha, Japan describe a filtration process for clarification, characterized by the addition of an inorganic coagulant, in an amount less than that with which formation of flocs is recognized by a coagulation test, to a suspended water having the turbidity of raw water lower than 500 ppm, maintaining the contacting time, as measured from the time that the coagulant is added to the suspension to the time that the suspension is supplied to a filter tank to within a few minutes, filtering the liquid at a filtration rate higher than 5 m^3/m^2 hr at all times during the process and during filtering, increasing the amount of the coagulant added by a small amount during the period after the beginning of filtration until the filtered water becomes stable, or increasing the filtration rate, or employing both of the two last-mentioned steps.

The purpose of the process is to make rapid and continuous filtration over a long period of time possible without deterioration of the quality of the filtered water by utilizing a filtering mechanism different from the conventional ones, wherein a minimum amount of coagulant is utilized at its maximum capability.

Figure 1.4a represents the relation between the turbidity of the raw water and the added amount of a coagulant [aluminum sulfate $Al_3(SO_4)_3 \cdot 18H_2O$] according to the process. The raw water was a lake water having pH 6.8 to 7.3, M-alkalinity 35 to 42 ppm, turbidity 4 to 100 ppm, and color 5 to 100 ppm, and the amount of the coagulant added was what was required to get a filtered water having a turbidity less than 0.5 ppm. The amount of aluminum sulfate added was determined with a filtration rate of 10 to 40 m^3/m^2 hr in winter, spring, summer and fall; the contacting time from adding aluminum sulfate to introducing the water into a filter was 40 seconds at a filtration rate of 10 m^3/m^2 hr, 20 seconds at 20 m^3/m^2 hr, 13 seconds at 30 m^3/m^2 hr, and 10 seconds at 40 m^3/m^2 hr. As clearly seen in this figure, in the range up to 100 ppm of the turbidity of the raw water, the turbidity has no correlative relation with the amount of aluminum sulfate added. A clarified water can be obtained by adding $Al_2(SO_4)_3{\cdot}18H_2O$ in amounts as little as 5 to 7 ppm. In this process the influence of the water temperature can be neglected, and the influence of the filtration rate is hardly recognized. A liquid can be treated with only one kind of coagulant in a very small amount. When adding 5 to 7 ppm of $Al_2(SO_4)_3{\cdot}18H_2O$, flocs cannot be recognized optically; the amount required to recognize flocs is higher than 12 ppm of $Al_2(SO_4)_3{\cdot}18H_2O$.

According to a conventional coagulation sedimentation filtering process, 12 to 30 ppm of aluminum sulfate is required for formation of flocs. When using 30 ppm of aluminum sulfate, flocs are hardly formed due to decrease of alkalinity, so an alkaline agent should be added. According to this process, however, as the amount of the aluminum sulfate added is so small, addition of a high polymer coagulant employed in a microfloc process is not necessitated, and clarification can be performed by adding a very small amount of aluminum sulfate. If the turbidity of raw water is 300 to 500 ppm, filtered water having a turbidity less than 1 ppm may be obtained by adding 15 ppm of aluminum sulfate where the turbidity of the raw water is 300 ppm, and 20 ppm of aluminum sulfate where the turbidity is 500 ppm.

When treating such high turbidity water with a usual coagulation sedimentation filtration process, it is necessary to add 50 to 60 ppm aluminum sulfate and several scores of ppm of an alkaline agent, but in this process, clarification can be carried out by adding only a small amount of aluminum sulfate.

FIGURE 1.4: FILTRATION PROCESS FOR CLARIFICATION

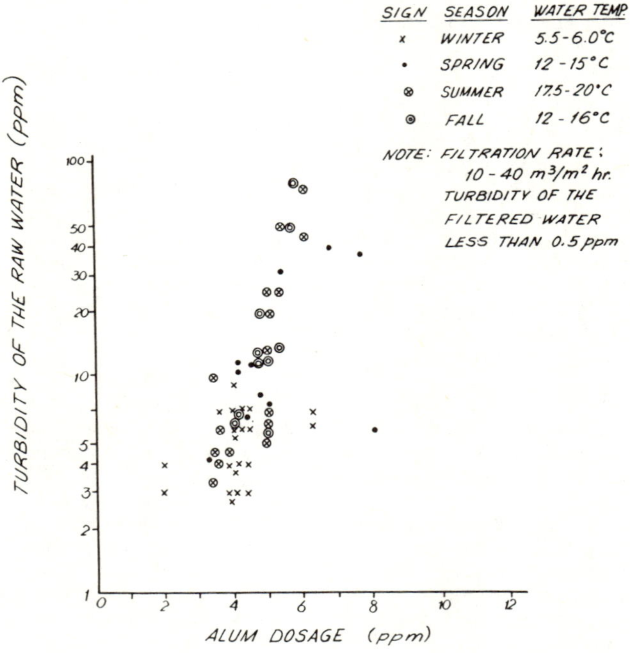

a.

Turbidity Versus Addition of Alum

(continued)

FIGURE 1.4: (continued)

b.

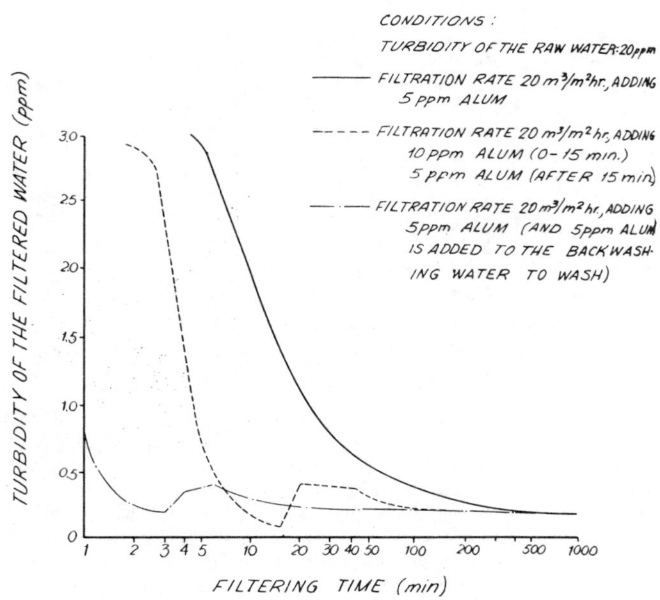

Turbidity Versus Filtering Time

Source: Y. Hayashi and A. Obara; U.S. Patent 3,719,592; March 6, 1973

When adding a small amount of a coagulant from the beginning of filtration as described above, the relation between clarification and the filtering time seen in Figure 1.4b is obtained. Scores of minutes or 1 hour after initiation of filtration, filtered water having about 1 ppm turbidity is obtained. Further precise experiments have been carried out to promote reduction of the turbidity at the beginning of filtration more rapidly. The resulting data are partially shown in the following table. The table shows the effects of the amount of aluminum sulfate added and the filtration rate on clarification of a liquid having a turbidity of 15 ppm. The values in the table are the time (minutes) required to get the filtered water having a turbidity of 0.5 ppm from beginning of filtration.

Filtration	----Added Amount of Aluminum Sulfate---		
Rate	5 ppm	7 ppm	10 ppm
10 m^3/m^2 hr	70	16.5	10.5
20 m^3/m^2 hr	70	10.5	6
30 m^3/m^2 hr	18	7.5	5
40 m^3/m^2 hr	18	6	2.5

The filter media used in each experiment was sand of 0.4 to 0.9 mm particle diameter and anthracite piled above the sand and having a 1.0 to 1.5 mm particle diameter. The sand and anthracite were packed in a filter of 0.8 m diameter to bed depths of 0.6 m and 0.3 m, respectively, the total bed depth being 0.9 m.

As seen in the above table, clear water can be produced just after initiation of filtration by adding 7 to 10 ppm of aluminum sulfate at the beginning of filtering and after 20 to

30 minutes, decreasing the aluminum sulfate to 5 ppm. Clear water can be also obtained in a short time by increasing the filtration rate temporarily at the beginning of filtration and then reducing the rate to the desired filtration rate. As seen in Figure 1.4b, though not shown in the table, clear water can be obtained in a short time by adding about 5 ppm of aluminum sulfate to the water which washes the filtering materials to regenerate same, and after sufficient washing, starting filtration. An economical filtering treatment with a lower water loss can be presented by combining the above described methods and the method of adding a very small amount of a coagulant, with which formation of flocs cannot be recognized optically, at the point in the liquid path from which the liquid can reach the filter within a few minutes.

Regeneration of Activated Carbon

A.C. Makrides and S.B. Brummer; U.S. Patent 3,730,885; May 1, 1973; assigned to Tyco Laboratories, Inc describe a method for controlling the adsorption-desorption behavior of activated carbon. Adsorption and desorption are controlled by controlling the potential at the interface between the carbon and the solution containing the species to be adsorbed or the solution into which the adsorbed species are to be desorbed.

Activated carbon is widely used in industry to remove impurities from fluids. By way of example, activated carbon is used to decolorize sugar solution, remove odors or flavors from drinking water, absorb gases and vapors, eliminate bacteria and toxic poisons from foods, facilitate separation of colloids and ultrafine particles from liquids by filtration or centrifugation, and remove organic materials from waste water. Usually the activated carbon adsorbent is employed in the so-called percolation method, in which the absorbent (in granular form) is held in place and the liquid or solution to be treated is caused to flow over and through it. Once the carbon has been saturated with adsorbate, it must be regenerated (i.e., revivified), which involves stripping it of the adsorbate.

The length of time which the activated carbon may be in service before it becomes saturated with foul products or impurities and must be regenerated will, of course, depend upon the amount of impurities passed through it per unit time. If properly regenerated, the carbon may be reused many times. Obviously, for industrial purposes, the regeneration process must be efficient and economical. Previously, the most commonly used methods of regeneration are (1) heating the carbon to a temperature at which the adsorbate is vaporized or decomposes into gaseous products and (2) subjecting the carbon to steam cleaning. It also has been suggested that regeneration may be accomplished by oxidizing the adsorbates with chemicals such as chlorine and hydrogen peroxide.

Regeneration by heating or steaming suffers from the limitation that the surface area of the carbon is degraded (often as much as 5 to 10%) during each regeneration. The losses in activity during regeneration are particularly critical in the case of activated charcoal which is more expensive than other commercially available activated carbons. Moreover the capital cost for regeneration can be substantial, particularly in water treatment plants. A further problem is that certain absorbates, notably organic materials, are difficult to desorb. Regeneration by oxidation with chemicals is unsatisfactory because regeneration is limited and tends to decrease rapidly after the first cycle. The following examples illustrate the process.

Example 1: A plurality of activated carbon adsorbing electrodes were made in the form of a hydrophilic rectangular fuel cell electrode from activated carbon and Teflon following the general procedure set forth in L.W. Niedmach and H.W. Alford, *Journal Electrochemistry Society* 112, 117 (1965). The carbon was Type BPL. It had a surface area of 1,000 square meters per gram. The carbon was ground to 325 mesh (average particle diameter is 0.044 mm), mixed with a Teflon-water dispersion (Type 30). The carbon and Teflon-water dispersion were mixed so as to provide by weight a ratio of 17 parts carbon to 3 parts Teflon. This mixture was pressed onto a gold plated tantalum screen and air dried at room temperature for 24 hours. The final electrodes had a geometric area of 12 cm^2. After fabrication, the electrodes were washed repeatedly in dilute HCl to remove acidextractables from the carbon, and then thoroughly washed in NaCl solution. Then the electrodes were gently evacuated to remove gases. Preliminary tests showed that the electrodes produced currents as high as 1 ma/g of carbon at potentials of possible interest. Accordingly, they were heavily oxidized by repeated galvanostatic charging using anodic currents of 100 to 300 ma which substantially reduced the magnitude of the currents passed at potentials of interest.

Adsorption-desorption studies were then made at room temperature with an aqueous solution containing 500 ppm of NaCl and 25 ppm of acetic acid. One liter of this solution was placed in a vessel and continuously flushed with nitrogen and stirred magnetically. A standard saturated calomel electrode (SCE) and one of the carbon-Teflon electrodes were immersed in the solution and coupled to a potentiostat capable of controlling the potential of the carbon electrode to ±1.0 millivolt. Then the potential of the carbon electrode was varied and at each selected potential level acetic acid analysis in solution was carried out by titration with 10^{-3} N NaOH. The results are shown in Figure 1.5a. A total adsorption of about 0.15 meq of acetic acid per gram of carbon was reached at about 0.30 volt versus SCE. No adsorption was found when the electrode was held at -0.30 volt versus SCE. By switching from 0.30 volt to -0.30 volt it was determined that the adsorption-desorption cycle was reversible.

The cycle was repeated four times and, within experimental error (1%), there was no degradation of the surface of the carbon electrode as a consequence of repeated adsorption and desorption. It was found that complete adsorption and readsorption was rapid (less than one hour) at potentials between about 0.4 and 0.75 volt versus SCE; it also was found that complete desorption occurred repeatedly and rapidly (less than about one hour) at potentials of -0.60 volt to -0.30 volt but that at a potential above -0.03 volt complete desorption took in excess of 18 hours. Some hysteresis in the potential region between -0.30 and +0.60 volt occured as shown in Figure 1.5a. This hysteresis is believed due to slow surface oxidation state changes in this region, perhaps changing the potential of zero charge.

FIGURE 1.5: ELECTROCHEMICAL CONTROL OF ADSORPTION-DESORPTION WITH ACTIVATED CARBON

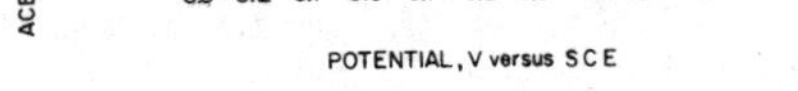

a. ADSORPTION OF ACETIC ACID ON ACTIVATED CARBON AS A FUNCTION OF POTENTIAL

b. ADSORPTION OF AMYL ALCOHOL ON ACTIVATED CARBON AS A FUNCTION OF POTENTIAL

Source: A.C. Makrides and S.B. Brummer; U.S. Patent 3,730,885; May 1, 1973

Example 2: An aqueous solution was prepared which consisted of 500 ppm of sodium chloride and 25 ppm of amyl alcohol. Then this solution was subjected to adsorption-desorption cycles using a carbon-Teflon electrode constructed as described in Example 1. The procedure of conducting the adsorption-desorption cycles was the same as in Example 1 except that the potential was varied between about -0.4 and +1.0 volt. The results are shown in Figure 1.5b. The total adsorption was greater than for acetic acid and was equal to about 20% of a molecule monolayer. Adsorption was maximum in the middle potential region (-0.4 to +0.2 volt) as expected for an uncharged or neutral species. The adsorption process was reversible but about half of the adsorbed alcohol was desorbed. The remainder was not desorbed and this is believed due to chemical bonding between the carbon surface and the alcohol.

Example 3: A carbon-Teflon electrode was prepared according to the procedure of Example 1 but using unregenerated carbon taken from a tertiary sewage treatment plant. The carbon was reported to have a loading of about 10 grams TOC (total oxidizable carbon in adsorbed organic compounds) per 100 grams of carbon. A second carbon-Teflon electrode was made in accordance with Example 1. These virgin and unregenerated carbon electrodes were immersed in separate 10^{-3} M Na_2SO_4 solutions which were vigorously stirred and each electrode was subjected to an applied potential of +0.85 volt versus SCE. No measureable COD was released by the virgin carbon electrode; however, the regenerated carbon electrode released 22.25 mg/l of COD into the solution. Then the regenerated carbon electrode was removed from solution and introduced into a fresh 10^{-2} M Na_2SO_4 solution where further desorption was carried out at -0.3 volt versus SCE. A further 17.44 mg/l of COD material was released, making a total of 39.69 mg/l which was equivalent to 3.97 grams of COD/100 g carbon.

Although adsorption and desorption according to the process can be practiced using thin fuel cell type electrodes as in the foregoing examples, the applications of the process on such scale are limited primarily to the laboratory. For large scale plants directed, for example, to tertiary treatment of sewage or treatment of drinking water, a more practical electrode arrangement is required. One possible configuration is a filter press arrangement. However, a preferred electrode configuration is a static bed similar to what has already been used in tertiary water treatment but with certain modifications which are described in detail.

J.R. Benzaria and C. Zundel; U.S. Patent 3,720,626; March 13, 1973; assigned to Degremont Societe Generale d'Epuration et d'Assainissement SA, France describes an improvement in the process for the regeneration of spent activated carbon combining an aqueous alkaline solution and an organic solvent, such as a lower alcohol.

The spent carbon is first contacted in a column under practically static conditions, and at between 25° and 90°C for 30 minutes to 2 hours with an aqueous alkaline solution, preferably 0.5 to 25% by weight sodium hydroxide. An alkaline treatment prepares the carbon for the elution action of the solvent. The solvent is a lower alcohol, preferably isopropanol combined with water. The solvent is then removed by steam and the carbon is regenerated by an acid treatment. Complete regeneration is obtained.

Other Filtration Techniques

K.S. Arvanitakis; U.S. Patent 3,705,648; December 12, 1972 describes an apparatus for clarifying a contaminated liquid by pumping unclarified liquid through a primary filter increasing the pressure drop across the filter until such pressure reaches a predetermined level. When the pressure reaches a predetermined level, the filtration operation is interrupted, the filtration system closed, and the filter cake accumulated on the primary filters is removed. The contents of the primary filter chamber are drained to a secondary filter chamber or desludger whereat the liquid bearing the dislodged filter cake suspension from the primary filters and additional filter aid medium is recirculated through a secondary filtration system forming a filter cake on the surface thereof and clarifying the contents of the primary filter chamber drawn thereinto.

The liquid in the primary filter chamber, clarified thereby, is pumped through the mixing chamber of a precoat feeder whereat various filter cake forming media are added thereto. The clarified liquid bearing the suspended filter aids is pumped through the primary filter chamber for recirculation through the primary filters to form a new filter cake on the surface thereof. When the new filter cake has been formed the filtration system is automatically placed on line to clarify the liquid passing therethrough.

While the primary filter is clarifying liquid passing through, the secondary filter of the secondary filtering chamber is cleansed of accumulated filter cake which settles to the bottom of the secondary filter chamber. The expended filter cake or sludge is conveyed downwardly by a central auger in the desludger chamber into a horizontal auger whereat the expended cake or sludge is conveyed to a vertical auger conveyer. The vertical auger is heat jacketed to expose the sludge to heat for solidifying the sludge to a desired consistency and expelling it into a container for disposal.

Q.L. Hampton; U.S. Patent 3,717,251; February 20, 1973 describes a method and apparatus for filtering suspended solids from a liquid suspension such as sewage, where the liquid to be filtered is forced upwardly through a filter medium. Upon leaving the filter medium, the filtered liquid flows over a weir and into an effluent trough, from where it is directed to an outlet conduit for suitable disposition. The solids entrapped in the filter medium during the filtering process are removed by a cleaning or washing process, which simply involves increasing the rate at which the liquid is passed through the filter medium.

During washing, the liquid, such as water, is forced through the filter at a rate of flow equal to several times the rate of flow the water moves during filtering, which acts to fluidize the entrapped solids removed from the filter medium, enters the effluent trough located above the filter medium and because the rate of flow is higher than that during the filtering process, is directed into a compartment spaced away from where the filtered water has been directed. The washwater may then be redirected through the treatment plant, or to other suitable disposal. The washing operation of the filter can either be preset to take place at specified time intervals, or can be initiated upon the establishment of a predetermined back pressure in the filter inlet resulting from clogging of the filter by the solids removed from the water being filtered.

T.L. Reece; U.S. Patent 3,692,178; September 19, 1972; assigned to Dover Corporation describes a pressure filtering method and apparatus in which the residual process liquid present in the pressure filter housing at the end of a filtration cycle is discharged from the housing, prior to the backwash cycle, and is sent to a recovery system and is treated so that it can be returned to the main stream of the process liquid.

OTHER SEPARATION TECHNIQUES

Electrostatic Field

According to a process described by *A.S. King; U.S. Patent 3,718,256; February 27, 1973* particles are removed from fluids, such as water, previously subjected to electrostatic fields. When subjected to an initial electrostatic treatment, the fluid undergoes a reduction in surface tension and neutralization of charged, dissolved particles. Subsequent field treatments effect nucleation or coagulation of dissolved impurities and break molecule clusters to aid the nucleation or coagulation action. Suspended and nucleated or coagulated matter is removed from the fluid after the electric field treatments by discharging the fluid against an electrically conductive surface at ground potential maintained at a temperature different from that of the fluid. This causes the suspended and nucleated or coagulated matter to be deposited from the fluid and thus readily permits separation and removal.

In Figure 1.6 the particle separator as shown comprises an upright, tubular, metallic housing **272** upon which a double-acting, pneumatic or hydraulic piston and cylinder assembly **274** is mounted. The piston rod within the assembly is axially aligned with the axis of the housing and is provided with a disc-shaped ram **276** at its lower end.

The housing contains a cylindrical, metallic inner wall member **278** spaced from the outer wall of the housing, a continuous, finned tube **280** of spiral configuration being disposed in the space between the two walls and communicating with an inlet **282** and an outlet **284**. A heating or cooling medium is pumped or flows by gravity into the inlet for flow downwardly through the spiral tube to the outlet. The inner wall member is thus either heated or cooled by the action of the medium flowing in the tube.

FIGURE 1.6: ELECTROSTATIC PARTICLE SEPARATOR

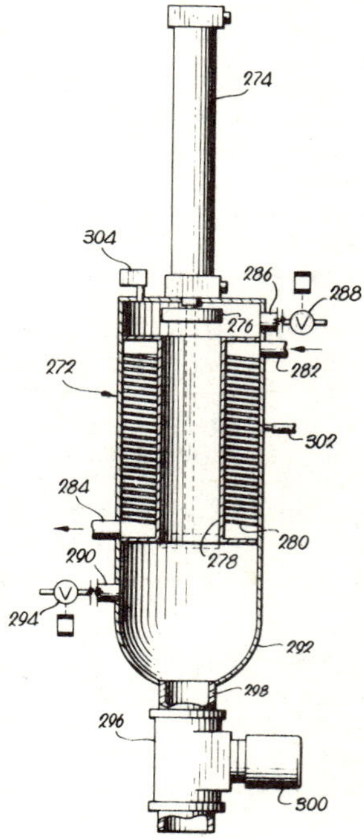

Source: A.S. King; U.S. Patent 3,718,256; February 27, 1973

A second inlet **286** communicates with the upper end of the housing above the inner wall member, flow of the previously treated fluid into the housing through inlet **286** being controlled by a solenoid valve **288**. A second outlet **290** communicates with a catch basin **292** formed by the lower portion of the housing, and is controlled by a solenoid valve **294**. A drain valve **296** communicates with a drain pipe **298** leading from the catch basin and is operated by a solenoid **300**. A ground connection is made to the housing as illustrated at **302**. A pressure switch **304** is sensitive to the pressure of fluid above the cylindrical inner wall for the purpose of activating the piston and the cylinder assembly when the pressure reaches a predetermined level.

During operation, the water containing suspended and nucleated or coagulated matter is directed into the inlet **286** of the particle separator of Figure 1.6, whereupon such water flows downwardly over the electrically conductive inner surface of the metallic wall member and into the catch basin for ultimate flow through outlet **290**. The temperature of the wall member is maintained a few degrees (10° to 15°F, for example) above or below the temperature of the incoming water, and the wall member is maintained at ground potential. It has been found that the matter in the water, due to an enhancement of nucleation or coagulation, is deposited in the catch basin and upon the inner surface of the wall member. This is a result of discharging the water, with attendant pressure drop, against an electrically conductive surface at ground potential maintained at a temperature a few degrees different from that of the water.

The pressure switch may be utilized in conjunction with suitable control circuitry (not shown) to activate the piston and the cylinder assembly when the pressure in the chamber above the wall member reaches a predetermined level. Such pressure buildup is caused by the gradual accumulation of deposited particles on the inner surface of the wall member. Activation of the assembly extends the ram to the broken line position illustrated and swabs the deposit from the internal surface of the wall member, whereupon such deposit falls into the catch basin.

Ultimately, when the accumulation of the deposit in the catch basin becomes excessive, the separator is cleaned by first closing valves **288** and **294** and then opening the drain valve, followed by activation of the assembly to drive the deposit from the catch basin. Sufficient water is trapped within the housing for the cleaning action. Thereafter, the inlet valve is opened to flush the housing; normal operation is resumed by reopening the outlet valve and reclosing the drain valve.

Multistage Process

D.L. Schmutzler and J.H. Schmutzler; U.S. Patent 3,674,145; July 4, 1972; assigned to Jadair, Incorporated describe an apparatus for continuously removing dusts, particles, aggregates and the like from liquids in a rapid, efficient manner to provide liquids clarified to meet projected pollution control requirements.

The process involves passing contaminated liquids into a two-stage chamber or tank in which heavier solids are removed in a first stage and suspended solids are settled out over a period of time and removed from the second stage.

Referring to Figure 1.7, there is shown a clarifying apparatus suitable for treating a variety of contaminated liquids. The apparatus comprises a generally rectangular elongated chamber or tank **10** consisting of sides **11**, rear end **12**, disposal ramp **13** and bottom **14**. A dividing plate **15** separates a first Stage A from second Stage B. The dividing plate may be positioned at any point between the sides, but generally is located in the upper forward portion of the dividing plate providing for the passage of liquid from one stage of the tank to the other. A drainboard **17** is an integral part of the disposal ramp and leads to a discharge station located below the upper edge of the ramp.

Endless chains **18** and **19** are mounted in Stages A and B on sprockets **20** and **21** carried on shafts **22** and **23** as shown in Figures 1.7a and 1.7b. Idlers **24** in Stages A and B hold down the endless chains for positioned movement.

A series of paddles **25** are affixed to the endless chains **18** and **19** respectively. The paddles are carried by the rotating endless chains so that they move forward while in position on the bottom of the tank. Stage A which removes the heavier solids employs paddles spaced at close intervals. Paddles used in Stage B are of the same design but spaced at longer intervals. Figure 1.7c shows a typical paddle. The paddle consists of lateral upper edges **26** connected to center edge **27** and side flanges **28**. Center flange **29** attached to the side flanges and the center edge has drain holes **30** located therein. A continuous hard rubber strip **31** is fastened to the flanges by bolts **33** and a metal backing strip. Ends **32** of the strip are tapered toward the edges of flanges **28**. A series of bolts or rivets connect the paddles to a link **34** forming part of the endless chains.

A series of baffles are positioned at several positions to direct the flow in the tank. Inlet baffle **35** is positioned adjacent to and parallel with the end of the tank. A rear baffle **36** is positioned parallel to the inlet baffle. The distance between the two baffles is approximately twice the distance from baffle **35** to the end but is dependent upon the influent flow rate, velocity and composition. The lower edge of baffle **35** extends below the lower edge of baffle **36**. Inclined baffle **37** is located in the forward portion of Stage B immediately behind the rear edge of the weir. This baffle extends down to close proximity with the bottom. A discharge baffle **38** is positioned at the rear of Stage B. The upper edge extends to a point slightly below the top edges of the sides. In operation, the water level is just above the upper edge of the discharge baffle.

FIGURE 1.7: CLARIFICATION OF LIQUIDS CONTAINING SOLIDS

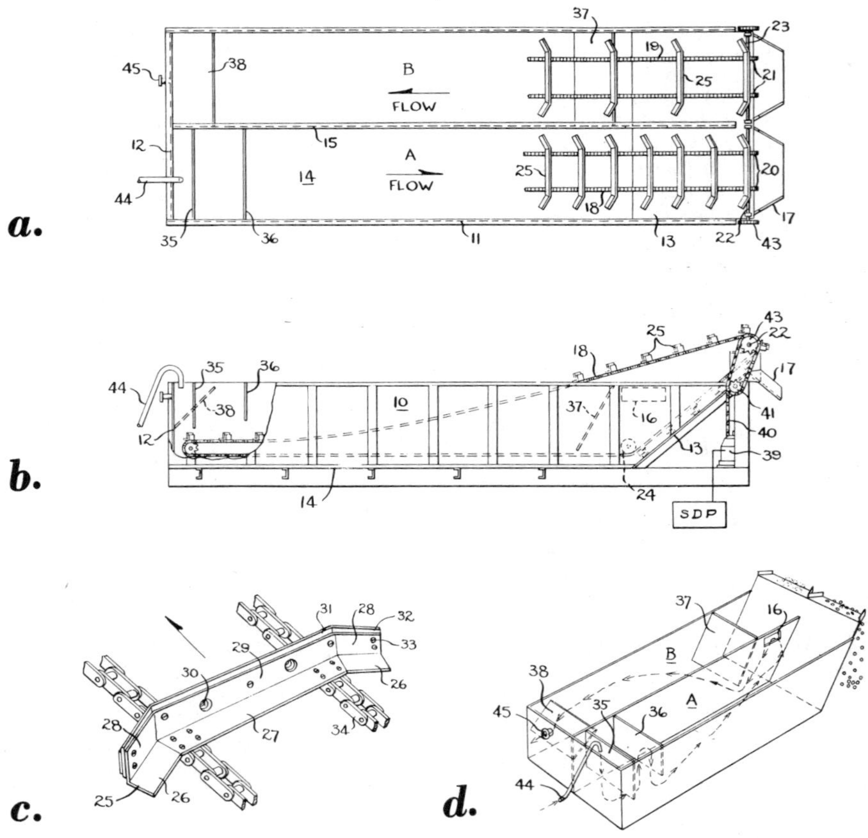

(a) Top View of Clarifying Apparatus
(b) Side View of Clarifying Apparatus
(c) Detail of Paddle
(d) Flow Pattern

Source: D.L. Schmutzler and J.H. Schmutzler; U.S. Patent 3,674,145; July 4, 1972

A drive chain **40** connects drive means **39** with take-off gear **43** at the forward portion of the tank. A connecting chain provides communication between the drive gear **41** and the take-off gear mounted on the shaft. An automatic variable control mounted on the drive means is actuated by a sensing device mounted at the outlet. The speed of the endless chains is varied according to the degree of clarity of the discharge. Programmed automatic controls such as a conventional stepping drum programmer and delay-off timers, designated schematically at **SDP** in Figure 1.7b, vary the periods of the conveyor's various operating velocities and provide for periods of nonmovement of the paddle and chain system.

The contaminated liquid enters through inlet **44** either by means of gravity flow or by forced flow. As shown in Figure 1.7a, the inlet is positioned midway between the side and the dividing plate. Actual flow is determined by the liquid requirements, residence time required and composition of the contaminated liquid. Clarified liquid is taken off either by gravity flow or pumps.

Referring to Figure 1.7d there is shown a schematic, perspective flow diagram. The contaminated liquid enters through the inlet as indicated by the arrow in a turbulent state. This turbulent condition is present in the area between the end and the inlet baffle. Very little settling takes place because of the unsettled state.

The incoming liquid passes under the lower edge of the inlet baffle into the area between baffle **35** and the rear baffle. A portion of the liquid which is still turbulent has higher energy or curl and moves upward until it strikes the rear baffle or mixes with other randomly turbulent liquid. This produces a calming effect due to mutual cancellation by the various turbulent components so that the liquid passing under the rear baffle is quelled and becomes a fluid dynamically controlled mass with a minimum amount of disturbing currents. As the liquid moves forward under the lower edge of the rear baffle, the separation commences with the larger particles rapidly settling to the bottom due to the small initial settling distance.

A natural density stratification begins to develop with the least dense, cleaner liquid rising upward and increasingly contaminated, more dense liquid appearing with increasing distance from the liquid surface. The paddles attached to endless chain **18** move the settled matter along the bottom toward the disposal ramp. Paddle and liquid velocities are less than the sediment transport speed for the particles being settled. Sediment transport speed is defined as the speed or velocity of a flowing stream which will move particles of a certain size, mass and density along the bottom or bed of the stream.

The cleanest liquid in the forward portion of Stage A is skimmed off by the weir and passes into the next stage. In the liquid remaining, the degree of contamination increases from top to bottom. Differential separation occurs in the area adjacent the moving paddles. The paddles are moved forward at a speed which allows maximum removal without producing turbulence which would cause the particles to rise into the contaminated liquid.

The hard rubber edge of the paddles provides a close sweeping action without the abrasiveness of metal on metal. As a paddle moves forward, it crosses the intersection of the disposal ramp with the bottom. There is a tendency for the ends of the lower edge of the rubber strip to drag at the intersection causing the middle of the paddle to rise releasing a portion of the particles. Tapering the end edges of the rubber strip as shown in Figure 1.7c minimizes the lifting action enabling a paddle to retain the contents in front of it until discharge.

As the paddle moves up the discharge ramp, a limited number of particles are washed through the drain holes. Near the upper edge of the discharge ramp the paddle emerges from the liquid and moves up to the lip of the ramp where the load is pushed over the edge to the discharge station below. During the interval from the time the paddle emerges from the liquid until discharge over the edge, excess liquid is released through the drain holes. This minimizes the amount of liquid loss and hence makeup required as well as producing a more solid discharge. The passage of particles through the drain holes at this stage is minimized due to the increased density of the sediment caused by the compacting against paddles.

While the larger particles have been removed in Stage A, the generally smaller particles remain to be separated in Stage B. As the cleaner liquid containing smaller suspended particles passes through the weir into Stage B, it is forced downward by the inclined baffle. Light turbulence which is created in the liquid during passage through the crossover weir dissipates during its residence in the chamber created by the baffle **37**, the side, the disposal ramp and the dividing plate. The partially contaminated liquid is smoothly introduced into the rear portion of Stage B through the opening between the lower edge of the inclined baffle and the bottom. Therefore, the particulate matter has a much smaller initial settling distance. The final clarification process for the liquid in Stage B is essentially identical to that which took place in Stage A. The conveyor velocity in Stage B can be set independent from conveyor **18** through the use of a split drive. The cleanest water appears near the surface at the rear of the unit. The baffle **38** causes this surface water to be skimmed off for discharge through outlet port **45**.

Continuous Split Loop Ion Exchange Resin Contactor

I.R. Higgins and J.D. Ferner; U.S. Patent 3,677,937; July 18, 1972; assigned to Chemical Separations Corporation describe a continuous cyclic process for contacting ion exchange resin with two or more liquids in a plurality of serially connected U-shaped columns. The process includes introducing a first liquid into a first U-shaped column loop whereupon a component is adsorbed on the ion exchange resin. A second liquid is introduced into a second column which contains a portion of the ion exchange resin previously contained in the first serially connected column loop, a portion of the ion exchange resin having adsorbed thereon the component from the first liquid.

The second liquid has a first component exchangeable with the component from the first liquid on the ion exchange resin and a second component combinable with the component from the first liquid which is released from the ion exchange resin during the exchange. A product which is the combination of the second component of the second liquid and the released component of the first liquid is then removed from the second column loop. An ion exchange apparatus for producing potassium phosphate is described in detail.

Separation Using Liquids of Different Densities

A.M. El-Hindi and R.G. Jaeggli; U.S. Patent 3,752,758; August 14, 1973 describe a method of removing solid contaminants from a liquid where the solid particles have a weight-to-buoyancy ratio approaching unity. The contaminated liquid is placed in a container with a significantly heavier or lighter liquid, forming an interface. The two liquids are mutually immiscible and are forced together to be thoroughly interspersed with one another and then allowed to separate again. The solid particles are thereby wetted to some extent by the initially clean liquid and tend to remain therein after the liquids separate, thus being removed from the initially contaminated liquid.

Referring to Figure 1.8, the initially contaminated liquid, e.g., water, is supplied through inlet **10** to tank **12**. A secondary liquid, having a specific gravity significantly lower than that of the contaminated liquid, e.g., kerosene, is admitted to the tank through inlet **14**. The light, initially clean liquid and the heavier, initially dirty liquid form an interface, indicated at **16**, in the tank.

A mixing unit comprising a pair of impellers **18** and **20**, carried on a common shaft **22**, is suitably mounted on the tank. The two impellers are arranged on opposite sides of the liquid interface, substantially equidistant therefrom, and have blades formed to direct liquid toward one another when rotated in the same direction. The shaft is journaled in conventional bearing means at **24** and **26**, and carries pulley **28** on its outer end. One or more belts **30** connect the pulley to the output shaft of electric motor **32**, thereby providing means for rotating the impellers at a desired speed. Conventional control means may be provided for selectively varying the speed of the motor, if desired. The blades of the impellers are preferably connected by rods or tubes **33** which assist in breaking up the liquids to achieve a more thorough mix.

The pumping action of the two impeller blades, one immersed in each of the two liquids, forces the two liquids to be mixed thoroughly, and they then flow outwardly from between the impellers, as indicated by the arrows, and separate again. When the liquids are mixed the particles are at least partially wetted by the lighter liquid, thus tending to assume the weight of the lighter liquid and reducing the weight-to-buoyancy ratio with respect to the heavy liquid. Therefore, once the liquids have been thoroughly mixed the particles will tend to remain in the lighter liquid since they are now light enough to float in the heavy liquid. The latter may then be allowed to drain off, passing under baffle **34** which extends from the top of the tank below the interface, and flowing over wall **36** which extends upwardly from the bottom of the tank to approximately the same height, but slightly below the level of the upper, lighter liquid. It may be desirable in some applications to make the height of the wall adjustable.

FIGURE 1.8: SEPARATION OF SOLIDS FROM LIQUID

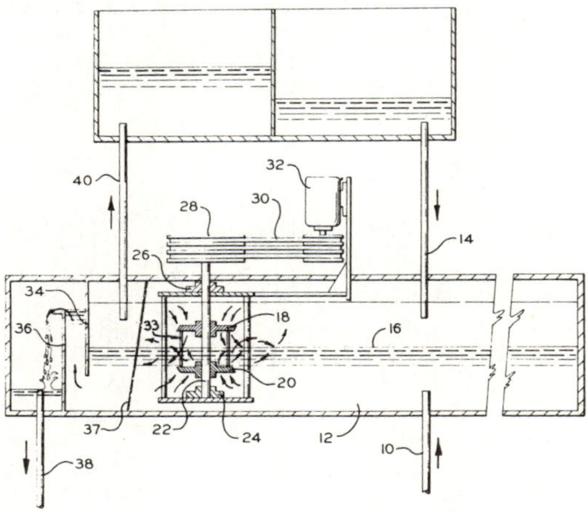

Source: A.M. El-Hindi and R.G. Jaeggli; U.S. Patent 3,752,758; August 14, 1973

A coagulation or coalescing screen, indicated by dotted lines **37**, may be included, if desired. This will assist in returning any lighter liquid which has been mixed into the heavy liquid back to the upper level. The heavy liquid, now essentially free of contaminating particles, is drawn off through outlet **38**, and the lighter liquid which now carries the particles will, of course, tend to settle in the lighter liquid and may therefore be removed by conventional means.

Controlled Rate of Settlement

P. Cahour, J. Decrop, J. Holder, and A. Ledac; U.S. Patent 3,666,419; May 30, 1972; assigned to Commissariat a l'Energie Atomique, France describe a method and device for automatically controlling the rate of settlement of a solid in suspension in a liquid medium, in particular of a precipitate formed in the liquid medium, as a consequence of a continuous chemical reaction.

The process consists mainly in periodically carrying out a cycle of operations which includes the successive steps of drawing off, preferably while effecting agitation, a part of the suspension to be studied and delivering it into a measurement cell, of arresting such delivery when the suspension reaches a set level in the cell, of stopping the agitation process, of measuring the settling time between two given levels in the cell of the interface (referred to as settling face) which forms in the body of the suspension at rest in the cell between the suspended precipitate undergoing settlement and the liquid floating on the surface, and of discharging this suspension from the cell.

Thus, by periodically measuring the rate of settlement of the precipitate, notably in cases of continuous production of a substance by precipitation, it is possible to detect fluctuations in the operation of the reaction, and accordingly, to adjust the different reaction parameters so as to maintain the physico-chemical properties of the precipitate within narrow limits.

The device for performing the above method includes a settlement cell equipped with stirring means and capable of being supplied with a suspension liquid, means for sensing the filling limit that are capable of arresting the supply and of arresting the stirring process when the suspension liquid has reached a set level at the top of the cell, means for detecting the passage of the settling face of the precipitate, within the suspension liquid at rest, to an upper and lower level in the cell, a device operated by the detecting means for measuring the time taken by the settling face to move from the upper level to the lower level, and a control unit controlled by the means for sensing the end of filling and by the detection means to periodically initiate the filling process, and, after the measurements have been effected, empty the cell.

Liquid Membrane Process

N.N. Li; U.S. Patent 3,740,329; June 19, 1973; assigned to Esso Research and Engineering Company has discovered that a dispersed phase can be removed from an external phase by contacting the dispersed phase with surfactant membrane coated droplets, the surfactant membrane coated droplets comprising an interior phase coated with a surfactant membrane. In the process the dispersed phase interacts with the surfactant membrane and is removed from the external phase by the removal of the surfactant membrane coated droplets.

This is a general procedure which can be applied in various diverse areas. The external phase is usually a liquid, although it can also be a gas. The dispersed phase may be a liquid which is nonmiscible with the external phase, a gas or a solid. Various dispersed phase-external phase systems which may be treated by the process include industrial and municipal wastewater streams. All of these streams may contain clay, oil, rust and algae as the dispersed phase. Industrial product streams which contain suspended matter may also be treated by the process.

A typical example of the process is as follows. To 100 ml of a suspension of calcium hydroxide in water (1 gram calcium hydroxide per 100 ml of water) are added 10 grams of an emulsion of 6 grams of normal hexane in 4 grams of a 0.2% saponin, 70% glycerol, and 29.8% water solution. In this system the saponin is the surfactant, the normal hexane is the interior phase, and the continuous phase of the emulsion is glycerol-water. The resultant mixture is stirred for 0.2 minutes at 50 rpm and then allowed to settle. The emulsion rises and is thus easily separated. The remaining water is clear and ready for recycling.

The above process may be carried out at any temperature above which the external phase, when the external phase is immiscible with the continuous phase, or the external phase-continuous phase mixture, when the phases are miscible, is fluid. For convenience, the process is usually carried out at ambient temperature, i.e., 15° to 30°C. It is, of course, critical that the surfactant membrane coated droplets must be stable at the process operating temperature chosen.

Example 1: Removal of Suspended Crude Oil from Separator Effluent — 500 grams of effluent from a petroleum processing separator, which is used to separate oil from petroleum processing wastewater and which contained 86 ppm of a heavy crude oil suspended in an aqueous phase, was mixed with 50 grams of an emulsion which comprised a 3% by weight sodium chloride in water-solution as the interior phase and a continuous phase comprising 2% by weight Span 80, 18% by weight ENJ3029, which is an amine with a molecular weight of about 2,000, and 80% S100N, which is a paraffinic solvent. The weight ratio of the interior phase to the continuous phase was 1:2. The mixture was stirred for 40 minutes and allowed to settle. The emulsion rose to the top taking the suspended hydrocarbons with it and was easily separated. The separator effluent which was now at the bottom was crystal-clear. Infrared analysis showed that the separator effluent contained only 4 ppm of suspended crude oil. The separator effluent was now of sufficient quality to reuse.

Example 2: Removal of Suspended Crude Oil from Separator Effluent – The above procedure was repeated except that the effluent contained 243 ppm of suspended heavy crude oil. The mixture was stirred for 20 minutes at 100 rpm and when the stirring was discontinued and the mixture allowed to settle for 5 minutes the aqueous effluent phase was found to contain only 73 ppm of suspended heavy crude oil. This treated effluent can be recycled to a second stage which may be substantially the same as the first stage for further reduction of suspended oil.

Example 3: Treatment of Wastewater Containing a Solid Dispersed Phase – 500 grams of an aqueous phase which contained 2% by weight of finely dispersed clay and iron rust (50/50 mixture) was mixed with 50 grams of an emulsion having hexane as the interior phase and a solution comprising 0.2% by weight saponin, 70% by weight glycerol and 29.8% by weight water as the continuous phase. The weight ratio of the interior phase to the continuous phase was 1:1. The mixture was stirred at 50 rpm for 0.2 of a minute. When the stirring was terminated the aqueous phase was clear with all suspended matter in the emulsion phase.

Example 4: Removal of a High Molecular Weight Insoluble Polymer from an Aqueous External Phase – 500 grams of a 20% by weight starch-in-water mixture were stirred at 50 rpm for 0.5 minute with a 50 gram emulsion comprising S100N (a paraffinic solvent) as the interior phase and a solution comprising 0.2% by weight saponin, 70% by weight glycerol, 0.5% by weight HCl and 29.3% by weight water with a weight ratio of the interior phase to the continuous phase being 1:1. The starch was concentrated in the emulsion phase and the water phase was clear.

Example 5: Treating a Nonaqueous External Phase – 500 grams of a 1% by weight ferric sulfate suspension in a paraffin oil were mixed with 50 grams of the emulsion described in Example 4. The mixture was stirred for 15 minutes at 100 rpm and allowed to settle after termination of the stirring for 5 minutes. The emulsion settled to the bottom carrying the ferric sulfate with it and leaving behind a sparkling clear oil phase. This example shows that nonaqueous external phases can be treated within the process.

Electromagnetic Field in Radio Frequency Range

J.J. Carnes, J.H. Green and J.H. Heller; U.S. Patent 3,715,305; February 6, 1973; assigned to The New England Institute, Inc. describe a process for increasing the sedimentation rate of heterogeneous colloidal suspensions by subjecting the colloidal suspension to an electromagnetic field in the radio frequency region. An enhanced settling rate is produced without the need for additional electrolytes as frequently utilized in the prior art.

Certain experiments were performed which proved the concept of increasing the sedimentation rate of heterogeneous colloidal suspensions. Colloidal suspensions were put into two glass cuvettes, each measuring 5 cm x 1 cm and compared for percent transmission (%T) at time zero in a Beckman DB spectrophotometer at 600 nanometers. (Light scattering measurements were not possible, and visual, qualitative observations showed more dramatic changes than direct light transmission.) One cuvette was then inserted between two copper plates measuring 5.4 cm x 4.3 cm (23.22 cm^2 area) spaced 1 cm apart.

The output of a RF-generator was applied across these plates and a 5 ohm Koolohm noninductive resistor was put in series with the generator so the RF current flow could be determined by reading the potential drop on a calibrated Boonton RF voltmeter, model 91H. The potential difference across the plates ranging from 100 to 10,000 volts peak-to-peak was read with a calibrated Tektronix 545 oscilloscope and Jennings divider. The

RF generator supplied a pulsed output with pulse widths 5 to 100 microseconds and pulse repetition rates of 4 to 1,000 pulses per second. All equipment was housed in an air-conditioned copper-shielded room.

After exposure in the field for times varying from 1 to 6 minutes, one cuvette was returned to the spectrophotometer and its %T relative to a control cuvette which was not subjected to the RF field was recorded at times ranging from 10 to 60 minutes after the beginning of the experiment. This provided the data on changes in sedimentation rate induced in the particle suspension due to exposure to the radio frequency field.

Stable suspensions representative of biological wastes were prepared from a protohemin (light) suspension containing 35 mg of protohemin in 20 ml distilled water and 3 ml pyridine. Effluents from metal working plants were principally studied and were representative of various stages of pickle liquor treatment before discharge to a river or waterway: raw dump to the settling lagoon, wet tank and clarifier samples. The samples were used directly or, as in the case of the dense raw dump, diluted four times with distilled water.

The composition of the dried sludge from the settling lagoons averages as follows: 13.8% water at 212°F, 2.0% carbonate ($CO_3^=$), 2.4% silica (SiO_2), 15.6% calcium oxide (CaO), 32.5% ferric oxide (Fe_2O_3), 28.6% sulfate ($SO_4^=$), 1.8% copper oxide (CuO), 1.2% zinc oxide (ZnO), and 0.9% alumina (Al_2O_3). The following examples illustrate the process.

Example 1: Results of a typical experiment with inorganic colloids are given in the table below. The wet tank samples from a stage of the pickle liquor treatment were thoroughly shaken up and poured into the control cuvette and the cuvette to be subjected to the RF field. The wet tank sample is a mixture of calcium sulfate and ferrous hydroxide colloidally suspended in water. The cuvettes were arranged in the spectrophotometer so that any change in the irradiated suspensoid leading to an increased sedimentation rate would be shown by a decrease from 100% in transmittance of irradiated sample (%T) with reference to the control over a period of time.

Time, min	%T with Reference to Unirradiated Sample	Δ%T
1	Sample in field	
4	Sample out of field	
5	107	0
6	94	13
6.15	90	17
6.45	85	22
7.45	90	17
8	85	22
9	90	17
16	95	12
18	94	13

The sample was subjected to the RF field for 3 minutes, frequency: 19 MHz, pulse repetition rate: 60 pps, pulse width: 5 microseconds, the potential drop across the 5 ohm resistor was 48 Vrms and the potential across the plates was 6,000 V.

Visual comparison of the irradiated and control samples showed much larger particles in the irradiated sample and these had all settled out within 13 minutes after removal from the RF field. At time 6.45 minutes almost complete settling was observed. The change in Δ%T is not linear with percent sedimentation.

The variation with time of the %T readings reflects three factors in the measurement: first, the light beam goes through the suspension about a third of the way up from the bottom of the cuvette; second, convection stirring caused by the field initially keeps the newly coagulated larger particles in the beam and the initial reading shows an apparent negative effect of the field; then there is a rapid settling, shown by a large change in %T, followed by a return to a value characteristic of dissolved ions absorbing at 600 nm. The negative effect results from the initial agglomeration forming larger particles which causes a decrease in the transmission. When the colloidal suspension is removed from the RF field, this initial negative effect was observed.

Example 2: Results of a typical experiment with a model biological colloid are given in the table below. This colloid or stable suspension is composed of a protohemin (light) suspension containing 35 mg of protohemin in 20 ml distilled water and 3 ml pyridine. This heterogeneous colloidal suspension is prepared by dissolving the protohemin, which is a porphyrin compound, in the pyridine, making a solution and rapidly mixing the solution with water. The colloidal suspension precipitates out as a finely divided colloid. Results are given in the following table and it is seen that the initial sedimentation rate was relatively slow, but that the final sedimentation was greater than that shown in the previous table. (Results in the table below are expressed at %T, reference air, and show an increase if the field exposure has been effective.) RF field treatment of a protohemin suspension, 19 MHz, 3 kv, 100 pps, 50 microseconds 7 Vrms across 5 ohms was measured.

Time, min	%T with Reference to Air*
0	9
1	Sample in field
4	Sample out of field
5	9
2	9
4	9
39	16
60	20

*Increased sedimentation is shown by increase in %T

Solidification of Waste Sludge

C.L. Smith and W.C. Webster; U.S. Patent 3,720,609; March 13, 1973; assigned to G. & W.H. Corson, Inc. describe a method for treating by-product sludges from chemical plants and treating facilities and the product of this treating method. In this method, residual reactive materials, including sulfate ions, aluminum ions, iron ions, calcium ions and magnesium ions or sources of these ions present in various forms in the sludge, are reacted with ion yielding aluminum compounds, lime and soluble sulfates, added when necessary to raise the concentrations of these materials in the sludge above certain minimum levels, to produce a composition which will harden over a period of time, usually on the order of days or weeks. The sludge may therefore be handled in solid form for disposition.

Alternatively, the hardenable composition may be used as a cementitious material either as a filler or base material in some other location or in the place of treatment, e.g., a settling pond. For the latter purpose an aggregate material may be included in the hardenable sludge mixture.

The sludge treatment process depends on the formation of cementitious materials generally known as calcium sulfo-aluminate hydrates or calcium alumino-sulfate hydrates. The basic reactants in this process are calcium ions, aluminum ions, and sulfate ions. Further, iron ions may be substituted for the aluminum ions and magnesium ions for calcium ions. Certain minimum concentrations of these materials must be present and available for reaction to produce the hardenable composition of the process. The dry weight percent of available sulfate ions should be not less than 0.5%; that of aluminum ions, expressed as aluminum oxide, 0.2%; and calcium ions, expressed as calcium oxide, 0.5%.

Test Procedure — Place 7.500 grams of the material under test in a small vial of approximately 14 ml capacity. (The materials which shall be tested are the cementitious nonaggregate portion of the composition.) Add sufficient water to fill the vial, stirring continuously so that a uniform mixture is maintained. Seal the vial with a screw cap and store at 38°C. Open the vial at the age of 7 days and transfer the liquid contents of the vial to a 250 ml beaker and the solid cake to a mortar. Grind the cake with the addition of some water until a uniform slurry is formed and no large lumps remain. Transfer this slurry to the same 250 ml beaker and adjust the volume to 100 ml.

Add 100 ml of HCl((1 to 4.5 dilution ratio to yield an overall final concentration of free HCl equivalent to a dilution volume ratio of 1:10). Allow the beaker and its contents to stand overnight after frequent stirring for the first several hours. Filter the liquid into a 250 ml flask and wash the residue thoroughly with hot water. Allow the liquid to cool and fill the flask to the mark. Aliquots of this solution shall be used for the determination of the acid-soluble constituents.

Available Al^{+3} and Fe^{+3} (reported as weight of molar equivalent amount of Al_2O_3), Ca^{+2} and Mg^{+2} (reported as CaO) and SO_4^{-2} is then determined in the resulting filtrate using atomic absorption spectrograph, x-ray spectrograph, or some other standard qualitative elemental analytical procedure.

The following are several examples in which the process has been used to produce solidified load-bearing products from aqueous chemical waste sludges from industrial chemical plants, or synthetic approximations thereof.

Example 1: Using a synthetic pickle liquor (approximately that which might be discharged by a steel plant) consisting of a water solution containing 10,000 ppm Fe^{++} and 150,000 ppm $SO_4^{=}$ ions, sludges were formed by neutralizing portions of this liquor to approximately pH 7 and in excess of pH 12 with dolomitic monohydrated lime. The two sludges so prepared were treated by addition of varying weight percentages of fly ash (by weight of sludge) as indicated in the table below. The blended materials were placed in segments of glass tubing 2 inches in diameter and 6 inches long which were capped at the bottom end with nylon cloth held in place with a rubber band. The filled test assemblies were placed in 1 inch of deionized water in aquaria.

At varying intervals as indicated in the table these specimens were tested for resistance to penetration (determined as resistance in psi required for penetration of 1 inch into the specimen by a 0.025 square inch rod). Values for penetration resistance of sludge/fly ash mixtures at designated time are below.

Sludge Type	Sludge pH	% Added fly ash	Deionized Water 2 weeks	4 weeks	8 weeks
Dol.	12+	0	0	0	0
Dol	12+	10	0	0	•
Dol.	12+	20	0•	•	•
Dol	12+	50	2200	3200	•
Dol	12+	75	3000	5600	8000+
Dol	12+	100	4400	8000	8000+
Dol.	7	0	0	0	0
Dol.	7	10	0	0	0
Dol.	7	20	0	0	0
Dol.	7	50	0	0	400
Dol.	7	75	0	400	1400
Dol.	7	100	Trace	1400	1840

• Indicates glass tube cracked (usually from expansive forces), no reading possible in most cases.

Example 2: Specimens were prepared as in Example 1 using dolomitic monohydrated lime to neutralize synthetic pickle liquor to approximately pH 7 and in excess of pH 12. These were placed in test assemblies as in Example 1 but were cured in sea water. Penetration resistance values of sludge/fly ash mixtures are given below.

Sludge Type	Sludge pH	% Added Fly Ash	Sea Water 2 weeks	4 weeks	8 weeks
Dol.	12+	0	0	0	0
Dol.	12+	10	0	400	•
Dol.	12+	20	400	2000	1000

(continued)

Sludge Type	Sludge pH	% Added Fly Ash	Sea Water 2 weeks	4 weeks	8 weeks
Dol.	12+	50	3200	2400	8000*
Dol.	12+	75	3600	4800	*
Dol.	12+	100	6200	8000+	*
Dol.	7	0	0	0	0
Dol.	7	10	0	0	0
Dol.	7	20	0	0	0
Dol.	7	50	0	0	440
Dol.	7	75	0	600	2400
Dol.	7	100	600	1400	2800

*Indicates glass tube cracked (usually from expansive forces), no reading possible in most cases.

Example 3: Specimens were prepared as in Example 1 using dolomitic monohydrated lime to neutralize synthetic pickle liquor to approximately pH 7 and in excess of pH 12. These were placed in test assemblies as in Example 1 but were cured in synthetic mine effluent consisting of a water solution containing 300 ppm Fe^{++} and 75 ppm Al^{+++}, adjusted to pH 3 with sulfuric acid. Penetration resistance values of sludge/fly ash mixtures are given as follows.

Sludge Type	Sludge pH	% Added Fly Ash	Mine Water 2 weeks	4 weeks	8 weeks
Dol.	12+	0	0	0	0
Dol.	12+	10	0	0	0
Dol.	12+	20	Trace	*	*
Dol.	12+	50	2400	4800	*
Dol.	12+	75	1200*	*	*
Dol.	12+	100	4800	*	*
Dol.	7	0	0	0	0
Dol.	7	10	0	0	0
Dol.	7	20	0	0	0
Dol.	7	50	0	0	Trace
Dol.	7	75	0	800	800
Dol.	7	100	Trace	1600	1600

*Indicates glass tube cracked (usually from expansive forces), no reading possible in most cases.

Silicon-Containing Aminomethyl Phosphonates

A process described by *D. Redmore and F.T. Welge; U.S. Patent 3,779,912; December 18, 1973; assigned to Petrolite Corporation* involves silicon-containing aminomethyl, or substituted aminomethyl phosphonic acids and derivatives. More particularly this process relates to such compounds containing –SiO– groups and polymers and copolymers. These compounds may be described as being N-methyl phosphonic acid derivatives of the following compositions

$$(RO)_3SiA\overset{\overset{H}{|}}{N}-Z$$

and polymers and copolymers; where R is for example a hydrocarbon group such as alkyl, aryl, etc., but preferably lower alkyl, i.e., methyl, ethyl, etc., and Z is for example hydrogen, hydrocarbon such as alkyl, aryl, etc.

$$-\left(A\overset{\overset{H}{|}}{N}\right)_n-R'$$

where R' is for example hydrogen, a hydrocarbon group such as alkyl, aryl, etc., where n is a number for example 1 to 25 or more, such as 1 to 10, but preferably 1 to 5. A is an alkylene group $(-CH_2-)_m$ where m is for example 1 to 10, such as 2 to 5, but preferably 2 to 3; carboxy containing; hydroxy-containing; ether containing, etc.

The aminomethyl or substituted aminomethyl phosphonic acids of this process may be prepared by reacting a siliconamino compound having at least one reactive N–H group with a carbonyl, such as an aldehyde or a ketone, and a phosphorous acid or a derivative, The following examples illustrate the process.

Example 1: (Trimethoxysilylpropyl) ethylene diamine (55.6 grams; 0.25 mol) was placed in a flask fitted with mechanical stirrer, condenser and thermometer. To this amine was added phosphorous acid (42 grams; 0.5 mol) dissolved in water (50 ml) and hydrochloric acid (50 ml). The mixture was heated to gentle reflux (105° to 110°C) and 40% formaldehyde (38 ml; 0.5 mol) was added dropwise during 1 hour. After the addition the mixture was heated under reflux for 1 hour to complete the reaction. The product is represented by the following formulas:

$$(MeO)_3Si(CH_2)_3NHCH_2CH_2N\left[CH_2\overset{O}{\overset{\|}{P}}(OH)_2\right]_2$$

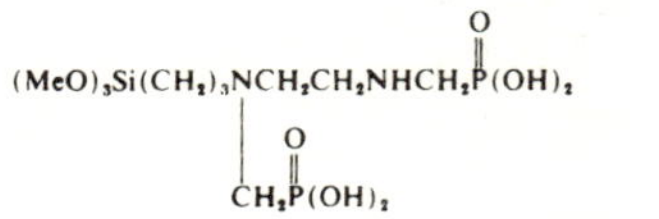

Example 2: Following the procedure of Example 1 (trimethoxysilylpropyl) ethylene diamine (55.6 grams; 0.25 mol) was reacted with 40% formaldehyde (19 ml; 0.25 mol) and phosphorous acid (21 grams; 0.25 mol). The product is represented by the following formula:

$$(MeO)_3Si(CH_2)_3NHCH_2CH_2NHCH_2\overset{O}{\overset{\|}{P}}(OH)_2$$

These phosphonates are useful, for example, as corrosion inhibitors, scale inhibitors and water clarifiers.

Ultrasonic Treatment

C.E. Farman, T.C. Green, H.D. Pepper, and S. Slone; U.S. Patent 3,686,115; August 22, 1972 describe a method and apparatus for the degassing of and continuous processing of liquid in a flow line to deactivate the shear susceptible chemical compounds and/or to kill the microorganisms. Polluted water containing shear susceptible chemicals and aerobic and/or anaerobic microorganisms is purified while in transit, by transferring energy into the moving liquid by means of an ultrasonic generator driving a transducer immersed in the liquid in a unique disposition, the impedance of the generator and immersed transducer per se being matched. A feature of the process is the immersion of the transducer in liquid to be degassed and/or in the column of liquid being processed and its axial disposition focused in the direction of flow, and which affects the entire column of liquid in the flow line.

Continuous Granulating Method

H. Suzuki, H. Arimitsu, and S. Tanaka; U.S. Patent 3,730,887; May 1, 1973; assigned to Ebara Infilco, KK, Japan describe a granulating method and apparatus for separating the suspended solids, in the form of well dehydratable granular solids, from a suspension containing suspended materials such as clays, metal minerals, metal hydroxides or organic materials.

The process comprises maintaining the concentration of the suspended solids in the suspension at a level preferably above 10 g/l, adding a high molecular compound in this suspension and treating it in a treating tank, giving forcible swirling to the solution to form an orientated flow while keeping the suspension concentration in the treating tank at above 30 g/l, shaping the suspended material in the suspension into well dehydratable granular solid masses which are fundamentally different in nature from the flock, and finally separating the masses from the solution.

One of the important features of the process resides in the obstacle-free interior construction of a tank into which the suspension liquid containing suspended material is flowed and separated into solid and liquid portions. The tank is also characterized by providing at its lower part the inlet ports for introducing the suspension solution and the high molecular compound, respectively, an outlet port provided at its upper part for discharging the solution in the tank, and a swirling flow generating means adapted to produce a swirling flow in the tank without forming any turbulent flow. The mixture solution of the suspension and the high molecular compound flows swirlingly in the tank and is discharged from the outlet in the form of slurry.

According to the process, the suspended solids are turned into the granular solid masses which are highly dehydratable, so that, compared with the flock-like mass obtained in the conventional processes, the filtering velocity is almost tripled and the produced solid mass can be easily and efficiently dehydrated by a roll press. Also, the suspended solids in the sludge can be shaped into a well dehydratable granular solid mass and directly extracted from the suspension liquid, so that the sludge may be dehydrated with extremely simple operation and apparatus and consequently an excellent cake is obtained. The following examples illustrate the process.

Example 1: A suspension having a clay-containing solid concentration of 100 g/l was added with polyacrylamide (molecular weight being about 5,000,000) in an amount of 0.5 kg/ton of the suspensoid and was introduced into a cylindrical tank equipped with a paddle-like impeller. A swirling flow was produced in this tank by using the impeller under the conditions that the impeller peripheral velocity is 1.0 m/sec, the apparent RT is 3 minutes and the concentration of granular solids at the granular solid generating section in the tank is 150 g/l. There were obtained the granular solids having an average grain diameter of 3.3 mm (maximum 6 mm). The produced solids were picked out of the liquid with a pincette and their water content was determined. It was 53%.

The granular solid slurry overflowing from the tank was piled on a flat concrete floor. When it was allowed to stand, water was drained off from the slurry mass. Just 5 minutes after suspension of the operation, the top portion of the pile (where the slurry was falling till 5 minutes before) was subjected to analysis to determine water content of the granular solids at that portion. It was 51%. In another test, the granular solid slurry was flowed out of the tank directly onto a plate so that it is piled up and then this pile was subjected to pressing by a roll. The resulting cake had a water content of 43%.

In still another test, the granular solid slurry (about 100 g/l) from the tank was vacuum-dehydrated under a given condition to know its filtering performance. It was found that the filtering rate was about 240 kg/m^2 hr.

With a view to make comparison with the products obtained from a conventional method, 4 pieces of baffle plates were mounted in an ordinary circular agitation tank and it was operated under otherwise the same conditions. As a result, the suspensoid solids were not granulated but formed into flock-like masses. This flock slurry, when poured onto a concrete floor, was not piled up but guttered away like mud. It was impossible to pick the flocks out of water with a pincette or to dehydrate them with a roll press.

In a vacuum dehydration test conducted on this flock slurry, it was found that its filtering rate is about ⅓ (70 kg/m^2 hr) of that attained in the process. The flocks were gathered from the slurry by using a 0.3 mm-mesh screen to determine their water content. It was 70%.

Example 2: A suspension having a clay-containing suspensoid concentration of 20 g/l was added with polyacrylamide (molecular weight being about 5,000,000) in an amount of

0.7 kg/ton of suspensoid, and the same operation as in Example 1 was repeated by using the same apparatus with no baffle plates; peripheral velocity is 0.7 m/sec). Resultant therefrom were the granular solids having an average grain diameter of about 2 mm. Their water content was 54%. The produced slurry could be piled up in a mountainous form. The concentration of granular solids in the tank at that time was 140 g/l. 100 cc of liquid separated from the granular solids was added to 300 cc of raw water and precipitation was effected in a cylinder. The suspensoid precipitated at a rate of more than 100 mm/min, thus remarkably promoting concentration of the raw water.

In another test, the shape of the impeller was deformed such as to hinder accumulation of the suspended solids in the tank, whereby the concentration of granular solids in the tank was reduced below 30 g/l and the solids overflowed from the tank were turned into flock-like masses which were sludgy on the floor.

Electrolysis Procedure Using Aluminum Anode

M. Ichiki and K. Nakade; U.S. Patent 3,816,276; June 11, 1974; assigned to Mitsui Mining & Smelting Co., Ltd., Japan describe a method for obtaining harmless and clean water by separating and eliminating metal ions, oils or suspended solids which are contained in the water.

This method comprises three serial processes. In the first process, water containing metals, oils, suspended solids and other contents is electrolyzed in the first electrolytic cell provided with an anode of aluminum or aluminum alloys while air is blown into the water, so that the contents are changed into compounds insoluble in water, and at the same time the compounds are caused to fully absorb and retain foams of air, or hydrogen gas or oxygen gas generated by the electrolysis so as to bring the specific gravity of the compounds to an extremely low level. In the second process, anionic water-soluble high molecular materials are added to the water obtained from the first process so that the compounds are converted into coagulums which by their structures are apt to ascend and float on the water surface.

Finally, in the third process, the water obtained from the second process is electrolyzed in the second electrolytic cell provided with an insoluble anode, so that all the coagulums are substantially caused to ascend to and float on the water surface, thus obtaining a water solution which scarcely contains harmful contents below the floated coagulums.

Example: Electrolysis was conducted with an electrolyte of industrial wastewater which had a pH of 7.5 containing 22.4 ppm cadmium (including 15.7 ppm in the dissolved state), 14.2 ppm lead, and 150 ppm emulsified oil; an anode of an aluminum alloy containing 5% zinc; a cathode of iron. The electrolysis lasted for 150 minutes during which 133 amp direct current was passed per 1,000 liter of the electrolyte, and air was blown in at a rate of 200 l/min per 1,000 liter of the electrolyte. As a result, generation of water-insoluble compounds was observed in the electrolyte. To this suspension, sodium polyacrylic acid was added at a ratio of 10 grams to 1,000 liter of the suspension, and after mixing and dissolving, the water-insoluble compounds became agglomerated to form coagulums of smaller bulk specific gravity.

Then another electrolysis was conducted, using the suspension containing the agglomerated coagulums as an electrolyte, graphite as an anode, and iron as a cathode, and applying DC current of 50 amp per 1,000 liter of the electrolyte. After 10 minutes electrolysis, almost all the coagulums ascended to the water surface to float there, leaving the solution under them clean. Analysis of this clean solution provided that the solution was a harmless water solution containing 0.00 ppm cadmium, 0.00 ppm lead and 1.2 ppm oil.

OIL-WATER SEPARATION PROCESSES

EQUIPMENT

Cyclone-Flotation Cell

According to a process described by *J.A. Garcia; U.S. Patent 3,784,468; January 8, 1974; assigned to Esso Production Research Company* a mixture of liquids capable of gravity separation, such as oil and water, is charged with gas or aerated in a cyclone. The aerated liquid mixture is then separated in a flotation chamber or cell. The lighter liquid component floats to the surface of the liquid mixture in the flotation cell aided by the rising gas in the liquid mixture. Several cyclone-flotation cells may be connected together in series in which arrangement the heavier separated component mixed with the lighter unseparated component is cycled from a particular flotation cell to the next succeeding cyclone-flotation cell.

As shown in Figure 2.1a, a liquid-liquid separation system indicated at **10** separates liquid components having different specific gravities from each other. A series of flotation chambers, cells or compartments and cyclone or vortex liquid-gas mixers are shown. A flotation compartment **20** has a common wall **21** with another flotation compartment **22**, which in turn has a common wall **23** with another flotation compartment **24**. Compartments **20**, **22** and **24** are covered by a gas-tight top **30**. A vortex or cyclone **40** is mounted above compartment **20** and a liquid delivery tube **41** extends from the upper end of cyclone **40** into compartment **20**.

Another similar cyclone **42** is mounted above compartment **22** and a liquid delivery tube **43** extends from the top of cyclone **42** into compartment **22**. A third cyclone **44** is mounted above compartment **24** and a liquid delivery tube **45** extends from cyclone **44** into compartment **24**. As shown, tubes **41**, **43** and **45** terminate below the surface of the liquid in their respective compartments. Positioned just above the outlet of tube **41** is a perforated distribution plate **46**. Each of the other tubes **43** and **45** is similarly provided with a perforated liquid distribution plate **47** and **48**, respectively.

Gas is supplied to the compartments from a source indicated at **50**. The gas is located between the top **30** of the compartments and the level of the liquid indicated at **80** (Figure 2.1c). The bottom of each cyclone is immersed in a gas layer as shown. Each of the compartments **20**, **22** and **24** overflow into a common trough **52** which extends the full length of the unit and which discharges into a discharge duct or conduit **56** through opening **54** in the end wall of the unit. The upturned lip of trough **52** determines the maximum liquid level in the compartments. Duct **56** extends downwardly into an outer

FIGURE 2.1: GRAVITY SEPARATION USING CYCLONE-FLOTATION CELLS

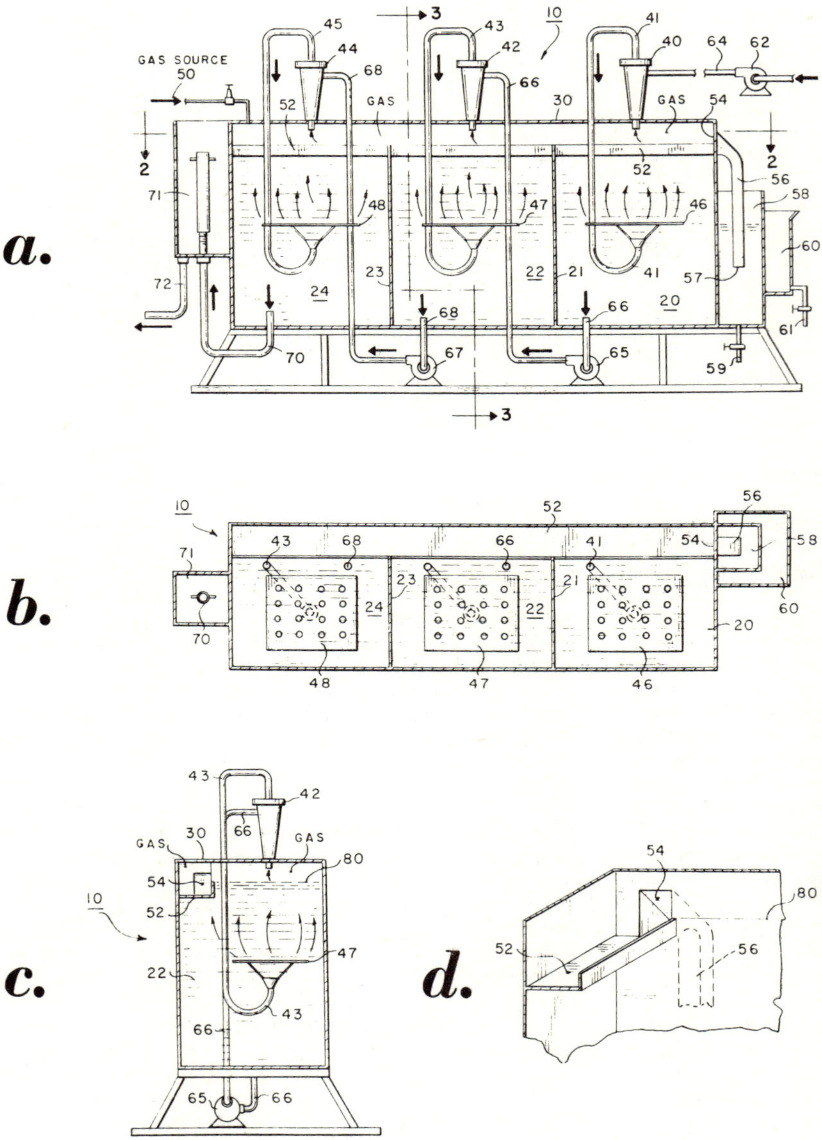

(a) Liquid-Liquid Separator
(b) View Along Lines **2—2** of Figure 2.1a
(c) Sectional View Along Lines **3—3** of Figure 2.1a
(d) Isometric View of Overflow Trough

Source: J.A. Garcia; U.S. Patent 3,784,468; January 8, 1974

compartment **58** with its open end **57** submerged in liquid so as to maintain a gas head above the liquid in the duct and also above the liquid in the compartments **20**, **22** and **24**. Any solids in the overflow into compartment **58** deposit on the bottom of the compartment and may be drawn off at suitable intervals through conduit **59**. As oil fills the compartment **58** it overflows into a collection chamber **60** from where it may be discharged by way of conduit **61** to a sump.

A pump **62** pumps the liquid mixture to be separated through a conduit **64** into cyclone **40**. Another pump **65** located in a conduit **66** pumps liquid from the bottom of compartment **20** into cyclone **42** and still another pump **67** located in a conduit **68** pumps liquid from the bottom of compartment **22** into cyclone **44**. A conduit **70** extends from the bottom of compartment **24** into a small compartment **71**. Conduit **70** is adjustable in height in compartment **71** to establish the liquid level in compartments **20**, **22** and **24**. Clean water is discharged from compartment **71** through line **72**.

Operation: In operation of the system a mixture of liquids and possibly some solids, such as water, oil and some sand, is delivered through conduit **64** to cyclone **40** by means of pump **62**. The bottom of cyclone **40** exists in the gas environment. The vortex formed in the center of cyclone **40** sucks gas up which mixes with the swirling liquid under pressure. The aerated liquid mixture from cyclone **40** is discharged into compartment **20** through tube **41** and distribution plate **46**. The rising gas charged in the liquid mixture in cyclone **40** floats oil and possibly light solids to the surface of the liquid. Heavier solids are separated in the cyclone **40** and dropped from the base and remain suspended by the gaseous froth on the surface of the liquid.

The floated oil and solids overflow into trough **52** and through duct **56** into compartment **58**. When compartment **58** fills the oil overflows into compartment **60**. Any solids carried with the oil into compartment **58** may be discharged through conduit **59**. The heavier separated liquid along with the unseparated lighter liquid component is pumped by means of pump **65** through conduit **66** into cyclone **42** where the mixing and charging with gas is repeated. The liquids charged with gas are discharged through conduit **43** into compartment **22** through distribution plate **47**.

Similarly, the heavier liquid component along with the unseparated lighter liquid component in compartment **22** is pumped by means of pump **67** through conduit **68** into cyclone **44** where the liquids are again charged with gas and are discharged through tube **45** into compartment **24** through distribution plate **48**. The heavier component is then overflowed through conduit **70** into compartment **71** where it is discharged from the system through conduit **72**. The action of the liquids in the cyclone causes the gas to be sucked up the center of the vortex formed by the swirling liquid mixture in the cyclone and beaten into the liquid to form small gas bubbles in the liquid.

The vortex or cyclone separators **40**, **42** and **44** are of conventional design and may be suitably of the type discussed and shown on pp 3932-3935 of the *Composite Catalog of Oil Field Equipment of Services,* 1966-67 ed, published by World Oil. The inner wall of each cyclone is preferably lined with a replaceable sleeve of material such as polyurethane or butyl rubber which is durable. The gas may be methane (natural gas) or any suitable nonoxidizing, inert gas. Where oxidation is not a problem, air may be used.

Vortex Separator

A.E. Mesing; U.S. Patent 3,780,862; December 25, 1973; assigned to United Aircraft Corp. describes a vortex separator for separating a mixture of two or more materials of different densities into two separate components. In particular, this process provides an improved separator for immiscible liquids such as oil and water, and results in a component containing primarily oil and a component of substantially pure water. A specific application of the process is for separating ship's ballast or bilge water into an oil component which may be stored in a limited volume, and a water component which may be dumped overboard without contaminating the ocean water. The separator may be used for other applications such as separating the liquid waste from restaurants or other business establishments into components, for example grease and water, for ease of disposal.

In the process there is provided an improved oil-water separator where an oil-water mixture is injected tangentially into a circumferential chamber and a vortex is generated within the chamber forcing the lighter oil to form a core along the axis of the chamber. The oil is removed from the chamber through a duct in the center of the end wall of the chamber opposite the end at which the oil-water mixture is injected. The water which surrounds the oil core is removed from the chamber through a port in the same end wall in which the oil duct is located, the water removal port being concentric with and of larger radius than the oil removal duct.

In accordance with another aspect of the process, clean or recirculated water is injected in a tangential direction into the circumferential chamber adjacent the end wall from which the oil and water are removed from the chamber. The addition of clean or recirculated water assists in maintaining the vortex in the chamber, and results in improved operation. An annular baffle plate having a circular aperture in the center may be inserted in the vortex chamber parallel to the end walls and located between the end walls. Its function is to reduce the number of small oil particles which reach the end wall and may be exhausted with the water.

Two or more vortex separators may be connected in series if the water exhausted from the first vortex separator contains an undesirably high oil content so that the remaining oil may be fully removed before the water is dumped back into the ocean. If minimum oil storage volume is desired, the vortex separator of the type described in U.S. Patent 3,743,102 may be connected in series with these modified vortex separators, the first vortex separator of the series producing relatively clean water which may be dumped into the ocean, and the second vortex separator being fed the oil output from the first separator to produce substantially pure oil. The water output from the second separator may then be recirculated back into the first separator to remove any oil contained in the water from the second vortex separator. Complete details of the construction of the modified vortex separator are provided.

Alteration of Flow Diameters in Extractor Tube

A.L. Markel; U.S. Patent 3,660,285; May 2, 1972; assigned to Reynolds Sub-Marine Services Corporation describes methods of separating fluids having different densities where the fluids are first separated by passing them through an axial flow pump. Thereafter, the flow diameters of the fluids are altered so that the flow diameter of one of the fluids is substantially equal to the internal diameter of a suitable extractor tube.

In one form of the process, a mandrel having a tapered end is inserted into one of the fluids after it has been separated in a manner to increase its flow diameter to that of the internal diameter of an extractor tube. In another example, an additional fluid, such as air, is injected into the fluids being monitored, thereby increasing the flow diameter of the fluid to be collected to that of the internal diameter of an extractor tube. In all cases it is preferred to use a section of conduit downstream from the axial flow pump that is at least translucent whereby visual inspection permits the extractor tube to be positioned properly for maximum withdrawal efficiency.

Referring to Figure 2.2a, there is shown a fluid separating system, indicated generally at **10**. This fluid separating system **10** has an inlet **12** for receiving fluids of different densities, such as oil and water. These fluids are introduced into a fluid separating means **14** which may by way of example be an axial flow pump such as is disclosed in U.S. Patent 3,595,392. The outlet of such an axial flow is a plurality of substantially concentric streams with a fluid having a lower density being positioned centrally of fluids having greater densities. In a specific example, oil and water are separated in an outlet section **16** in which an extractor tube means, such as a Pitot tube **18** having a constant diameter is also positioned and which is reciprocable axially therein.

FIGURE 2.2: SEPARATION OF FLUIDS OF DIFFERENT DENSITIES

a.

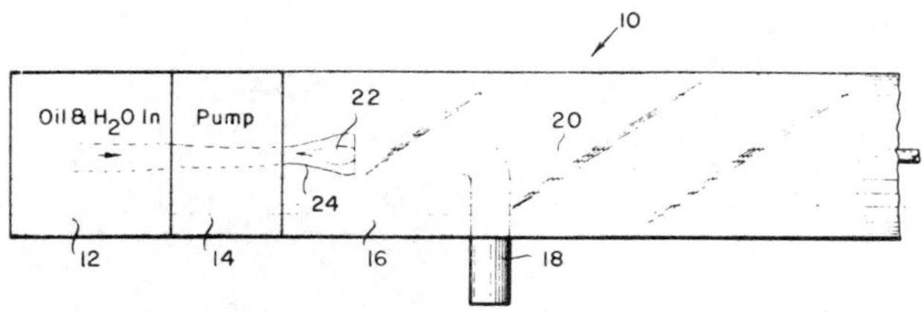

b.

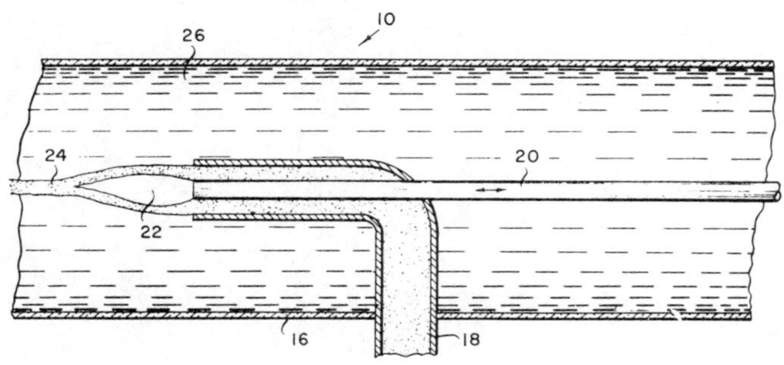

c.

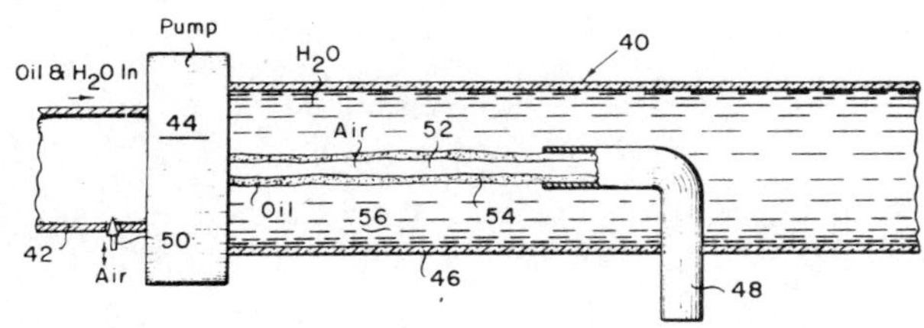

Source: A.L. Markel; U.S. Patent 3,660,285; May 2, 1972

Associated with Pitot tube **18** and extending therethrough is a mandrel means **20** having a tapered, substantially conical end portion **22**. As is indicated in Figure 2.2b, mandrel means **20** is reciprocable axially within tube **18** in order to reach a desired position in the outlet section **16**. Since the material of the outlet section **16** contains at least a portion which is transparent or translucent, and for purposes of this process must be at least translucent, it is possible to select any desired position for the mandrel means **20**. The fluids such as oil **24** and water **26** are readily observable in such an arrangement insofar as their flow diameters are concerned because the oil is customarily much darker.

Thus it is relatively simple to determine when the flow diameter of the oil **24** has been split by means of the tapered, substantially conical end **22** of mandrel means **20**, so that its flow diameter substantially matches the internal diameter of extractor tube means **18**. In this manner, the flow of fluid into extractor tube means **18** is substantially only the fluid which is desired to be collected, namely oil.

Referring to Figure 2.2c, there is illustrated another example of the process where a fluid separating system is indicated generally at **40**. An inlet section is provided at **42** and a fluid separating means, such as an axial flow pump **44**, is employed as discussed previously. The outlet section from the axial flow pump **44** is shown at **46**, and at least a portion is made at least translucent whereby the position of an extractor tube means **48** may be determined with precision.

In this form of the process, it is preferred to introduce another fluid, preferably in the intake section **42** of the fluid separating means **44**. Thus there is indicated a throttle valve **50** through which air is introduced into the inlet **42** of the fluid separating means **44**. The outlet in outlet section **46** is shown to comprise a plurality of substantially concentric streams indicated to be air **52**, oil **54**, and water **56**. In this form of the process, by observing the streams as they flow toward extractor tube means **48**, the amount of air admitted through air throttle valve **50** is adjusted so that the diameter of the oil stream **54** substantially matches the internal diameter of the extractor tube means **48**. Thus only the desired liquid, namely oil, coupled with air is withdrawn by the extractor tube means **48**.

Roller-Doctor Blade Technique

M.H. Van Stavern, H.D. Moorer, G.H. Miller and K.M. Gunn; U.S. Patent 3,685,653; August 22, 1972; assigned to Texaco, Inc. describe a method and apparatus for recovering oil slicks from marine surfaces at a reasonable rate of recovery with a limited pickup of water with the oil. When oil is recovered by a roller which dips down into the layer of oil and carries it up to a doctor blade, the oil tends to be accompanied by a film of water. At higher rates of roller rotation and recovery there is usually greater water contamination. For this and other reasons, it may be beneficial to decrease the quantity of water which is recovered with the oil.

The process involves means and steps for accelerating the breaking of an emulsion, an objective which is of particular significance with today's demand for clarification of increased quantities of water. This is accomplished by picking up on the surface of a rotating roll, a coating of emulsion which is then brought into contact with a supplemental or so-called transfer roll located above and out of contact with the body of emulsion. The transfer roll is spaced closely from the first named roll or pickup roll a critical distance in the range of 1 to 6 mils, effective to cause the minute particles of dispersed oil to coalesce and segregate.

Referring to Figure 2.3, **10** represents the frame of the device which is carried by a pair of floats **12** which may take the form of any suitable marine vessel. Where it is desired to arrange the oil pickup device over a fixed channel, the frame can be simply mounted on fixed supports over the channel. A pickup roller, comprising cylinder **14**, is journaled as at **16** in frame **10**. Roll **14** is so arranged that its lower extremity dips into the water surface **18**, as at **20**. A supplemental or transfer roller comprises cylinder **22**, likewise journaled at **24** on the frame **10** and located in a position substantially above and out of contact with the oil slick **38**.

Pickup troughs **26** and **28** are arranged respectively adjacent the pickup roll **14** and transfer roll **22** to receive film removed by scrapers or doctor blades **30** and **32**, which bear against the respective rollers. The rollers may be rotated in opposite rotational directions by drive motor **34** and suitable belt and gear drive, so that the rollers move, at their point of close juxtaposition, at approximately the same speed and the same direction.

The roller **14** rotates in a counterclockwise direction. This is a good arrangement where the direction of movement of the vessel is toward the left as viewed in the figure, or where a flow of oil is from the left. In this arrangement the rotating surface which has just passed the doctor blade first contacts oil layer **38** and carries the adherent oil downwardly and, if the oil layer is thin, through the water surface. Thereafter it emerges and proceeds up to the bight between rolls **14** and **22**, at which point the surface of roll **22**, rotating in the opposite angular direction, contacts the oil layer.

FIGURE 2.3: SEPARATION PROCESS UTILIZING ROLLER-DOCTOR BLADE PROCESS

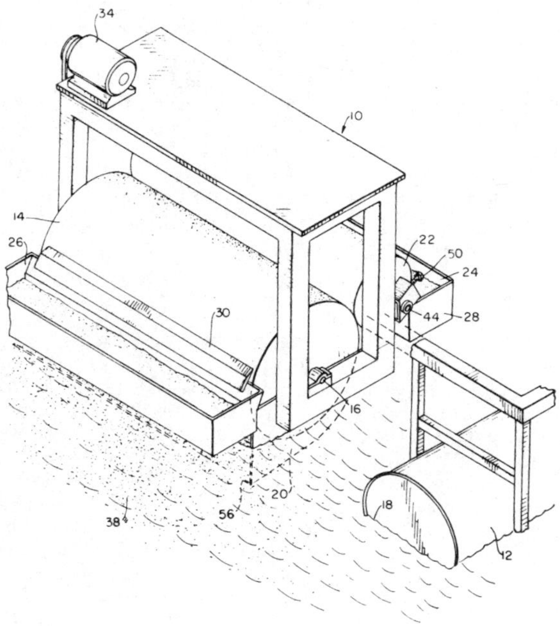

Source: M.H. Van Stavern, H.D. Moorer, G.H. Miller and K.M. Gunn; U.S. Patent 3,685,653; August 22, 1972

By way of exemplifying the process, a series of tests were run using a laboratory scale piece of roller pickup and transfer device identical to that noted in Figure 2.3, but modified to provide a relatively small trough, closely fitted under the pickup or main roll so that the emulsion is continuously fed into contact with the roller until essentially all of the feed stream has been picked up.

One emulsion employed for the tests comprises a 2% water emulsified in a continuous phase of Code 1562 Capell oil, wax free Grade D. The separation properties of the emulsion during quiescent holding show about 72% of the water settled out of the emulsion at 90 hours of standing time. The data derived from the operation of the dual roller pickup and transfer device were as follows, using rollers having a Teflon surface rotated at 80 rpm in each instance.

		Water Collected, cc			
Roll Spacing	No. Cycles	Charge Trough	Product Pan	Accumulated Total	Percent of Water Removed
1 mill	1	8.0	5.0	13.0	66.6
1-1.5 mill	1	13.5	2.1	15.6	80.0
	2	1.2	0.2	17.0	87.3

Apart from the significant effect of this treatment in expediting separation of the soft emulsion, it is to be especially observed that one is here concerned with breaking the emulsion, rather than separating the streams. Stated in another way, the amount of product delivered to trough **26** and **28** is of no particular significance since the material which is discharged downwardly from between the bight of the rollers **14** and **22** is also de-emulsified product. In other words, it is to be noted that the demulsification figures above represent not only product (both oil and water) recovery in troughs **26** and **28** but also the product which has been returned to the charge trough.

This has many beneficial possibilities. First, the troughs **26** and **28** may be omitted so that the entire product is continuously returned to the original reservoir. This may be particularly advantageous and economical in treating the contents of a large settlement basin or reservoir where the separated oil and water proceed to stratify below the upper layer of emulsion. This equipment, arranged in a settling pond of this type, does not require any alteration of the system but simply accelerates the separation. As a result, the treatment capacity is correspondingly increased.

Magnetic Separation Using Ferrofluids

A process described by *R. Kaiser; U.S. Patent 3,796,660; March 12, 1974; assigned to Avco Corporation* is largely based on the existence of a recently discovered class of materials known as ferrofluids. Briefly stated, a ferrofluid is a stable suspension of finely divided magnetic particles in a carrier liquid. Characteristically the suspension or colloidal dispersion does not settle out under the influence of gravity and no interaction between the particles occurs even in a magnetic field. Within the context of the process the finely divided magnetic particles include the materials usually recognized as being ferro or ferrimagnetic, such as magnetite, gamma iron oxide, metallic iron, etc, and more generally any material having a saturation magnetization at room temperature in excess of 1,000 gauss. This class of materials includes also cobalt, nickel, gadolinium, samarium-cobalt and similar alloys, chromium dioxide, ferrites, such as manganese-zinc ferrite, manganese ferrite, nickel ferrite, and many other materials. The preferred materials for the process are magnetite and gamma iron oxide.

The particles suspended in ferrofluids are very small. The particle size is normally less than about 1000 A and can be as small as 25 A. For magnetite, for example, the particle size will average about 50 to 150 A. The range will differ with other materials. It has been found that a surprisingly low differential concentration of magnetic particles in one phase of a mixture makes the mixture amenable to magnetic concentration of that phase, i.e., creaming or actual separation. A single magnetic particle in a 10-micron droplet and a strong magnetic field gradient is adequate to overcome the dispersing effect of Brownian movement on the droplet.

For logistic and economic reasons, it is manifestly desirable to add as little ferrofluid as reasonably possible to whatever mixture is being treated. The minimum quantity of ferrofluid necessary to achieve significant result can be just enough to provide about 0.0001% by volume of magnetic particles in the phase, where the ferrofluid is preferentially soluble. Such a small concentration does, of course, require powerful magnetic forces. As a practical matter, if the expense of incorporating ferrofluid in a particular mixture is justified by the circumstance, then enough should be added for facile separation by the readily available magnetic forces. The general range of magnetic particles contemplated for preferred practice of the process is 0.001 to 0.5% by volume of magnetic particles in the final mixture.

Example 1: A series of oil based ferrofluids were formed by ball milling finely divided magnetite, carrier liquid and stabilizing agent (surfactant). Details of the individual ferrofluids are tabulated on the following page (Tables 1, 2 and 3). With all the ferrofluids described in Table 1, the drop did not spread on distilled water, passing the spreading test, and indicating thereby the presence of very little free surface active material in solution. Essentially all of the surfactant added to stabilize the colloidal suspension is absorbed on the surface of the magnetite particles and not free to interact at the oil/water interface. Ferrofluid 1109 (Run 1-6) contains 11% oleic acid by volume, equivalent to a concentration of 0.35 mol/liter of solution, yet passing the spreading test indicates a concentration of free oleic acid less than 0.01 mol/liter, and probably far less.

TABLE 1: OIL BASED FERROFLUIDS

	Overall composition						Principal physical properties			
									Magnetic properties	
Run	Magnetite size, A.	Volume, percent	Stabilizing agent	Volume, percent	Carrier liquid	Volume, percent	Density, gm./cm.³ (25° C.)	Kinematic viscosity, cs., 30° C.	Initial suscept-ibility	Saturation magneti-zation, gauss
1-1a (1102)	88	1.4	PIBSA	5.3	Kerosene	93	.844	4.1	0.10	43
1-1b (1103)	88	2.6	PIBSA	9.9	do	87	.878	6.0	0.19	80
1-1c (1104)	88	5.7	PIBSA	27.1	do	73	1.01	53	0.40	172
1-2 (1105)	84	6.7	PIBSA	17.1	Methyl cyclohexane	75	1.096	54	0.47	195
1-3 (1106)	84	5.0	PIBSA	12.8	Decahydronaphthalene	82	1.068	32	0.35	150
1-4 (1107)	84	4.6	PIBSA	11.7	Tetradecane	84	0.962	32	0.32	138
1-5 (1108)	85	3.0	PIBA	7.5	Kerosene	90	0.915	6.8	0.18	80
1-6 (1109)	80	4.8	Oleic acid	11.0	Tetradecane	84	0.965	3.6	0.17	131
1-7 (1110)	70	1.1	Dodecyl amine	2.0	Decahydronaphthalene	97	0.897	5.6	.03	28
1-8 (1111)	84	3.4	PIBSA	8.6	Kerosene	88	0.972	5.1	0.24	100

TABLE 2: PROPERTIES OF STABILIZING AGENTS USED TO PREPARE TEST FERROFLUIDS

	Oleic acid (purified)	Polyisobutene acid (PIBA)	Polyisobutene succinic acid (PIBSA)	Dodecyl amine (DA)
Source	Fisher Scientific Co.	Enjay Chemical Co.	Enjay Chemical Co.	Fisher Scientific.
Molecular weight	282	1,000 peak	1,000 peak	185.
Acid number (ASTM D-664) (where applicable)	198	30-45	105-20	N.A.

TABLE 3: DESCRIPTION AND PROPERTIES OF CARRIER LIQUIDS USED TO PREPARE TEST FERROFLUIDS

	Refined kerosene	n-Tetradecane	Methyl cyclohexane	Decahydronaphthalene
Source	Humble Oil & Refining	Eastman Kodak	Fisher Scientific Co.	Fisher Scientific Co.
Category	Mised paraffins	Paraffin	Monocyclic branched naphthene	Bicyclic naphthene.
Density (25° C.) gr./cm.³	0.786	0.759	0.765	0.877.
Viscosity (30° C.) cs	1.9	2.3	1.1	2.1.
Boiling point, ° C.	205-254	254	100	187-195.
Spreading coefficient (of sample on distilled water at room temp.).	S<0	S<0	S<0	S<0.

To demonstrate that the proportion of carrier may be varied greatly without affecting ferrofluid stability, ferrofluids 1102 and 1103 were obtained from ferrofluid 1101 by vacuum distillation removal of carrier liquid. In another instance, a ferrofluid originally formulated with a substantial excess of surfactant was adjusted down in spreading coefficient to the 20 dynes/cm level by contacting the ferrofluid with activated charcoal. This ferrofluid later became ferrofluid 1109.

The above ferrofluids were, of course, formulated so that all exhibited a spreading coefficient less than 20 dynes/cm against distilled water at room temperature. However, determination of whether a candidate ferrofluid meets this criterion can be done by the relatively simple pass-fail test outlined below.

An aliquot of ferrofluid is dissolved in a nonspreading oil of known negative spreading coefficient using a dilution ratio which depends on the coefficient of the oil (20/S.o. where S.o. is the spreading coefficient of the oil). In tests on the above ferrofluids, one volume of ferrofluid was diluted with 12.5 volumes of mixed cis and trans decahydronaphthalene which measured a spreading coefficient of -1.4 dynes/cm against distilled water. A drop of the mixture is placed on water surface in a glass vessel that was carefully cleaned in chromic acid and rinsed with distilled water. If the drop does not spread, the original ferrofluid has a sufficiently low spreading coefficient for the process.

Example 2: Some available ferrofluids were tested by the above-described spreading coefficient test to determine suitability for ballast water separation. None passed. These ferrofluids are described in Table 4 on the following page.

TABLE 4: FERROFLUIDS NOT APPLICABLE

	Examples		
	2-1 (1112)	2-2 (1088)	2-3 (1059)
Magnetite:			
Size, A	88	87	80.
Concentration [1]	12.2	3.6	2.0.
Stabilizing agent	PBISA	Tenlo 70	Oleic acid.
Concentration [1]	46.5	8.7	7.0.
Carrier	Kerosene	Kerosene	Kerosene.
Concentration [1]	40	88	91.
Density, gm./cm.³ (25° C.)	1.27	0.894	0.875.
Kinematic viscosity, cs. (30° C.)	6,800	2.5	1.7.
Initial susceptibility	0.78	0.27	0.07.
Saturation magnetization, gauss.	368	108	55.
Spreading coefficient against distilled water.	Passes test	Does not pass test.	Does not pass test.
Average particle diameter, angstroms.	88	87	80.

[1] Volume percent.

Ferrofluid 1059 differs from ferrofluid 1109 principally in the concentration of free oleic acid. Percolation of ferrofluid 1059 through a column filled with activated charcoal granules removed virtually all of the free oleic acid and the ferrofluid was then employed to prepare ferrofluid 1109.

Ferrofluid 1088 shows that the surfactant must be firmly attached to the magnetic particle surfaces, reversible adsorption is not desirable. The surfactant used in this instance, Tenlo-70, is an oil-soluble, nonionic condensation product of an amino ester of a fatty acid. With this surfactant (and a kerosene carrier) a substantial free surfactant content has been required to maintain a stable dispersion of colloidal magnetite.

Aside from failing the spreading coefficient test, ferrofluid 1088 exhibited a related disability. Specifically, substantial dilution induced flocculation of the magnetite particles. In terms of emulsion treatment, considerable dilution of the ferrofluid (in the hydrocarbon oil, for example) is contemplated and the ferrofluid should be dilutable to almost any degree by a liquid miscible therewith without flocculating.

Ferrofluid 1112 is included to demonstrate that the particle content can be increased to what is an essentially nonusable level as a practical matter without destroying colloid stability or upsetting the spreading coefficient test. Ferrofluid 1112 was prepared by vacuum evaporation of the carrier liquid from ferrofluid 1102.

Example 3: This example demonstrates qualitatively that emulsions can be separated (or at lease one phase concentrated) by application of magnetic forces.

Test 1 — Emulsions of crude oil (mixed sweet crude, vis 4 cp at 30°C, 0.83 g/cm³ 25°C) magnetized with a kerosene based ferrofluid (3.5% magnetite, 10.0% PIBA by volume) were pumped through a pair of beds of 50 to 80-mesh cobalt powder, equal quantities of oil passing through each bed. One bed was held between the poles of a permanent magnet, the second bed being unmagnetized. Initially, the beds were filled with distilled water.

Emulsion Composition A: 99 ml water, 1 ml 10 gauss crude oil (10:1 oil to ferrofluid) and 0.2 ml Tween 80.

Results

Volume effluent, ml.	Percent oil in effluent	
	Magnetized bed	Unmagnetized bed
3	0	0
6	0.001	0.012
9	0.028	0.054
12	0.018	0.110
15	0.052	0.150

Emulsion Composition B: 99 ml water, 1 ml 1.0 gauss crude oil (100:1 oil to ferrofluid) and 0.2 ml Tween 80.

Results

Volume effluent, ml.	Percent oil in effluent	
	Magnetized bed	Unmagnetized bed
3	0	0
6	0.006	0.006
9	0.024	0.003
12	0.040	0.053
15	0.053	0.087
18	0.067	0.116

Emulsion Composition C: 99 ml water, 1 ml 0.1 gauss crude oil, and 0.2 ml Tween 80.

Results

Volume effluent, ml.	Percent oil in effluent	
	Magnetized bed	Unmagnetized bed
3	0	0
6	0	0.010
9	0.031	0.081
12	0.059	0.270
15	0.091	0.33
18	0.128	0.36

Conclusions: (1) The magnetized beds are more efficient demulsifiers than unmagnetized beds. (2) Even at very low magnetization of the oil, the demulsification efficiency is 2 to 3 times as high as it is in the absence of the magnetic field.

Test 2 — Emulsions of fluorocarbon/water were formed using the fluorocarbon ferrofluid 2 described in the *Journal of Applied Physics,* vol 41, no 3, 1964–1072 containing 2.7% magnetite, 6.8% Freon-E-3, HFPO hexamer acid 90.5%. Again, the emulsion was passed through a pair of cobalt powder beds, one between the poles of a permanent magnet (having an air field of 3,000 gauss). Initially, the beds were filled with distilled water.

Emulsion Composition: 98 ml water, 1.6 ml aqua ammonia, 1.0 ml 83 gauss fluorocarbon, and 0.2 ml HFPO acid.

Results

Volume effluent, ml.	Percent oil in effluent	
	Magnetized bed	Unmagnetized bed
3		
6		
9		
12	0	0.018
15	0.0022	0.050
18	0.0030	0.099

Conclusions: The magnetic fields aid the coalescence of a magnetized fluorocarbon in water emulsion.

Test 3 — This test illustrates separation of an emulsion containing two internal phases, one internal phase being magnetized.

Procedure: An emulsion of the crude oil in water was mixed with some of the Test 2 emulsion of magnetized fluorocarbon in water. The mixed emulsion was passed through two cobalt powder beds. One bed was held between the pole pieces of a magnet having an air field of 3,000 gauss. The second bed was not magnetized. Both beds were filled with water prior to introducing the emulsion.

Emulsion Composition A: 99 ml water, 0.4 ml Tween 80 and 1 ml crude oil.

Emulsion Composition B: 99 ml water, 0.4 ml Tween 80 and 1 ml 8.3 gauss fluorocarbon.

Results

Volume effluent, ml.	Oil from magnetized bed/oil from unmagnetized bed	Fluorocarbon from magnetized bed/ fluorocarbon from unmagnetized bed
3	----	----
6	----	----
9	0.42	0.23
12	0.64	0.00
15	0.75	0.24
18	0.94	----

Conclusions: The magnetized droplets are removed much more effectively than the nonmagnetized are, in the magnetized bed relative to the nonmagnetized bed.

Ultraviolet Radiation and Magnetic Separation of Ions

A process described by *R.R. Lucero; U.S. Patent 3,767,545; October 23, 1973; assigned to Interface Development Co., Inc.* relates to the separation of ions from gases and liquids including seawater, petroleum hydrocarbons, molten metals, and gases discharged into the atmosphere.

It is known that hydrocarbons, and other substances including water and gases, if subjected to an ionizing potential so that they are ionized or broken into charged submolecular fragments, then directed through a magnetic field, will separate according to mass. Each species of molecule under this influence produces a unique mass-distribution pattern.

With the advent of cryogenics and thus superconductivity, it is now possible to design magnets of extremely high field strength for continuous operation. This makes possible the use of the above-mentioned principle on a larger scale than ever before. The use of this method of separation to replace the distillation method of fractionation offers improvements such as economy of energy, lower maintenance cost, safety of operation, simplicity of operation, and an improved fractionation.

The process utilizes the principle that, if the material to be separated is first acted upon by electrical potential so that it, the material, is either ionized or broken into charged submolecular fragments and then directed through a magnetic field, each species of molecule, i.e., each component will manifest a unique distribution pattern in the magnetic field, depending on its mass. The process would then draw off these components or fractions by vents placed according to the distribution pattern. Therefore, an improvement in fractionation itself would be achieved by the use of molecular separation, since it is a finer and more discriminate separation than boiling range.

It is known that the conductivity of certain hydrocarbons can be increased up to 10^8 by exposure to intense ultraviolet radiation. This is considered as useful to this process to aid in ionization. Also laser application of radiation is considered as an ionization aid.

A solenoid-type magnet is believed to be of best use to the process, since its magnetic flux lines arrange in concentric formation through the bore of the magnet; thus arranging concentric separation rings from which the fractions of material could be drawn off as a continuous flow by the properly arranged flow path. Also, a spinning magnet and/or spinning magnetic flux would cause the magnetic flux lines to pass through more material in a given amount of time as well as add a slight centrifugal force to the remainder of the material. A rotation of magnetic flux alone can be achieved by an arrangement of switches.

In the process for removing like substances from a fluid there is radiated into the fluid electromagnetic energy of wavelengths substantially within the spectrum of ultraviolet and up to the visible range to make like ions of material in the fluid more susceptible to a magnetic field. A vortex motion is induced in the fluid to recycle the ionized particles. The radiated fluid, while undergoing a vortex motion, is passed through a substantially cylindrical magnetic field so that like ionized particles converge at a focal point within the magnetic field. Fluid is then removed at the focal point of the ions whereby fluid of higher ion concentration is separated from fluid of lower ion concentration. The process may be carried out by an apparatus which comprises means for establishing a substantially cylindrical magnetic field such as that established by a solenoid magnet and means for flowing a fluid through the magnetic field. There are means for radiating the fluid within the cylindrical magnetic field with energy of wavelengths substantially within the spectrum of ultraviolet and up to the visible range to make like ions in the fluid more susceptible to a magnetic field. Means are provided for inducing a vortex motion in the radiated fluid whereby the ions converge at a focal point within the magnetic field. At the focal point there is provided means for removing fluid having a higher ion concentration.

The fluid may be flowed through an enclosure and a cylindrical magnet may be positioned around the enclosure or within the enclosure to establish the cylindrical magnetic field therein. The fluid may also be in the form of gases which are passed through a tubular member, such as a stack or exhaust, with the stack containing an enlarged chamber in which the fluid is moved in a vortex motion through a magnetic field while being radiated with electromagnetic energy.

Three-Dimensional Anolyte Stream

A process described by *H.T. Anderson; U.S. Patent 3,816,274; June 11, 1974; assigned to Swift & Company* involves the treatment of effluent having oil-water emulsion and comprises impressing direct electrical current through the effluent in two or more zones so as to create a three-dimensional anolyte stream whereby the emulsion breaks allowing the fat to float to the surface where it is separated from the clarified effluent.

The process concerns the separation of greasy matter from aqueous liquid found in water-oil systems by forming a surface layer of greasy material which can be removed. The process comprises the steps of conveying wastewater containing fat-water emulsion to a first zone of a substantially rectangular container possessing a plurality of essentially equally spaced apart anodes so as to give the emulsion a rapid pH change, conveying the solution to a second zone having rows of horizontal, parallel, essentially equally spaced apart anodes so as to form a three-dimensional anolyte stream whereby the fat particles separate cleanly from the water, float to the surface and are subsequently separated from the wastewater.

In order to carry out the process, apparatus for treating water-oil emulsions has been devised. Such an apparatus comprises a substantially rectangular-shaped container having an entry conduit and an exit conduit and having a first anolyte stream zone at the entry end and a second anolyte stream zone at the discharge end of the container. In the first anolyte zone are rows of a plurality of spaced apart anodes transverse the longitudinal axis of the container while the second anolyte zone has rows of a plurality of anodes spaced apart and substantially parallel to the longitudinal axis of the container. The longitudinal anodes are positioned so as to be immersed in effluent and cause a three-dimensional anolyte stream.

The following example is set forth using a two-dimensional anolyte stream.

Example 1: A steel tank about 3 feet wide, 6 feet long, and filled to a liquid level of 5 feet 6 inches was used as a test cell. The wastewater consisted of packinghouse wastewater having fat, proteins and some cellulose. The test unit had a baffle across the width of the tank about 26 inches from the inlet end. The baffle, perpendicular to the floor of the tank and parallel to the inlet end, was 28 inches high. It also contained an 8-inch upper section which was inclined at a 45° angle toward the discharge end.

Two anodes were suspended in the wastewater by means of insulated support members. Two anodes were placed in a transverse alignment and were in a vertical plane parallel to each other and also parallel to the wall of the inlet section as well as the baffle. Two anodes were installed parallel to each other and to side walls in a 45° diagonal position. The upper end of the anodes was approximately 6 inches from the discharge end. The lower end of the anodes was about 6 inches from the bottom of the baffle. The anodes were approximately 11 inches from side walls and 5 feet in length. Two pairs of additional anodes parallel to each other, tank walls and surface were also installed approximately 11 inches from the side walls. The upper pair of anodes was 8 inches below liquid level. The lower pair cleared the baffles separating the inlet and discharge end. Discharged effluent entered the water box through a 2 inch wide opening across the width of the tank. Discharged effluent from the cell entered the water box which extended across the width of the flotation unit. Effluent discharge from this unit was of poor quality the first day. Some improvement was produced the third day.

Current requirements were fairly high. Using a rectified DC current to energize the anodes, a rectifier setting was at 14 volts causing 48 amperes of current to flow. Current flow dropped slightly the second day and on the third day current flow was at 42 amperes. Calcial deposits on the tank walls were rather heavy after the third day. These deposits existed on the tank walls parallel to the anodes. Heavy deposits of sludge adhered to the underside of the inclined upper portion of the baffle. Floc settled in the effluent water discharge box. Considerable floc remained transit and was carried by a final effluent discharge.

The diagonal anodes were removed and replaced with pairs of parallel anodes parallel to the tank bottom and to the upper anodes along the longitudinal axis of the tank for use in the next example.

Example 2: Three pairs of parallel anodes were installed at three different levels in a tank of the same size as recited in Example 1. The lower pair of anodes was spaced 14 inches from the floor. A second pair of transverse anodes was mounted 28 inches above the floor level and a third pair of transverse anodes was placed so as to be 3 inches below liquid level. Using a rectified DC current to energize the anodes, a setting of 10 volts was used. The anodes were positioned such that at any point between adjacent anodes, the lowest potential was between about 0.85 and 3 volts.

Wastewater containing emulsions of fat and water was flowed through the system. Approximately 80 to 90% of the floc rose to the surface in the area directly above the space between the inlet end and the baffle. It was possible to see clearly most of the anodes in the discharge zone. Additional floc rose to the surface in the discharge zone and all floc was skimmed off.

ABSORBENTS AND FILTRATION

Polypropylene Oil Mop Structure

H.M. Rhodes; U.S. Patent 3,794,583; February 26, 1974; assigned to Oil Mop, Inc. describes a method and apparatus for separating oil from an oil and water mixture by first subjecting the mixture to passage through a chamber containing an oil mop structure made from fibrillated strips of polypropylene secured to a polypropylene line so that the oil will be attracted by the polypropylene strips and the water will pass through the mop structure. The mixture is then subjected to passage through a fabric membrane having a 2% fluorocarbon solution impregnating the fabric which will permit passage of the water through the membrane rejecting the oil and permitting the water to pass on and be drawn off separately from the oil which may likewise be drawn separately from the chamber.

Referring to Figures 2.4a and 2.4b, **10** designates a cylindrical container having a concentric tube **11** and a second concentric tube **12**. Located within the tube **11** is an oil mop structure **13** similar to the oil mop structure shown in U.S. Patent 3,668,118. At the upper portion of the container **10** is a helical serpentine winding of oil mop structure **14** which is retained by foraminous plates **15** and **16** and is provided with an oil outlet **17**. The second concentric tube **12** defines between the container **10** and the concentric tube **12** a second chamber **17** which is defined with a fabric membrane **18** which has been impregnated with a 2% fluorocarbon solution which will pass water but reject oil.

FIGURE 2.4: OIL AND WATER SEPARATION PROCESS USING POLYPROPYLENE MOP

a.

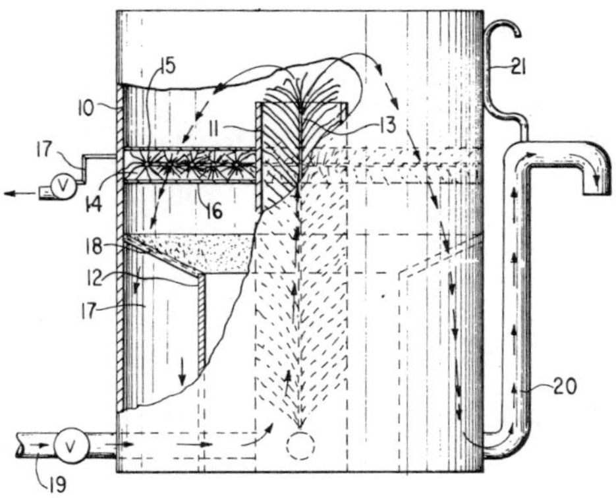

b.

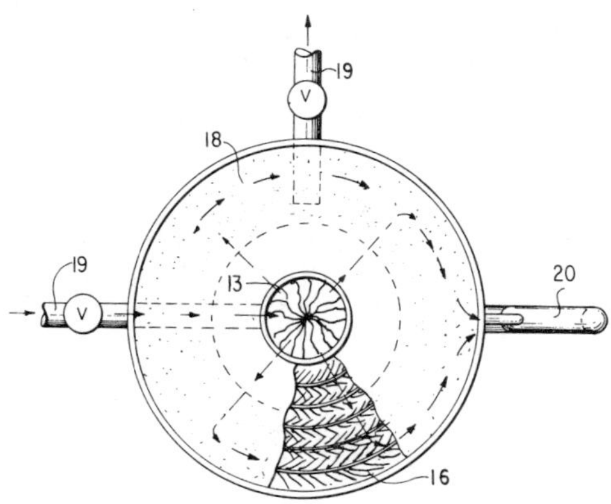

(a) Side View of Apparatus
(b) Top View of Apparatus

Source: H.M. Rhodes; U.S. Patent 3,794,583; February 26, 1974

The inlet or supply side of the container **10** has an inlet pipe **19** controlled by a valve for supplying a mixture of oil and water to the inner concentric tube **11** through which oil and water will pass upwardly through the mop structure **13** and over into the upper portion of the upper chamber across the foraminous plates **15** and **16** and over the helically wound oil mop structure **14** where a major portion of the oil in the comixture will be trapped. Thereafter the water will pass through the mop structure **14** and come into contact with the fabric membrane **18** and oil will be separated and the water passed into the lower chamber where it will be drawn off through pipe arrangement **20** having an air vent **21**.

In related work *H.M. Rhodes; U.S. Patent 3,689,407; September 5, 1972; assigned to Oil Mop, Inc.* describes a method and apparatus for separating oil from a mixture of oil and water by pouring the mixture over two oil-attracting media (polypropylene mops), each having a different physical characteristic and placing the second oil-attracting media between two floats having different specific gravities so that the second oil-attracting media is wrung out due to its compression between the two floats to release the oil and to permit it to rise above the water level in the apparatus. Thus the oil may be drawn off separately from the water.

Polypropylene Webs

A process described by *W.E. Sohl; U.S. Patent 3,764,527; October 9, 1973; assigned to Minnesota Mining and Manufacturing Company* relates to separating oils from mixtures of oil in an aqueous medium by means of contacting the mixture with certain oleophilic fibrous webs comprising large numbers of interconnected, interstitial spaces for reception and retention of the oil to the exclusion of the water.

Generally speaking, fiber-forming materials which are effective as fibrous webs in carrying out the process are those which are preferentially wetted by the oil of an oil-water mixture to the extent that water will be rejected from the internal voids of the web, while the oil will be retained. The operable fiber-forming materials may be defined broadly as those materials (normally polymeric) which when immersed in an oil-water mixture, allow the oil to displace water from a smooth surface and to form a contact angle with the surface of less than 90°. Generally, the effectiveness of the fiber-forming material has an inverse relationship to the size of the above-described contact angle. Thus, polypropylene and polyethylene surfaces allow oil to assume a contact angle of substantially zero degrees and therefore fibers of these polymers provide the preferred webs for the process.

Webs may be formed employing any one of several processes known in the art. Particularly useful webs may be formed according to the methods of U.S. Patent 2,571,457 to Ladisch, and U.S. Patent 3,231,639 to Mabru. The following examples illustrate the process.

Example 1: 18.95 g of mineral oil was mixed with approximately 1,500 ml of water at room temperature in a pyrex dish having a depth of 100 mm and a diameter of 190 mm. The oil had a viscosity of 11.71 to 14.38 cs at 100°F. A web of polypropylene microfibers measuring 4" x 4" and weighing 1.39 g was then placed on the surface of the oil-water mixture for 1.5 minutes, after which it was noted to be saturated with oil. The saturated web was then removed from the mixture and the amount of oil contained therein was calculated by weighing. The amount of isolated oil was noted to be 18.0 grams or 95% of the original amount. The remaining oil in the mixture was substantially totally removed with a second clean web. The 18.0 g of oil removed in the first operation represented a 1,295% removal of oil based on the weight of the fibers.

Example 2: 200 ml of mineral oil identical to that used in Example 1 was poured onto a surface of approximately 1,500 ml of water at room temperature. A web of nylon microfibers measuring 4" x 4" and weighing 0.60 g was placed on the surface of the mixture. The web was observed to have been saturated within approximately one second. Upon removal of weighing, the web was observed to contain a 6.88 g of oil, or 1145% based on the weight of the dry web.

Coal and Polyethylene Matrix

C.O. Bunn; U.S. Patent 3,798,158; March 19, 1974; assigned to Col-Mont Corporation describes a method for the removal of oil and other organic contaminants from polluted effluent, particularly oil refinery effluent. The oil is sorbed by carbonaceous material either in the form of –200 mesh or finer carbon particles in the form of coal, fly ash or activated carbon, or a porous matrix comprised of such finely divided carbon particles and a powdered plastic such as polyethylene, or a combination of powdered carbon particles and matrix. The sorbent is stripped of the oil, thereby conditioning the same for reuse in the system, or withdrawal from the system, in the case of matrix formation, for use independent of the system.

It has been found that the sorption material sorbs substantially all the oil impurities including oil emulsions, the treatment of which has plagued the industry for years. By properly controlling the system through the provision of predetermined quantities of coal, based on the estimated or known amounts of oil contamination in the effluent, and the proper mixing of the oil with the coal to assure intimate contact of all of the oil with the coal, substantially all of the oil can be removed from the effluent.

An experiment which illustrates the reduced retention time of the hybrid system is described below.

A dosage of 3 g per 200 ml of –200 mesh sub-bituminous coal was added to refinery sewage effluent containing emulsified oil in the amount of 15 to 30 ppm. To this was added 3g/200 ml of 4/10 mesh coal matrix, prepared from polyethylene and –200 mesh sub-bituminous coal. The mixture was shaken vigorously 10 minutes, followed by gravity filtration in a conventional manner and without vacuum. The filtrate was substantially purified and without oil odor. There were no difficulties in the gravity filtration of the sorbent material from the liquid.

The filter cake was loose and the coal powder did not separate from a tight hard cake on the filter paper. The filter cake was readily washed from the filter paper. It was apparent that the powdered coal agglomerated substantially with the coal matrix. It was also apparent that the coal powder facilitated the intimate intermixing of coal and the emulsified oil, and that the coal matrix facilitated the filtration of the enriched coal powder as well as assisting the sorption process.

This particular water contained 20 ppm total organics as an emulsion consisting of a mixture of highly oxygenated aliphatic and aromatic hydrocarbon compounds. The oxygenated compounds were both carboxylic acid and ester types. It was likely from the spectrum that a complexity of individual molecules made up the organic portion, not just one or two or three individuals.

After treatment with the combined coal powder and coal matrix as described above, the original 20 ppm total organics had been reduced to 3 ppm. The organic material consisted of highly oxygenated aliphatic hydrocarbons without the apparent presence of any aromatics which were present in the untreated sample. Also, the location of the carbonyl absorption band near 5.78 microns indicates that the oxygenated material was mostly esters rather than a mixture of esters and acids, as was the case for the untreated sample. The absence of bands near 3.3, 6.23, and 6.65 microns in the treated sample and their presence in the untreated sample indicate the total sorption of the aromatic hydrocarbons. There is strong evidence that the organic acids also were sorbed.

Compost

According to a process described by *J.P. Harnett; U.S. Patent 3,771,653; November 13, 1973; assigned to International Minerals & Chemical Corporation* an oil film is removed from the surface of water by contacting the same with a compost prepared by the aerobic digestion of organic waste material. The compost with the absorbed oil floats on the water and can be easily collected to recover the oil. The oil is then separated from the compost, which may then again be used for recovering more oil. Alternatively, the compost is weighted to such a degree that it will sink to the bottom of the water after absorbing oil from the layer of oil on top of the water. The following example illustrates the process.

Example: A quantity of 33° API crude oil was added to water to form a film. Compost obtained under the tradename of Alive was added to the water while it was being agitated to simulate wave action. The compost completely absorbed the oil, picking up about five times its weight in oil. The compost containing the absorbed oil was removed from the oil and finally squeezed at 1,000 psi to effect a 90% recovery of the absorbed oil. A briquette which may be used as fuel remained after the oil was removed by the compression.

Gypsum and Lime

A process described by *C.J. Carmichael and A.G. Franklin; U.S. Patent 3,781,201; Dec. 25, 1973* is directed to separating oil from a mixture of oil and wastewater on offshore rigs where the water cannot merely be allowed to wash overboard containing oil which pollutes the environment and endangers marine life. The mixture of oil and water is first passed through a primary skimmer tank where oil that quickly separates from water is skimmed off into a blow case and water from the bottom of the skimmer tank is fed to a mixing tank where gypsum and lime are added. The gypsum-oil and excess gypsum and other pollutants are mechanically separated from the water.

Pumice Stones

J. Cincotta; U.S. Patent 3,752,762; August 14, 1973 describes apparatus for using pumice stones for removing oil from oil-contaminated water and for enabling saturated pumice stones to be cleaned so that they can be reused.

According to the process, the pumice stones can be placed in a container having an open top, a removable bottom, and porous opposed first and second side walls. The pumice stones are placed in the container through the top and oil-contaminated water, such as from an oil spill, etc, is directed through the first porous side wall, through the container and out the second side wall. The pumice stones in the container absorb the oil from the water, and when they become saturated the bottom of the container is removed allowing the pumice stones to drop on a conveyor. The conveyor transports the pumice stones to a steam-cleaning area which removes the oil. The pumice stones are then recycled back to the container for further use.

The process can be used industrially for removing oil from water that is used for cooling and other purposes so that the water can be recycled as required.

Wax-Filled Bed

According to a process described by *J. Yamamoto, K. Minakawa, H. Nishikado and S. Imon; U.S. Patent 3,770,628; November 6, 1973; assigned to Nippon Oil Company, Limited, Japan* oil-containing contaminated drainage can be effectively clarified by passing the same through a wax-filled bed. In the preferred case, the treatment of oil-containing contaminated drainage is carried out by the use of combination process selected from: (a) a wax-filled bed, an active carbon bed process; (b) a sandfilter, a wax-filled bed process; and (c) a sandfilter, a wax-filled bed and an active carbon bed process.

The following examples illustrate the process.

Examples 1 and 2: Removal of the oil content by absorption was conducted on drainage containing oil at a concentration of 12 ppm by passing the same through a vessel filled with globular wax particles (MP 60°C and oil content 0.4%) 1 mm in diameter and pelleted wax particles (MP 55°C and oil content 0.4%) 3 mm x 5 mm in diameter at ordinary temperature. The results are shown in the following table in comparison with those in the case where a commercially available oil absorbent was employed.

	Oil content in the drainage—			
	Before treatment (p.p.m.)	After treatment (p.p.m.)[1]	Ratio of removal	L.H.S.V. (liter hr.)[2]
Example:				
1	12	0.6	95	7.9
2	12	0.8	93	7.9
Reference example:				
1[3]	12	10.1	16	7.9
2[4]	12	2.5	79	7.9
3[5]	12	8.6	29	7.9

[1] Oil content in the aqueous layer after separation by settlement.
[2] 1 m^3 of the drainage supplied per m^3 of the wax per hour.
[3] Commercially available product composed mainly of fibrous polyurethane.
[4] Commercially available product composed mainly of fibrous polypropylene.
[5] Commercially available product composed mainly of fibrous polyolefins.

As is evidently seen from these results, the method is highly effective for removal of the oil in oil-containing drainage.

Example 3: In a vertical drum 1 m in diameter and 2 m in height was placed globular microwax 1 mm in diameter (MP 70°C and oil content 0.3%) to a wax layer of 1.2 m^3. Oil-containing drainage was passed at ordinary temperature through the wax layer at an LHSV of 30 vol/vol/hr. The contacted drainage was then passed to a settlement tank from the bottom of which the treated water was drawn as the drainage. The experiment was contained for 30 days. The results are shown below.

	Oil content in the drainage—	
Progress	Before treatment (p.p.m.)	After treatment (p.p.m.)
1 day	6.4	1.3
5 days	6.0	0.8
10 days	2.0	0.4
20 days	10.3	1.4
30 days	2.2	0.5

As is evident from the results shown above, the method is excellent in view of the fact that no reduction in ratio of the oil removed occurred after progress of 30 days at such a high rate of the treated drainage at an LHSV of 30.

Cellulosic Filter

According to a process described by *M. Ohta; U.S. Patent 3,689,406; September 5, 1972* oil is removed from water, e.g. wastewater, by means of a fibrous filter composed essentially of a hydrophilic cellulose material such as regenerated cellulose, mercerized cellulose and acetylated cellulose. The following examples illustrate the process.

Example 1: Cotton fiber subjected in a usual manner to bleaching, exoleation and mercerization was spun and woven to form a thick cloth. The cloth was mounted to a filter press through which an aqueous emulsion containing oils (the diameter of oil droplets was about 0.1 mm) in an amount of 5,000 ppm was filtered. As the result of measurement, it was found that the quantity of oils contained in the filtrate was only 0.3 ppm and, moreover, the oils were water-soluble.

Example 2: Commercially available bleached cotton fabrics were treated according to a usual manner with caustic soda to effect mercerization. The hydrophilic cellulose thus obtained was mounted to a filter press as described in Example 1, through which the aqueous emulsion identical with that used in Example 1 was filtered. Oily substances were scarcely found in the filtrate.

Example 3: Cellulose triacetate was prepared by treating cotton fiber according to a known manner with acetic anhydride and acetic acid in the presence of concentrated sulfuric acid. The resulting cellulose triacetate was dissolved into acetone and the dope thus obtained was injected in hot air to form a filament which was then woven to a thick cloth. This cloth was mounted to a filter press similar to that used in Example 1, and the aqueous emulsion used in Example 1 was filtered through. Oily substances were scarcely found in the filtrate.

Ion Exchange Resins with Fixed Organic Ions

A. Abadie, A. Gazost and H. Roques; U.S. Patent 3,729,410; April 24, 1973; assigned to Anvar, France describe a process for deoiling polluted waters which comprises contacting the oil-containing water with an oleophilic resin having a high surface area, for example, in the form of a powder or small spherules or granules, and removing the treated water from the resin after several stages, as necessary, if a particular degree of deoiling is desired. The oleophilic resin can be an ion exchange resin with the oleophilic properties imparted by organic ions fixed thereon. Accordingly, an ion exchange operation can be conducted concurrently with the deoiling process, if desired.

The oleophilic resins to be employed in accordance with this process are conventional ion exchange resins upon which organic ions are ionically fixed. As such resins, use is made, for example, of two commercial products which can be obtained either in the form of a powder or spherules having a diameter of approximately 10 ml. These are: (a) a cationic resin of a strong type (a copolymer of sulfonated styrene divinylbenzene); and (b) an anionic resin of a strong type (a copolymer of styrene divinylbenzene) upon which there are fixed or attached ionic functions such as, for example, $—N^+(CH_3)_3Cl^-$.

As organic ions, the following may be utilized: (a) various alkyl pyridinium cations such as lauryl pyridinium chloride

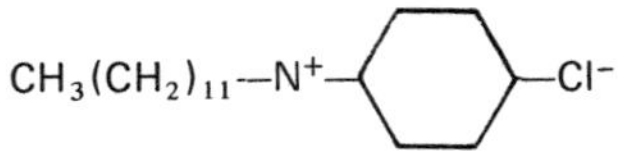

and cetyl pyridinium chloride

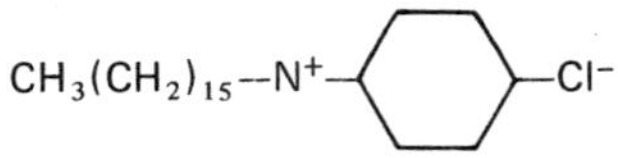

(b) anionic sulfates of fatty alcohols such as $CH_3(CH_2)_{11}OSO_3^-\ Na^+$.

Water containing the following specific contaminants may be deoiled by using this process.

1. Hydrocarbons dissolved naturally without an emulsifier, e.g., benzene—Approximately 55% of the initial content of benzene (1.6 g/l) was eliminated on the first theoretical plateau when treated on a column of oleophilic resin. With a sufficient number of plateaus, it is possible to continue the elimination or exhaustion of the benzene as far as may be desired.

2. Emulsified hydrocarbons (without an emulsifier), e.g., n-decane—On the first theoretical plateau, about 90% of the initial quantity (0.5 g/l) can be eliminated with an oleophilic resin contained on a column. It is not necessary in this case to treat the emulsion any further than in case 1 above.

3. Oils "solubilized" by means of an ionic emulsifying agent—The preliminary operation consists in eliminating the surface-active material by adsorption on a standard resin. The hydrocarbons are then freed from the solution by using an oleophilic resin in a mixed bed.

The advantages of the process are as follows:

(1) The integral regeneration of the oleophile properties is rendered easy either by means of a humid hot air current or an organic solvent, such as acetone, methanol, etc.

(2) The oleophilic resins may be employed, without danger to their specific properties, with rather concentrated saline solutions (e.g., 2 g/l of sodium chloride) the maximum acceptable salinity conditions being determined in each case. For example, a strong cationic resin of the type RSO_3^- having 50% in the form of Na^+ and 50% in the form of $CH_3(CH_2)_{15}N^+$ cannot be used in solutions where the concentration of NaCl is greater than 5.5 g/l.

(3) The nonorganic ions of the resin will remain exchangeable, with the result that an ion exchange operation can be accomplished at the same time that the deoiling operation is carried out. The regeneration of the mineral portion of the resin, without affecting the oleophilic properties of the resin, can be carried out easily under the condition that the permissible limit concentration of the regenerating agent is known.

Permeable Polyurethane Foam

According to a process described by *M.D. Gregory; U.S. Patent 3,779,908; December 18, 1973; assigned to Continental Oil Company* a dispersion of water and an oleophilic liquid is passed through a permeable foam of polyurethane which is equilibrated with respect to sorption of the oleophilic liquid to coalesce the oleophilic liquid and the water into separate layers which readily separate by gravity. The oleophilic liquid layer and the water layer can then be readily separated from each other.

Preferably the oleophilic liquid is a liquid petroleum hydrocarbon which is a liquid at a temperature in the range of 0° to 100°C. The process is believed to be particularly applicable to removing dispersions of crude oil from production waters.

Example 1: A dispersion of water in oleophilic liquid was prepared by injecting 50 ml of 5 weight percent sodium chloride brine into the intake of a centrifugal pump which was continuously circulating kerosene in a 5-gallon reservoir. A small amount of Tretolite F-46 surfactant was added in producing the dispersion to produce a more stable dispersion. Upon completion of formation of the resulting dispersion, it was allowed to settle for 30 minutes. Thereupon, the water content of the water in oleophilic liquid dispersion was determined to be 1,000 ppm.

A flexible open-cell polyurethane foam having a density of 1.55 pounds per cubic foot and having a high content of ether linkages in the polyurethane polymer was cut into cylinders 3½" in diameter by 2" in thickness. This polyurethane foam had a compression set maximum of 10%, a tensile strength of 15 lb/in^2 minimum, an elongation of 180% minimum, and a resilience of 40% minimum. The cylinders were packed into an elongated cylinder of Lucite plastic having an inside diameter of 3½" and a height of 16½" to form a foam bed having a depth of 11½". The long axis of the coalescer cylinder was vertical. The foam employed in the bed weighed 59.3 grams. A perforated Lucite plastic disc supported the foam bed above a void space of 5". An inlet was provided at the bottom of the coalescer cylinder and an outlet was provided at the top of the coalescer cylinder with a line leading to a water trap.

Water-kerosene dispersion prepared as specified was pumped through the foam bed in the coalescer at a rate of about 40 ml per hour. This flow rate gave a residence time of about 30 minutes in the foam bed. The effluent from the coalescer was passed to a water trap and the water concentration in the kerosene downstream was measured after the foam bed had become equilibrated with respect to the oleophilic liquid and coalescence had begun.

The dispersion was at a temperature of about 25°C. The water concentration in the effluent downstream of the water trap was monitored over the period of the run where 39 liters of dispersion were passed through the coalescer. The water concentration of the effluent ranged from about 40 to 100 ppm. This constituted a reduction of over 90% of the water dispersed in the kerosene by passage through the coalescer and the water separator.

Example 2: An elongated cylindrical coalescer fabricated of Lucite plastic having an internal diameter of 7.5", a long axis in a vertical plane, and an inlet at the bottom, was set up. It had a bed depth of 12" of the polyurethane foam of Example 1, a separation reservoir above the polyurethane foam bed, with an aqueous liquid inlet near the bottom of the separation reservoir, and an oil outlet near the top of the separation reservoir.

In the South Ponca field of Oklahoma is a dispersion of South Ponca crude oil (43° API gravity) in water having a temperature of 90°F and an oil concentration ranging from about 100 to 200 ppm. The oil in water dispersion of the South Ponca field was passed to the coalescer and then upward through the foam bed. After sufficient of the dispersion had passed through the foam bed to equilibrate the polyurethane form, coalescence was initiated.

Continued upward passage of the dispersion, coalescence thereby, and separation of the oil, which was drawn from the top of the separation reservoir, and water which was drawn from the center of the separation reservoir situated above the foam bed effected separation into oil and water components. The following table presents data for the run where the oil was coalesced from the oil-in-water dispersion.

Flow rate bbl/day, ft.² of foam surface	Press. drop across the bed ΔP (psi)	Influent oil conc. (ppm) of dispersion	Effluent oil conc. (ppm) of aqueous phase	Total hrs. operated
229	2.0	116	1.8	1
229	2.0	120	1.8	2½
229	2.0	174	2.0	3½
229	2.0	126	1.5	4½
229	2.0	113	1.2	5
229	2.0	180	1.3	6
229	2.7	(*)	1.2	6¾
229	(*)	194	1.3	7¾
(*)	(*)	(*)	1.1	8¾
(*)	4.0	214	2.4	9¼

* Data not taken.

This example demonstrates the effective coalescence of crude oil from a crude oil-in-water dispersion according to the process.

Highly Dispersed Melamine-Formaldehyde Resin

A. Renner; U.S. Patent 3,716,483; February 13, 1973; assigned to Ciba-Geigy, Switzerland describes a process for removing dissolved, emulsified or suspended organic substances, for example, fat, resin, oil or dyestuff, from water in which the contaminated water is brought into contact with a highly disperse, solid, water-insoluble organic polymer such as melamine-formaldehyde resin or polyacrylonitrile of average molecular weight greater than 1,000, and a specific surface area greater than 5 m^2/g, and the polymer charged with the contamination is separated from the water. In the examples which follow, parts denote parts by weight and percentages denote percentages by weight.

Example 1: 140 parts of highly disperse polymer, 140 parts of kieselguhr, 108 parts of calcium formate and 12 parts of silicone oil are well mixed in a Lödige mixer. As the highly disperse polymer, a highly disperse cross-linked melamine-formaldehyde resin, manufactured according to the data of Example 12 of French Patent 1,453,570, is used in Experiment 1.

In Experiment 2, a cross-linked urea-formaldehyde resin, manufactured according to Example 1 of Swiss Application 4,169/68 (Case 6,418) is used as the highly disperse polymer. In Experiment 3, a polyacrylonitrile of specific surface area 65 m^2/g is used as the highly disperse polymer. For comparison, a blank experiment (Experiment 4) was carried out with 140 parts of kieselguhr alone (specific surface area 28 m^2/g), and also a comparison experiment with 140 parts of pyrogenic silica aerogel (Aerosil) which has a specific surface area of 220 m^2/g. and 140 parts of kieselguhr (Experiment 5).

The oil adsorption test for the assessment of various highly disperse polymers was carried out as follows: 8 to 16 parts of spent motor oil were dispersed in 200 parts of water in a beaker by means of a vibro-mixer, 4 parts of the above mixture are added and the vibration continued for some minutes. The way in which the oil detaches from the wall of the container and is taken up by the polymer is observed. The polymer is filtered off and the filtrate is examined for freedom from oil. In a series of experiments with increasing amounts of oil, the amount of oil which still just gives an oil-free filtrate is determined. The results are presented in the following table.

Experiment No.	Highly disperse material	Specific Surface area m²/g	Maximum oil absorption part	%
1	Melamine-formaldehyde resin	150	14	350
2	Urea-formaldehyde resin	75	15	375
3	Polyacrylonitrile	65	10	250
4 (blank experiment	(Kieselguhr alone)	28	<8	<200
5 (comparison experiment)	Silica aerogel ("Aerosil")	220	12	300

Example 2: A dyestuff solution contains 1 part of the pyrazolone dyestuff of formula:

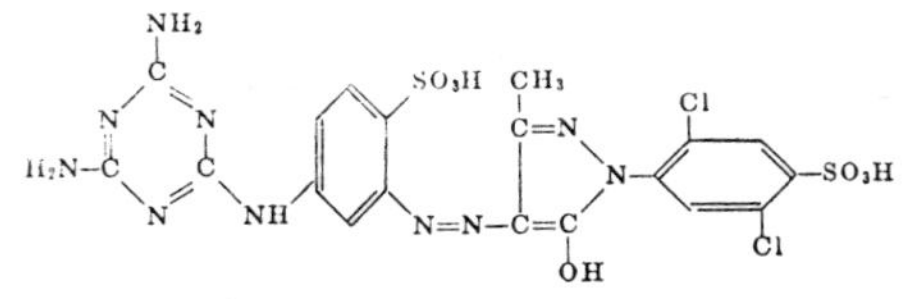

per 1,000 parts. It is a deep yellow color. It is acidified to pH 2 to 3, 7.5 parts of highly disperse melamine-formaldehyde polymer (manufactured according to Example 1 of French Patent 1,453,570) are added, and the mixture heated to 90°C and vibrated for 3 minutes. Thereafter all the dyestuff has been taken up by the polymer (13.3%). The polymer charged with dyestuff settles out well. The supernatant, absolutely colorless solution, can easily be decanted.

Example 3: A sample of effluent contains 2.5% by weight of an emulsified lacquer resin of the following composition:

	Parts
Castor oil alkyd resin (Beckosol 241)	30
Melamine-formaldehyde resin etherified with butanol (Cibamine M 86)	16.5
Silicone oil (1% in toluene)	0.5
Titanium dioxide	28
Xylene	23
Butanol	2

If 200 parts of the above emulsion are stirred with 1 part of a highly disperse urea-formaldehyde polymer according to Example 1 of Swiss Application 4,169/68 and with 0.5 part of aluminum sulfate, a clear effluent is obtained after filtration. Its light transmission at wavelengths of 460, 530 and 650 mμ is 99.5%, which indicates that the water is free of the constituents of the lacquer resin.

Magnetite Particles

According to a process described by *W.F. Lorenc and J.A. Hyde; U.S. Patent 3,767,571; October 23, 1973; assigned to Nalco Chemical Company* an aqueous liquid which contains suspended or dispersed oil or oily organic contaminants can be treated with magnetite particles. The process of treating aqueous liquids for the purpose of clarification comprises the following steps:

(1) Adding the magnetite particles to the aqueous liquids.
(2) Agitating the aqueous liquid and the magnetite particles thoroughly.
(3) Contacting the oil and the magnetite so as to provide adsorption.
(4) Removing the magnetite particles containing the adsorbed oil.

There are many convenient methods of removing the magnetite particles after adsorption. These methods may be filtration, magnetic means, settling, etc. The magnetite can be regenerated using different solvents to extract the oil or other organic contaminants. The organic contaminants could also be removed by burning. The magnetite particles would lose some activity but could be returned to the aqueous liquid for further adsorption.

A preferred magnetite particulate substance which is readily available and economical are those oxides of iron produced as a by-product in the oxygen blowing of iron and in the oxygen openhearth process to produce steel. These by-products are commercially available from many steel mills and are supplied as particulates which contain 80% magnetite. A preferred group of these materials are fine iron oxide powders which contain approximately 80% by weight of magnetite and which have an average particle size ranging from 20 to 60 microns.

Example 1: A wastewater containing 500 ppm of oil (an anionic, emulsified oil) was rapidly mixed at turbulent conditions with magnetite at a ratio of 40:1 by weight, of magnetite to oil, for a period of 10 minutes. Aliquot samples were removed each minute and analyzed for total carbon on a Total Carbon Analyzer. The carbon content remained constant at 16.5 ppm at 2 to 10 minutes. A hexane extractable test was also run and showed only 2 ppm oil remaining in the water.

The hexane extractable test consists of mixing a known amount of hexane with an aliquot sample of the wastewater. The liquids are allowed to separate. The hexane layer containing the oil is heated in a tared beaker to drive off hexane. The oil content of the original sample can then be calculated.

Example 2: A wastewater containing 500 ppm oil (an anionic, emulsified oil) required a ratio of 20:1 by weight of magnetite to oil. The carbon content was reduced to less than 10 ppm. Likewise, a cationic oil was also tested and required only a ratio of 5:1 by weight of magnetite to oil. The carbon content as determined by a Total Carbon Analyzer was less than 10 ppm. The ratio of magnetite to oil can vary from 1:1 to 100:1 by weight.

A convenient method of removing the magnetite particles is by the use of a magnetic field of sufficient flux density to attract the particles. The aqueous liquid so treated is then removed from the area of the magnetic field, thereby providing a relatively clear liquid, or a liquid containing substantially less oil than was present before the treatment.

Acid Treatment of Tar Sands Middlings Water

R. Schutte; U.S. Patent 3,816,305; June 11, 1974; assigned to Gulf Oil Canada Limited, Canada-Cities Service, Ltd., Imperial Oil, Limited and Richfield Canada Ltd. describes a process for the clarification of tailings and middlings waters produced in conjunction with the hot water treatment on tar sands. In general, the process comprises the introduction of sulfuric acid into the water to reduce the pH to within the range of 5.5 to 7.0 to cause flocculation and coagulation of the clays and small solids. The product is allowed to stand for several days to allow the pH to recover to the substantially neutral condition and the solids to settle from the middlings water.

The process may further comprise stirring the treated tailings or middlings water until a pH equal to about 7.0 is obtained where the water may be returned to the native water source subsequent to clarification and neutralization.

Thus it has been found that flocculation or coagulation of clay in small solid particles occurs when sulfuric acid is added to the waters containing up to about 10% solids. The amount of acid required is generally between about 2.0 and 5.75×10^{-4} grams per gram of solid material contained within the middlings water. In general, the acid concentration is adjusted so that immediately after addition, the pH of the mixture of treated waters is within the range of 5.5 to 6.5. The solids contained within the middlings water, which is treated with acid, coagulate through flocculation and settle rapidly. In general, the treated middlings water acts to reduce the sludge settling pond area required. The area of the sludge settling pond is greatly reduced through the introduction of acid such that retention times of middlings water in the sludge settling pond is curtailed to a small percentage of the total time necessary without treatment. The following examples illustrate the process.

Example 1: To investigate the time dependence of pH, two sets of examples of middlings, and a tailings water treatment were studied. Various amounts of sulfuric acid were added to study the clarification rate. The initial set was allowed to settle naturally with the second set (those being the even number samples in Table 1) being stirred after each pH measurement. Table 1 relates the amount of acid added and the pH of the supernatant fluid as a function of time for each sample. After 7 days settling, the pH of all the samples was found to have a pH over the neutral value.

Table 2 represents middlings samples obtained from a semicontinuous unit run in which 0.03 weight percent sodium hydroxide, an often used processing chemical, was first added in the process. It can be seen that after five days' settling, the pHs of the clarified portion of all samples were also returned to a neutral value. Table 3 represents a set of tailings water samples containing about 40% solids obtained from a non-chemical run in which 1.0 to 5.82×10^{-4} grams of sulfuric acid per gram of solids was utilized in order to depict a time dependent rise of pH to the basic side of the pH scale. A more rapid return of the pH to a basic value was found in the stirred rather than the nonstirred samples. The stirred samples also showed that the amount of change in the middlings and tailings water brought about with the addition of acid was permanent, even when the pH was initially basic excellent clarification was experienced.

Example 2: The clarified portions of the samples of Table 1 were centrifuged after 170 hours. The specific conductance, ultraviolet spectrum, surface tension and pH were measured. The results are given in Table 4. Color of the samples was light yellow (Sample M1) to clear (Samples M7 and M8). An increase in acid gave a small increase (20%) in conductance, a decrease in UV absorbance by a factor of 2.5 and an increase in surface tension by about 6%.

The supernatant liquid from several sludge settling experiments was combined to give 6 liters of clarified water of a yellowish appearance and a pH of 5.5. This water was used in small batch extraction tests to show that the acid-clarified water did not adversely affect oil recovery and solids content of the froth, but increased the water content of the froth. The latter was found to be corrected by addition of 0.02 weight percent NaOH to the slurry.

TABLE 1

VARIATION OF pH WITH TIME

Sample No.	H_2SO_4 ml	g/g Solids $\times 10^{-4}$	Hours after addition of acid 2	5	12	23	29	47	76	167
	(0.13N)	Middlings from a non-chemical run. 5.6 wt. % Solids. 500 ml Samples								
M1	1		6.90	6.90	7.30	7.14	7.00	7.10	7.35	7.44
M2	1	1.92	6.93	6.65	7.11	7.17	7.07	7.20	7.23	7.76
M3	2		6.67	6.52	6.92	6.93	6.98	6.82	7.08.	7.52
M4	2	3.91	6.78	6.60	6.97	6.99	7.00	7.18	7.10	7.80
M5	4		6.53	6.51	6.79	6.68	6.70	6.81	7.00	7.18
M6	4	7.48	6.56	6.50	6.79	6.90	6.80	6.98	6.92	7.30
M7	8		6.10	6.28	6.43	6.23	6.48	6.40	6.75	6.97
M8	8	15.0	6.14	6.22	6.60	6.63	6.29	6.58	6.85	7.11
Blank	—	—	7.30	7.10	7.48	7.40	7.32	7.42	7.44	7.81

TABLE 2

VARIATION OF pH WITH TIME

Sample No.	H_2SO_4 ml	g/g Solids $\times 10^{-4}$	pH: Hours after addition of acid 1	9	19	25	43	49	121	140	547
	(0.382N)		Middlings contained .03 wt % NaOH. 12.5 wt % Solids								
1	2		8.05	8.0	8.0	7.87	8.23	8.05	8.07	8.05	8.00
2	2	5.64	8.18	8.0	8.09	7.72	8.20	7.88	8.10	8.10	
3	4		7.66	7.52	7.70	7.83	8.04	7.77	7.20	8.07	8.00
4	4	11.19	7.72	7.61	7.52	7.57	8.07	7.90	7.50	8.10	
5	6		7.09	7.1	7.60	7.59	7.70	7.70	7.09	7.70	7.80
6	6	16.94	7.10	7.0	6.94	7.0	7.80	7.70	7.17	7.80	
7	8		6.60	7.13	7.17	7.20	6.83	6.80	7.48	7.30	7.60
8	8	22.04	6.66	6.52	6.65	6.65	7.35	7.00	7.30	7.32	
Blank			8.79	8.70	8.49	8.38	8.17	8.02	8.26	8.20	

TABLE 3

VARIATIONS OF pH WITH TIME

Sample No.	H_2SO_4 ml	g/g Solids $\times 10^{-4}$	pH: Hours after addition of acid 2	5	12	23	29	47	76	167	318
			Tailings from a non-chemical run. 40 wt % Solids. 500 ml Samples								
	(1. ON)										
T1	1		6.27	6.10	6.49	6.68	6.97	6.88	6.80	7.38	7.70
T2	1	1.56	6.33	6.18	6.48	6.54	6.81	6.64	6.58	7.47	7.80
T3	2		5.80	5.81	6.03	6.01	6.18	6.13	6.41	6.94	7.24
T4	2	3.08	5.90	5.98	6.17	6.20	6.30	6.35	6.42	7.19	7.54
T5	4		5.52	5.57	5.40	5.21	5.39	4.90	5.11	5.50	6.20
T6	4	5.74	5.53	5.57	5.54	5.52	5.71	5.19	5.40	5.81	6.56
T7	8		5.13	5.16	5.08	4.70	4.80	4.31	4.53	4.42	3.98
T8	8	11.81	5.14	5.22	5.30	5.01	5.21	4.58	4.64	4.43	4.30
Blank			6.88	7.10	7.33	7.20	7.13	7.23	6.97	7.50	7.62

TABLE 4

PROPERTIES OF CENTRIFUGED CLARIFIED WATER

Sample	Specific Conductance X 10^{-4}	UV Absorbance at 250 nm	Surface Tension (relative values)	pH
M1	3.98	0.79	68.1	7.55
M2	4.32	0.49	68.3	7.58
M3	4.35	0.55	69.7	7.52
M4	4.41	0.40	70.9	7.50
M5	4.75	0.42	72.2	7.44
M6	4.88	0.36	70.5	7.56
M7	5.12	0.33	70.7	7.45
M8	5.44	0.29	72.7	7.60
Blank	4.27	1.69	68.7	7.58
Distilled Water	1.66×10^{-6}	—	74.4	—

METALS

HEAVY METALS–GENERAL

Mercapto-s-Triazines

Y. Nakamua, A. Umehara and I. Yamada; U.S. Patent 3,778,368; December 11, 1973; assigned to Sankyo Kasei Company, Limited, Japan have found that when a mercapto-s-triazine or water-soluble salt thereof is added to a wastewater containing a metal ion having a normal electrode potential of at least -0.440 v (at 25°C), the metal ion can be removed from the waste with high efficiency and with a simple procedure. In fact, the metal ion can be effectively removed according to the process with efficiency as high as several ten times that attained by the conventional neutralizing method. By the treatment the metal content in the waste is reduced to substantially negligible amount of less than 0.1 ppm by weight, whereby the waste thus treated can be discharged without pollution with the metal.

Further, the treatment can effectively be conducted in a wide pH range, so that the waste can be subjected to this treatment as it is without any preliminary treatment. The sludges produced by the process, moreover, never allow the metal contained therein to be dissolved in water again and are easily subjected to post-treatment. For a better understanding of the process examples are given below, in which ppm, ppb and percent are all by weight and metal ion concentration was determined by atomic absorption analysis disclosed in JIS 0102-1971.

Example 1: To four kinds of 200 ml of wastewater samples containing different metal ions was added monosodium salt of 2,4,6-trimercapto-s-triazine trihydrate

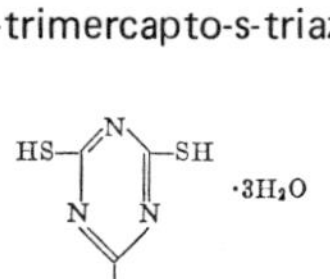

and each sample was stirred for 15 minutes. The samples thus treated were filtered with suction and the concentration of metal ions in the filtrates and the amount of the precipitates were determined. The results are shown in Table 1.

For comparison in Table 1 are also shown the results according to the conventional neutralizing method in which the same wastewater samples were respectively adjusted to pH 7.5 with a 10% aqueous solution of sodium hydroxide and then filtered in the same manner.

TABLE 1

	Waste water sample: Sample No.	Kind of metal ion	Metal ion concentration (p.p.m.)	pH	Amount of triazine comp. added (g.)	Metal ion concentration in filtrate (p.p.m.)	Amount of precipitate (g.)
According to the process	1	Cu^{++}	495	4	0.68	<0.03	1.6
	2	Cd^{++}	480	5	0.39	<0.02	1.3
	3	Pb^{++}	480	2	0.20	<0.1	0.6
	4	Hg^{++}	1,960	2	0.80	<0.001	3.6
According to the conventional neutralizing method	1'	Cu^{++}	495	4		1.9	3.0
	2'	Cd^{++}	480	5		1.3	2.5
	3'	Pb^{++}	480	2		6.1	2.2
	4'	Hg^{++}	1,960	2		0.1	9.2

From Table 1 above, it is evident that the method displays excellent effect in removing metal ion from the wastewater, giving only a small amount of the precipitate. The precipitates formed in Example 1 were added to water to aqueous solutions of various acids and left to stand for 3 days, to determine stability of the precipitates by measuring metal ions dissolved in the water and the acid solutions. The results are given in Table 2.

TABLE 2

Precipitate No.*	Water (p.p.m.)	2% aqueous solution of citric acid	Aqueous solution of nitric acid (pH 2)	Aqueous solution of sulphuric acid (pH 1.5)
1	None	None	None	None.
2	do	do	do	Do.
3	do	do	do	Do.
4	do	do	Trace	Do.
1'	2.9			
2'	1.8			
3'	7.1			
4'	Trace			

*The reference number given in Table 2 represents a precipitate obtained from the waste sample referred to by the same number.

From Table 2 above, it is apparent that the precipitates formed are very stable and will not release metal ions.

Example 2: To 200 ml of a waste fixing solution of pH 8 containing 500 ppm of Ag^+ was added 0.1 gram of monosodium salt of 2,4,6-trimercapto-s-triazine hydrate and after being stirred for 15 minutes, the wastewater was filtered. The concentration of Ag^+ in the filtrate was less than 0.02 ppm.

Example 3: To 200 ml of wastewater samples of pH 4 containing 500 ppm of Cd^{++} were respectively added 0.5 gram of monosodium salts of 4,6-dimercapto-s-triazines represented by the formula in the table below. After being stirred for 15 minutes, each sample was filtered. The concentration of Cd^{++} in each filtrate is shown below.

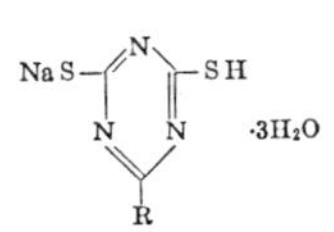

−R	Metal ion concentration in filtrate (p.p.m.)
$-C_2H_5$	<0.05
$-C_3H_7$	<0.05
$-C_4H_9$	<0.05
$-C_6H_{11}$	<0.05
$-SC_3H_7$	<0.05
$-SC_4H_9$	<0.05
$-N(C_3H_7)_2$	<0.05
$-N(C_4H_9)_2$	<0.05
$-N(C_6H_{11})_2$	<0.05
$-NHC_6H_{11}$	<0.05
$-NHC_6H_5$	<0.05
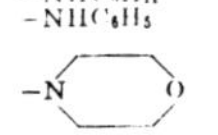	<0.05 <0.05

Example 4: To 500 ml of wastewater of pH 8.6 which was discharged from an electroplating plant and contained $Cu_2P_2O_7$ (78.5 ppm as Cu^{++}) was added 6 ml of 3% aqueous solution of monosodium salt of 2,4,6-trimercapto-s-triazine. After the wastewater was stirred for 15 minutes and left to stand for 30 minutes, reddish brown flocks were precipitated. The concentration of Cu^{++} in the supernatant was less than 0.03 ppm.

Precipitation of Heavy Metal Sulfides

J.R. Anderson and C.O. Weiss; U.S. Patent 3,740,331; June 19, 1973; assigned to Sybron Corporation describe a process for removing heavy metal pollutants from aqueous solutions by precipitation.

Soluble sulfides such as sodium slufide have been used in the past to precipitate metal values from solutions. With proper controls, such separations are generally substantially quantitatively complete, however, prior art sulfide precipitation processes have certain disadvantages which have prevented their wide-spread adoption.

The most serious problem has been the formation of noxious quantities of hydrogen sulfide. Excess sulfide is usually added to insure complete precipitation of the heavy metal pollutant. The excess sulfide forms hydrogen sulfide gas, whose vapors are detectible even above highly dilute alkaline solutions.

In some cases, the use of excess sulfide, instead of leading to complete precipitation of the heavy metal pollutant, tends to form complex sulfide ions which remained in solution. Among the complex ions which may be formed in this manner are $HgS_2^{=}$, $SnS_3^{=}$ and AgS^-, which are formed according to the following equations:

$$HgS + S^{=} \longrightarrow HgS_2^{=}$$

$$SnS_2 + S^{=} \longrightarrow SnS_3^{=}$$

$$Ag + S^{=} \longrightarrow AgS^-$$

Due to these undesirable secondary effects of sulfide precipitation processes, hydroxide precipitation has been used more widely than sulfide precipitation for the removal of heavy metals. Unfortunately, hydroxide precipitation is generally no longer as complete as sulfide precipitation because most metal bearing waste streams now contain complexing agents which have been used in conjunction with the heavy metal pollutants and which in large measure stabilize the pollutant ions against hydroxide but not against sulfide precipitation.

In this process, sulfide ion and a heavy metal ion that forms a sulfide having a higher equilibrium sulfide ion concentration than the sulfide of the heavy metal pollutant to be removed are added to the aqueous solution to be treated. Since the sulfide of the added heavy metal is more soluble than the sulfide of the pollutant, the heavy metal pollutant ions are precipitated in preference to the added heavy metal ions.

However, the added heavy metal ions act as a scavenger for excess sulfide, thereby preventing the formation of noxious amounts of hydrogen sulfide and complex sulfide ions. In addition, in certain cases, the added heavy metal promotes coprecipitation of insoluble mixed metal sulfide compounds that lead to lower pollutant concentration in the effluent.

The following table shows the sulfide ion concentrations for several of the more common heavy metal sulfides as calculated from their solubility products. The values are arranged in descending order of equilibrium sulfide ion concentration. Each metal ion can, therefore, be used as a scavenger for excess sulfide during precipitation of those metal ions lower in the table.

The choice of the heavy metal salt to be added to the solution to be treated depends upon what ions are to be removed from the solution, the use to which the water is to be put after it has been treated and such practical considerations as cost and availability of the various heavy metal salts.

Metal ion:	Equilibrium sulfide ion concentration (moles/l.)
Mn^{++}	3.75×10^{-8}
Fe^{++}	6.1×10^{-10}
Zn^{++}	3.46×10^{-12}
Ni^{++}	1.18×10^{-12}
Sn^{++}	3.1×10^{-13}
Co^{++}	1.73×10^{-13}
Pb^{++}	1.84×10^{-14}
Cd^{++}	6.0×10^{-15}
Ag^{+}	3.4×10^{-17}
Bi^{+++}	4.8×10^{-21}
Cu^{++}	9.2×10^{-23}
Hg^{++}	4.5×10^{-25}

For most waste treatment applications it is desirable to use a heavy metal salt which is relatively nontoxic and generally innocuous since the water may subsequently be discharged to reservoirs where stringent water quality standards are observed. Salts of metals at the top of the foregoing table are preferred since they can precipitate a greater variety of pollutant ions. Manganese forms the most soluble of these sulfides, however, iron salts will frequently be preferable because they are generally cheaper than manganese salts.

After the added heavy metal salt is dissolved in the solution, a soluble sulfide is added to precipitate sulfides of the various metals. The process may be manipulated to selectively precipitate the ions at the bottom of the table by adding just enough sulfide to precipitate the pollutant or pollutants which form the least soluble sulfides.

Alternatively, a heavy metal salt which forms a sulfide of intermediate solubility; i.e., a sulfide that is more soluble than the sulfides to be precipitated but less soluble than the sulfides of some of the pollutants in the solution; may be used to achieve the same result. In general, however, a metal salt is chosen whose sulfide is more soluble than the sulfides of all of the pollutants and enough sulfide ion is added to precipitate all of these less soluble sulfides.

Under equilibrium conditions, the resulting sulfide ion concentration is somewhere between 10^{-8} to 10^{-25} mols per liter, depending upon the pH of the solution and the solubility of the added heavy metal sulfide. If the soluble sulfide is introduced below the surface of the solution and agitation is sufficient to prevent a localized concentration of sulfide ions, the formation of the sulfide of the added heavy metal instantaneously reduces the sulfide ion concentration below the level at which it can form noxious amounts of hydrogen sulfide or soluble complex sulfides, while simultaneously allowing several orders of magnitude more sulfide ion than that required to precipitate the metallic pollutants in the solution. As a result, no objectionable odors are created and, since soluble complex sulfide ions are not formed, substantially all of the metallic pollutants in the solution are precipitated. The following are representative examples of the precipitation of heavy metal ions according to the process.

Example 1A: In this test, the waste sample was a composite of all the discharges from an electronics plant. Due to the complex nature of the material, analysis for all metallic impurities was regarded as impractical. However, following sulfide and hydroxide treatment respectively the waste was analyzed for metals known to be initially present that are generally regarded as serious pollutants. These data are shown in the following table.

Prior to sulfide treatment, the waste was hypochlorinated at a pH of about 8.5 to destroy NH_3. Using a 1,700 ml sample of hypochlorinated waste, the excess ClO^- was destroyed by the addition of 0.8 gram of sodium sulfite (Na_2SO_3) and after five minutes of agitation, the waste was further treated with 0.58 g of manganous sulfate monohydrate ($MnSO_4 \cdot H_2O$) followed by 0.75 gram of sodium sulfide nonahydrate ($Na_2S \cdot 9H_2O$). The treated waste was agitated for five minutes and then allowed to stand for one hour to allow the precipitates to settle. No hydrogen sulfide odors were detected during either the five minute agitation period or the one hour settling period. At the end of the one hour settling period, a sample of the supernatant was reserved for analysis.

Example 1B: In order to compare this process with hydroxide precipitation, a sample whose composition was similar to the composition of the sample used in Example 1A was treated as follows. A soluble hypochlorite was added to the water, which had a pH of 11.0, in order to convert cyanide ions known to be present to cyanate at which point metal hydroxides were precipitated. Sulfuric acid was then added to reduce the pH to 8.5 and convert the cyanate to nitrogen and bicarbonate. The material was allowed to stand for one hour after which the liquid was analyzed. The results, which are listed in the following table, show that precipitation by this method is not as complete as precipitation by this process.

Example 2A: In this test, waste from a printed circuit manufacturing operation containing 100 ppm Cu, 7.7 ppm Ni and 475 ppm NH_3 was treated via the sulfide process. The pH of a 100 ml sample was reduced to 7.7 with dilute H_2SO_4. A 0.123 gram dosage of ferrous ammonium sulfate was added immediately prior to 0.75 gram of $Na_2S{\cdot}9H_2O$. The sample was agitated for five minutes and allowed to stand for one hour. No hydrogen sulfide odors were detected during either the five minute agitation period or the one hour settling period. At the end of the one hour settling period, the effluent was analyzed for copper and found to contain 1.8 ppm.

Example 2B: The same waste used in Example 2A was adjusted to pH 7.7, the sample agitated and settled as described in Example 2A. It was found to contain 95.8 ppm Cu and 5.9 ppm Ni. This demonstrates the complexing powder of NH_3 and the problems it presents in hydroxide precipitation processes.

Examples 3A and 3B: The tests performed in Examples 2A and 2B were repeated except that 40 ppm $Fe_2(SO_4)_3$ and 8 ppm of an anionic polyacrylamide were added to coagulate the precipitates and the sample used in the hydroxide test was chlorinated to destroy the NH_3. As can be seen from the following table, the results for sulfide precipitation, even in the presence of NH_3, were as good as the results produced by hydroxide precipitation after chlorination.

Example 4A: The waste used in this experiment contained 4.8 ppm of Cr^{+6} and 3.5 ppm of Zn. The pH was adjusted to 7.7. Then 382 mg of Fe_2SO_4 and 333 mg of $Na_2S{\cdot}9HO_2$ were dissolved in small amounts of water and added simultaneously from separate pipettes. The sample was agitated for one minute, 100 ppm of clay and 2 ppm of an anionic polyacrylamide were added and slow agitation continued for an additional minute. The sample was then allowed to stand for one minute. Again no hydrogen sulfide odors were detected during agitation or settling. The effluent contained negligible amounts of Cr^{+6} and 0.3 ppm Zn.

Example 4B: A one liter sample of the same waste as in Example 4A containing 4.8 ppm of Cr^{+6} was adjusted to pH 2.0 with dilute sulfuric acid. A 0.0195 gram dosage of Na_2SO_3 was added with vigorous agitation. The pH of the reduced chromium solution was raised to 8.0 with dilute NaOH and the sludge was allowed to settle for 1 hour. A sample of the supernatant analyzed 0.05 ppm residual Cr^{+6} and 2.0 ppm residual Zn^{++}. Thus, it can be seen that precipitation of chromium according to this process is superior to conventional hydroxide precipitation and may be accomplished more simply. The results obtained in the foregoing sulfide and hydroxide precipitation processes are compared in the following table.

Example	Waste Composition in ppm	Sulfide Precipitation in ppm	Hydroxide Precipitation in ppm
1A-1B	Unknown	Cu, 0.1	0.8
	- -	Zn, negligible	2.0
2A-2B	Cu, 100	Cu, 1.8	95.8
	Ni, 7.7	- -	5.9
	NH_3, 475	- -	- -
3A-3B	NH_3, 475	Cu, 0.4	1.0
	- -	Ni, 2.0	2.0
4A-4B	Cr^{+6}, 4.8	Cr^{+6}, negligible	0.05
	Zn, 3.5	Zn, 0.03	2.0

Sulfur Compound and Diffusion Through Peat

J.-M. Lalancette; U.S. Patent 3,790,370; February 5, 1974; assigned to Universite de Sherbrooke, Canada describes a process for removing metal ions such as mercury, copper, cadmium, zinc, lead, nickel or silver from polluted waters which includes causing the metal polluted water and a sulfur-containing precipitating agent to diffuse through peat whereby a water-insoluble sulfide of the metal is formed and is adsorbed on the peat and the metal-free water is recovered. The process also relates to recovering the metal either as metal oxide or as metal in the case of silver, copper and mercury by burning the metal sulfide containing peat. The apparatus used in this process is described fully on pages 94 through 96.

Example 1: A solution (1 liter) of 2% of mercury (as $HgCl_2$) was stirred in the presence of metallic mercury. After allowing the solution to settle, it was flowed over five pads of peat, 1 inch thick and 5 inches in diameter. Each pad weighed approximately 55 grams and had been impregnated with a 1% solution of Na_2S. The effluent water had the same volume as the starting solution, since the peat was saturated with moisture. The concentration of mercury was lower than 1 ppb as determined with a "dial Atom" apparatus (Jarrell Ash) with a Hg hollow cathode.

After this treatment, the peat was wrung and left to dry for a few days. The moisture content of the peat was then decreased from 85% to a value of the order of 50%, the weight being reduced accordingly. This dry peat was then burned at 400°C using 0.1 cc of air per gram of peat per minute. The effluent gases were heated at 600°C for approximately 30 seconds, before leaving the furnace. The condensate was collected and the mercury was recovered, by decantation. Starting with a solution containing 2.71 grams of $HgCl_2$, 2.0 grams of mercury was recovered, a quantitative yield. The cooled gases were passed through a 1 x 24 inch tube containing wet limestone chips about one-fourth inch in diameter mixed with 5% in weight of sulfur. The removal of SO_2 was quantitative and no mercury vapors could be found in the effluent gases.

Example 2: By proceeding in the same manner as in Example 1 and using an acid solution of mercury acidified to pH 3 with HCl, the results were identical to Example 1.

Example 3: The procedure of Example 1 was used using a basic solution of mercury made alkaline to pH 12 with sodium hydroxide. The results were identical to Example 1.

Example 4: A solution of 1.83 grams of cadmium chloride in 1 liter of water was treated as in Example 1. The effluent solution contained no cadmium (test with H_2S) which was recovered as a partly sublimed oxide in the furnace. It has not been possible to obtain an exact weight of cadmium oxide.

Example 5: A solution of 1.61 grams of zinc sulfate in 1 liter of water was treated as in Example 1. The effluent solution contained no zinc (test with H_2S) which was recovered as oxide and difficult to weigh (0.4 gram) under the experimental conditions used.

N-Acylamino Acids

S. Inazuka, M. Takehara and R. Yoshida; U.S. Patent 3,755,158; August 28, 1973; assigned to Ajinomoto Co., Inc., Japan describe a process for capturing metal cations which comprises adding as a capturing agent N-higher aliphatic acylamino acid or a related water-soluble salt to an aqueous solution containing metal cations. The following examples illustrate the process.

Example 1: Nine aqueous solutions of cadmium sulfate containing 5 to 10,000 ppm as cadmium ion were prepared, and 1.2 equivalent of disodium N-hydrogenated tallow oil fatty acid acyl-L-glutamate was added as a 2.0% aqueous solution to the respective solutions. The stirring was continued at room temperature and thereby a large amount of insoluble cadmium salt crystallized out. The crystals were filtered out with a filter paper (Toyo No. 5A). The residual cadmium ion concentrations in the mother solution were measured by using an atomic adsorption photometer. The results are shown in the following table.

The Cd⁺⁺ conc. in the original solution	The Cd⁺⁺ conc. in the mother solution	The pH of the solution treated
10,000 ppm	0.100 ppm	6.40
5,000	0.020	6.30
1,000	0.050	6.45
500	0.009	6.72
100	0.012	6.43
50	0.009	6.35
20	0.007	6.25
10	0.005	6.20
5	0.008	6.40

Example 2: 2.0% aqueous solutions of various N-higher aliphatic acylamino acid salts are shown in the following table were prepared and the solutions were respectively added dropwise to aqueous solutions containing 100 ppm of Hg^{2+}, Al^{3+}, Sn^{4+}, Pb^{4+}, Co^{3+}, Ni^{3+} and U^{3+} under the same conditions as in Example 1. The resultant slurry in every case was filtered out by the following two different methods: (1) Treated with activated charcoal and then filtered out using the above described filter paper; and (2) Filtered out with a sand filter compacted with sands of various sizes and shapes.

The residual metal ion concentration in the respective filtrates was measured by using an atomic adsorption photometer. The results are shown in the following table.

Capturing agents	Metal ions							
	Hg^{2+}	Al^{3+}	Sn^{4+}	Pb^{4+}	Fe^{3+}	Co^{3+}	Ni^{3+}	U^{3+}
Disodium N-hydrogenated tallow oil fatty acid acyl-L-glutamate	0.005	0.010	0.009	0.006	0.010	0.012	0.010	[1] 0.055
	0.010	0.012	0.025	0.021	0.705	0.630	0.500	[2] 0.300
Disodium N-oletyl-L-glutamate	0.010	0.005	0.007	0.005	0.008	0.010	0.015	0.030
	0.020	0.018	0.023	0.014	0.170	0.150	0.230	0.200
Sodium N-lauroyl-L-phenylalaninate	0.002	0.007	0.002	0.004	0.009	0.008	0.005	0.010
	0.006	0.030	0.005	0.017	0.500	0.500	0.250	0.120
Potassium N-lauroyl-DL-methionate	0.008	0.015	0.010	0.005	0.030	0.035	0.020	0.010
	0.015	0.015	0.036	0.030	0.099	0.150	0.120	0.070
Sodium N-lauroyl-L-valinate	0.005	0.010	0.019	0.020	0.120	0.030	0.052	0.015
	0.012	0.085	0.025	0.029	0.860	0.750	0.820	0.080
Triethanolamine N-coconut oil fatty acid acyl glycinate	0.004	0.012	0.012	0.021	0.070	0.035	0.090	0.020
	0.005	0.008	0.040	0.088	0.500	0.500	0.650	0.100
Sodium N-lauroyl sarcosinate	0.010	0.017	0.020	0.010	0.130	0.250	0.085	2.90
	0.105	0.750	0.052	0.035	1.200	1.000	0.950	20.5
Sodium Nᵅ-lauroyl-L-lysinate	0.002	0.005	0.010	0.010	0.055	0.075	0.050	0.10
	0.007	0.080	0.050	0.038	0.910	0.850	0.050	1.50
Sodium Nᵅ, Nᵚ-dilauroyl-L-lysinate	0.012	0.002	0.001	0.002	0.002	0.010	0.030	0.020
	0.010	0.004	0.008	0.005	0.019	0.050	0.200	0.100
Control:								
Sodium lauryl sulfate	32.7	27.3	11.6	50.0	31.6	25.5	27.5	20.5
	49.9	56.4	59.8	69.1	70.5	59.5	43.5	25.8
Sodium dodecyl benzene sulfonate	39.7	32.4	27.4	21.2	31.2	35.0	32.0	15.7
	50.1	59.2	51.3	39.5	70.0	65.5	50.0	30.5
Fatty acid soap material	10.3	9.08	27.5	42.5	1.00	2.10	1.30	50.7
	9.2	13.7	48.3	61.7	3.05	12.5	10.7	25.7

[1] Upper step: treated with activated charcoal.
[2] Lower step: passed through a sand bed.

NOTE.—The concentration of metal ions in the original solution: 100 p.p.m.

Example 3: 1.0 ml of 2% aqueous solution of disodium N-hydrogenated tallow oil fatty acid acyl-L-glutamate was added dropwise to an aqueous solution containing 100 ppm of hexamminecobalt (III) ions as a complex ion. The precipitated solid was filtered off and the residual complex ion concentration in the mother solution was measured with an atomic adsorption photometer. The result was 0.48 ppm. Thus, N-higher aliphatic acylamino acid salt has satisfactory capturing activity to complex ions.

Ion Selective Membrane

J.S.H. Wu and B.L. Baker; U.S. Patent 3,718,583; February 27, 1973; assigned to U.S. Atomic Energy Commission describe a method for the concentration and recovery of heavy metal cations in aqueous solutions using an ion selective membrane system.

Heavy metal cations, in particular cations of actinides and lanthanides, previously have been concentrated, separated and recovered from aqueous solutions containing small quantities of these cations in ion selective membrane systems where the driving force for the concentration is provided by the difference in chemical potentials of the various ion species in the respective solutions on each side of the membrane. This concentration has been successfully accomplished without the use of an applied electrical potential across the membrane. Particular examples of a successful ion selective membrane concentration method, including the concentration and separation of actinides and lanthanides, are described in detail in U.S. Patent 3,454,490. This patent describes a method where ion species in aqueous solutions are selectively concentrated and separated by contacting a first aqueous solution containing the first ion species that is to be concentrated with one side of an ion selective membrane having polar selective permeability for the desired ion species while simultaneously contacting a second aqueous solution containing a second ion species having the same polarity as the first ion species with the opposite side of the membrane.

The compositions of both solutions are controlled to maintain a difference in the chemical potentials of the ion species and these solutions are maintained in contact with their respective side of the membrane for sufficient time to produce the transfer of a substantial portion of the first ion species and a portion of the second ion species through the membrane whereby the first ion species is concentrated in the second aqueous solution. The desired first ion species is then recovered from the second solution by conventional separation means.

This process is particularly useful and efficient for concentration, separation and recovery of heavy metal cations, such as uranyl ions, strontium ions, lanthanum ions and other valuable heavy metal cations which previously have been separated by conventional ion exchange resin techniques. The control of this separation method using difference in chemical potentials of the various ions across the membrane is accomplished by maintaining a difference in charge on the ions, by the use of complexing agents on opposite sides of the membrane, or by a difference in the ionic activities in the respective solutions.

Although the method described in the above noted patent has been very successful for the separation and recovery of heavy metal cations, it has been recognized that one of the difficulties inherent in this process is that the diffusion of water across the membrane by osmotic pressure tends to dilute the concentrating solution (strip) and thereby diminish the concentration of the desired cations. Some of the difficulty of diffusion of water by osmotic pressure can be overcome to a limited extent by careful selection of a membrane that will minimize this osmotic effect.

In addition, the effects of osmotic pressure can be partially controlled by careful continuous control of the chemical potentials (partial molar-free energies of the various ions) of the solutions and by maintaining a suitable solution contact time with the membrane. However, none of these attempts to control the osmotic pressure across the membrane have been sufficient to substantially reduce the detrimental effects of osmotic pressure.

Accordingly, it is an object of this process to provide an improved method for the concentration and separation of heavy metal ions in aqueous solutions using ion selective membranes. Thus, it is an object of this process to provide a method of suppressing the inherent osmotic diffusion of water across cation selective membranes for the improved concentration and separation of heavy metal cations from aqueous solutions.

This and other objects are accomplished in the process by removing from the ion selective membrane separation system a portion of the solution being concentrated in heavy metal cations (extract solution) and recirculating or recycling that portion of the extract solution into contact with the membrane. It has been found that recirculating a portion of the solution that is being concentrated in heavy metal cations (extract) (i.e., returning it to the initial strip solution) tends to suppress the osmotic diffusion of water across the membrane thereby increasing the concentration of the heavy metal cations in the strip solution without increasing the loss of heavy metal cations to the raffinate.

Thus, in a method for the concentration and recovery of heavy metal cations using an ion selective membrane which includes continuously contacting an aqueous feed solution containing the heavy metal cations with one side of the cation selective membrane and countercurrently contacting an aqueous strip solution with the opposite side of the membrane to effect a transfer and concentration of heavy metal cations in the strip solution, separating a portion of the solution that is being concentrated in heavy metal cations and recirculating that portion into contact with the membrane tends to suppress the osmotic diffusion of water across the membrane.

It will be apparent that this recirculation or recycle of a portion of the extract stream is different from a conventional reflux operation to enhance the concentration in a chemical process system because this recycle returns some of the extract to the initial strip solution rather than to an enriched stream. This process is particularly significant because it is the converse of conventional reflux in that a portion of an enriched stream is returned to a dilute stream.

The reduction in osmotic flow across the membrane may be caused by the substantial conversion of the membrane to the heavy metal cation form over its entire length. It is known that membrane in the hydrogen form swells more than in the uranyl form, thereby providing larger pores that can permit more transfer of water across the membrane with less resistance. It is believed that by providing more heavy metal cations in the strip solution the membrane is maintained in the heavy metal cation state and the effect is to suppress the transfer of water across the membrane.

Suppression of the inherent osmosis of the membrane using this method results in a significantly enhanced concentration of heavy metal cations without an increased loss of these same desired cations to the raffinate solution. The method has been found to be particularly useful for the separation of heavy metal cations, such as actinide and lanthanide cations. Also, it has been found that a suitable ratio of the recirculating portion to the initial strip solution is between 1 to 1 and 5 to 1.

At least three significant advantages result from this recirculation: (1) Better mixing of the solutions in the stripper solution, (2) avoidance of high acid concentration differences between the raffinate and the stripper stream, and (3) maintenance of the entire membrane length in the most suitable cation form for efficient concentration.

The cation selective membrane concentrator shown in Figure 3.1a is a parallel flow countercurrent concentrator that illustrates the single pass countercurrent concentration of cations in aqueous solutions. The concentrator **10**, which is arranged similar to conventional membrane systems, in stacks or cells, comprises a single feed channel **12** between adjacent strip channels are physically separated by semipermeable membranes **22** and **24** consisting of suitable cation selective membrane material having cation selective permeability for the cation species to be concentrated. The channels with the membranes are assembled between two end plates **18** and **20** with suitable spacers and membrane supports.

Although suitable spacers and membrane supports will be familiar to those skilled in the art, it has been found that satisfactory membrane support is provided by a stainless steel screen disposed between each channel. The stainless steel screen prevents sticking and collapse of a membrane due to the thin channel thickness and pressure difference between the two sides of the screens.

Feed and strip stream input and outlet means are provided at each end of the concentrator for feed and strip solution flow into the respective channels. Feed solution input is provided into feed channel **12** by feed inlet means **28** and the flow is controlled by feed metering pump **30**. After passing through channel **12** the depleted feed solution emerges through raffinate outlet means **32**. Strip solution is introduced through strip inlet means **34** and the strip solution flow is controlled by strip metering pump **36**. After flowing through strip channels **14** and **16**, the extract emerges through extract outlet means **38**. To provide for a recirculation or recycle flow, the extract is divided into product stream **42** and recycle stream **40**. The recycle flow **40** is controlled by a recycle metering pump **44**. The recycle is returned to the initial strip inlet means **34** to be combined with the strip solution.

FIGURE 3.1: CONCENTRATION OF HEAVY METAL CATIONS USING ION SELECTIVE MEMBRANE

a.

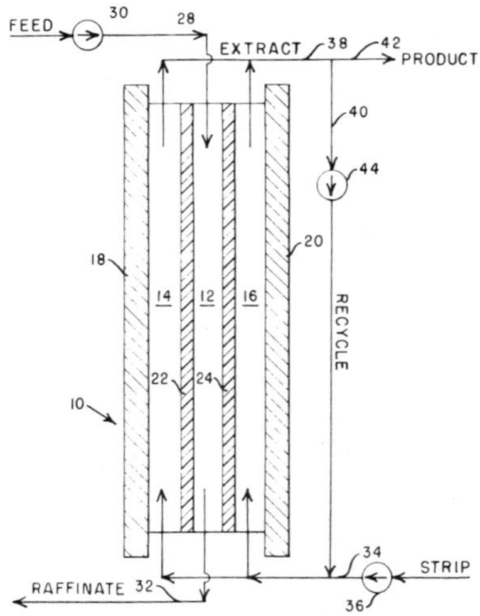

Single Pass Concentrator

b.

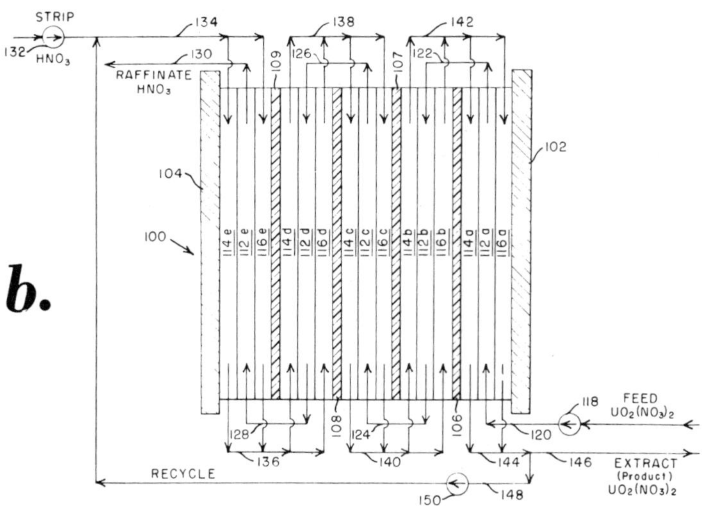

Multiple Pass Concentrator

Source: J.S.H. Wu and B.L. Baker; U.S. Patent 3,718,583; February 27, 1973

Figure 3.1b shows the schematic cross-section of a multiple pass cation exchange membrane concentrator assembly for the concentration of the uranyl ions from aqueous nitrate solution. The cation selective membrane concentrator assembly **100** of Figure 3.1b comprises five compartments each divided into three channels, one feed channel **112** and two strip channels **114** and **116**. Each concentrator compartment is separated from the adjacent compartment by suitable partitions **106, 107, 108** and **109.** The stack of membrane compartments and partitions is also arranged and disposed in conventional manner in a suitable frame between a pair of end plates **102** and **104.** Each of the channels is provided with a suitable inlet and outlet means and each compartment is interconnected, as illustrated, for continuous solution flow through each of the feed and strip compartments, respectively.

Feed solution inlet is provided through feed metering pump **118** and feed inlet means **120.** Feed solution then passes through **112a** to channel **112b** through conduit **122,** from channel **112b** to channel **112c** through conduit **124,** from channel **112c** to channel **112d** through conduit **126,** from channel **112d** to channel **112e** through conduit **128,** and the depleted feed solution finally emerges through raffinate outlet means **130.**

Strip solution is introduced countercurrent to the feed solution through strip solution metering pump **132** and strip inlet means **134** into channels **114e** and **116e,** and passes through the respective strip manifolds **136, 138, 140** and **142** until it passes through channels **114a** and **116a** and emerges as extract at extract outlet means **144.** As in Figure 3.1a, the extract at this point is divided into a product stream **146** and a recycle stream **148** is controlled by recycle metering pump **150** and returned to the initial strip inlet means **134** for recycle through the stripping system.

A wide variety of cation selective membranes may be satisfactorily employed in the method. However, membranes that have been found to be particularly effective are those membranes prepared by copolymerization processes. In these processes, hydrophobic films, such as polyethylene and polychlorotrifluoroethylene, are impregnated with styrene or styrene-divinylbenzene mixtures and are then polymerized by chemical means or by exposure to gamma radiation from a cobalt-60 source. Copolymerization of the styrene and divinylbenzene into the film base provides a suitable membrane base.

To obtain a strong acid cation selective membrane the polymerized membrane is then sulfonated by well-known conventional means. Strong acid cation selective membranes prepared by this process are between about 0.15 and 0.30 mm in thickness and have dry cation exchange capacities between about 0.6 and 1.6 meq/g. Detailed descriptions of suitable membranes, including methods of their preparation, are thoroggһly described in U.S. Patents 3,247,133; 3,257,334 and 3,113,889.

Membranes prepared as described are available commercially (AMFion). Suitable strong acid cation members are AMFion C-60, AMFion C-103 and AMFion C-113. In the example strong acid cation selective membrane refers to a membrane that has a polyethylene-styrene matrix with sulfonic acid ionic groups that is prepared by a polymerization process, such as the membrane AMFion C-103. The following example illustrates the process.

Example: Multiple Pass Membrane Concentration of Uranyl Nitrate — A multiple pass cation selective membrane concentrator was assembled from five compartments, each having three flow channels as shown in Figure 3.1b. A series folded arrangement provided a total of five feed channels each disposed between ten countercurrent flow strip channels. Each compartment contains two strong acid cation selective membranes 20 inches long and four inches wide disposed to form channels fifteen mils thick using a 12 mil screen for support. This provided a total effective feed channel length of 100 inches. The total membrane area exposed to feed solution was 800 square inches. Synthetic feed solutions containing 0.01 mol per liter uranyl nitrate were fed through feed channels of the membrane assembly at the flow rates indicated in the following table.

The feed rates were maintained by a controlled volume pump. Feed agitation was not required because of the narrow channel thickness between the membranes. A strip solution containing 3.0 mols per liter of nitric acid was fed into the strip channels adjacent each feed channel at the flow rates set forth below. The feed solution contained 0.01 M $UO_2(NO_3)_2$.

Multiple Pass Membrane Concentration of Uranyl Nitrate

					Analysis, M				
Run No.	Flow rate, ml./hour feed	Strip	θ[a]	α[b]	Product UO_2^{++}	Raffinate UO_2^{++}	Product H^+	Raffinate H^+	Osmosis moles in.², hr.
1	1,698	18.3	0.0	1.62	0.251	4.69×10^{-4}	0.108	2.87×10^{-2}	3.34×10^{-3}
2	1,698	18.4	0.0	1.62	0.244	4.37×10^{-4}	0.121	2.55×10^{-2}	3.37×10^{-3}
3	1,691	16.1	1.0	1.43	0.300	6.18×10^{-4}	0.048	2.76×10^{-2}	2.51×10^{-3}
4	1,689	16.0	1.0	1.42	0.304	5.80×10^{-4}	0.058	2.69×10^{-2}	2.56×10^{-3}
5	1,669	16.3	2.0	1.47	0.317	3.97×10^{-4}	0.088	2.71×10^{-2}	2.47×10^{-3}
6	1,678	16.3	2.0	1.46	0.322	4.50×10^{-4}	0.079	2.61×10^{-2}	2.50×10^{-3}
7	1,664	16.3	3.0	1.47	0.314	4.30×10^{-4}	0.106	2.66×10^{-2}	2.38×10^{-3}
8	1,664	16.8	3.0	1.52	0.313	3.50×10^{-4}	0.118	2.65×10^{-2}	2.38×10^{-3}
9	1,681	16.1	3.0	1.44	0.318	3.87×10^{-4}	0.096	2.69×10^{-2}	2.40×10^{-3}
10	1,676	18.4	3.0	1.65	0.2860	2.56×10^{-4}	0.185	2.78×10^{-2}	2.61×10^{-3}

[a] θ=The ratio of the flow rate of the recycling extract solution to the flow rate of the make-up strip solution.
[b] α=Equivalents of hydrogen ion per unit of time entering in the strip/equivalents of uranyl ion per unit of time in the feed.

Runs 1 and 2 were made without any recycle of the extract stream and subsequent runs (3 to 10) were made with increasing recycle of a portion of the extract stream to the initial strip stream to determine the effect of the recycle on product recovery and osmotic flux across the membrane. The amount of recycle entering the strip channels is in addition to the initial amount of strip solution. The table shows that with increasing recycle there is a significant decrease in osmosis across the membrane while at the same time uranyl ion product concentration is significantly enhanced.

This suppression osmotic diffusion of water across the membrane and improved recovery of uranyl ion can be accomplished while maintaining a low uranyl concentration in the raffinate. Runs 7 and 8 in particular with a recycle ratio of $\theta = 3.0$ have a very high uranyl ion product concentration and low uranyl ion loss to the raffinate.

In general any amount of recycle from a ratio of about $\theta = 1$ will result in a suppression of osmotic flux across the membrane and those skilled in the art will recognize that the value of θ can be increased to a value as high as would economically be feasible. Increasing the recycle ratio above a value of about $\theta = 5$ does not increase the concentration efficiency.

The ratio α, defined in the table, is a useful operating parameter for membrane concentration of uranium because it is a measure of the stoichiometric saturation of the product solution with respect to uranium when all of the uranium is removed from the feed solution. Thus, the value of the ratio α is an index to the approach to conditions that will yield maximum concentration. In general, optimum stoichiometric saturation of the product solution with respect to uranium for this system is obtained by maintaining the ratio α in the range of from about $\alpha = 1.45$ to $\alpha = 1.50$. A continuous process for the economic concentration and recovery of the uranyl ions, including the recycle of a portion of the extract stream, is also described.

Cation Exchange Resin

D.O. De Pree; U.S. Patent 3,725,259; April 3, 1973; assigned to Aerojet-General Corp. describes a process for the removal of Group Ib, IIb, III, IV, V, VI, VII and VIII metal salts from acidic waste streams such as acid mine drainage, pickling waste, and electroplating waste where the waste stream is highly acidic and contains these metal salts as pollutants. The process comprises contacting the stream in countercurrent fashion with a cation exchange bed containing a cation exchange resin either in the H^+ or alkali or alkaline earth substituted form to deposit metal ions from the salts, neutralizing the acidic effluent from the cation exchange resin, and regenerating the cation ion exchange resin with a solution of a chelating agent and precipitating the metal from the solution by treatment with a base or with a base and carbon dioxide.

According to the process, the chelating agent is subsequently recovered for recycle by precipitation of the transition metal with a base or with a base and carbon dioxide. If necessary, the acidity of the acid waste stream can be first adjusted to a pH of between about 3 to 5 using lime or calcium carbonate. The neutralization of the acidic effluent from the cation exchange resin is also preferably carried out with lime or calcium carbonate.

The most commonly occurring metals in acid waste streams are ferrous, ferric, aluminum, copper, zinc, nickel, chromium, cobalt and manganese salts, and it is the removal of these cations which is the prime objective of this process.

Figure 3.2a shows a simple block diagram of the process. Figure 3.2b shows the effectiveness of acetylacetone in removing ferrous ions from a cation exchange resin. Figure 3.2c shows the effectiveness of acetylacetone in removing ferric ions form a cation exchange resin. Figure 3.2d shows the effectiveness of acetylacetone in removing aluminum from a cation exchange resin.

The acidic waste stream which is adjusted, as required to a pH of from 3 to 5 with either lime or calcium carbonate is introduced through line **10** to the weak acid cation exchange bed **12** in a countercurrent pass. The cation exchange resin in the bed removes the metals from the stream leaving a transition metal-free acid stream passing through line **14** to the neutralization zone **16**. Lime [$Ca(OH)_2$] or calcium carbonate is continuously introduced via line **18** to the neutralization zone **16** to neutralize the remaining acidity in the water effluent metal-free acid stream. The purified product water is removed by line **20** in a condition suitable for discharge into streams or further industrial or other uses.

The spent resin from the bed is passed by line **22** to the resin regeneration zone **24**. The regenerated resin from zone **26** is returned to bed **12** by line **26**. The resin in the resin regeneration zone **24** is treated with a ligand or chelating solution introduced by line **28**. This solution chelates the metal present in the resin prior to regeneration to form a metal chelate solution which is removed via line **30** to the precipitation zone **32**.

In the precipitation zone **32** the spent ligand is continuously regenerated by the addition of a base or base and carbon dioxide via line **34**. The metal hydroxides formed within precipitation zone **32** are removed via line **36**. The regenerated ligand solution thus obtained is continuously recycled by line **28** to resin regeneration zone **24**.

In general, selective separation of the transition metal ion from other cations present in the acid waste stream is achieved by pH control of the input to the cation exchange resin so that only the strongly absorbed metal ions are deposited on the resin. In most acid waste streams a single metal predominates and hence this treatment is entirely adequate. The other cation present in most acid mine waste streams in calcium which is nontoxic and therefore does not provide an environmental hazard. Some unusual streams contain a variety of both the subject metal ions and nontoxic alkali and alkaline earth cations. Such streams may require final desalination treatment of the effluent to complete deionization. The following examples illustrate the process. In the examples the parts and percentages are by weight unless otherwise indicated.

Example 1: Effect of pH on the Deposition of Iron on Polymethacrylic Acid Resins — Both an aqueous solution of ferric sulfate (237 ppm), having a pH of 3.0 and a solution of ferric sulfate (237 ppm) in 0.1 N H_2SO_4, pH 1.5, were passed through Amberlite CG-50 resin and 20 ml fractions of eluate collected. At pH of 3.0 the first two fractions were free of iron as indicated by treatment with KCNS solution and fractions 3 and 4 showed traces of ferric ion. At the pH of 1.5 no iron was deposited on the resin as indicated by strong positive tests for iron in all eluate fractions. Thus, it may be concluded that at pHs of 3.0 or higher ferric ion can be removed from aqueous solution. Samples of CG resin were individually loaded by treatment with 1 N $Fe_2(SO_4)_3$ solution (pH ~3.0), 1 N $FeSO_4$ solution (pH ~3.0) and $Al_2(SO_4)_3$ solution (pH ~3.0), for evaluation in regeneration experiments. The resins were thoroughly washed with distilled water to remove any undeposited cations and then with methanol and dried under vacuum.

FIGURE 3.2: RECOVERY OF MINERAL POLLUTANTS FROM ACIDIC WASTE STREAMS

a.

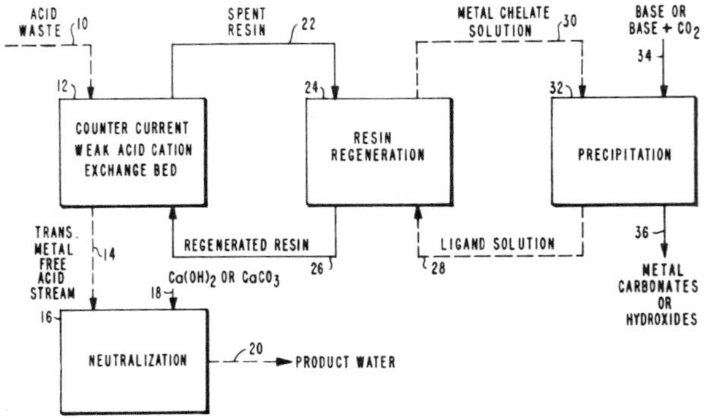

Process Flow Sheet

b.

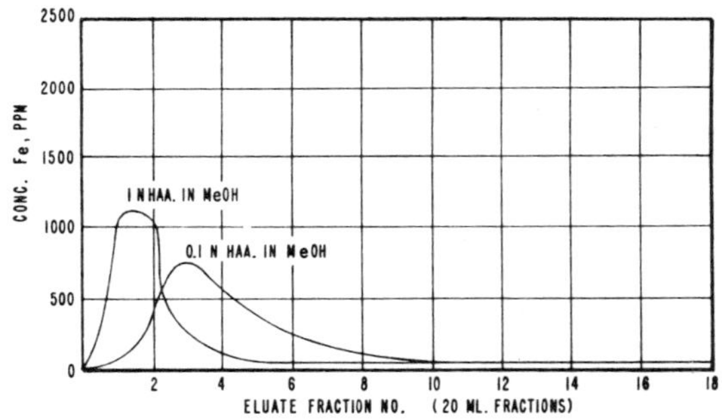

Effect of Acetylacetone Concentration on the Removal of Iron from Amberlite CG-50 Resin, Methanol Solution (Resin Loaded from $FeSO_4$ Solution, pH ~3.0)

c.

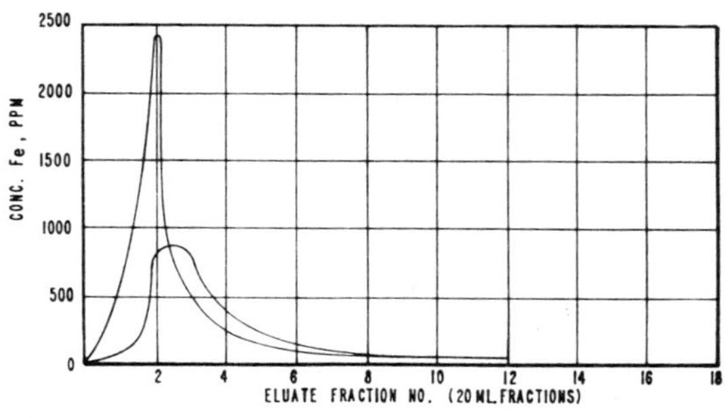

Effect of Acetylacetone Concentration on the Removal of Iron from Amberlite CG-50 Resin, Methanol Solution [Resin Loaded from $Fe_2(SO_4)_3$ Solution, pH ~3.0]

(continued)

FIGURE 3.2: (continued)

d.

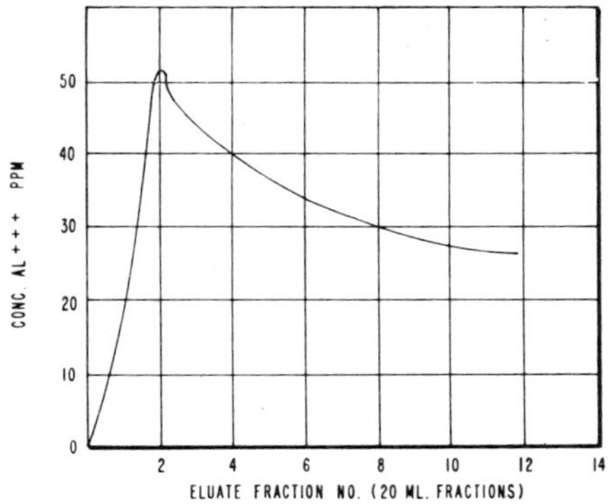

Removal of Aluminum from Amberlite CG-50 Resin
(Regenerating Sol = 0.1 N Acetylacetone in Methanol)

Source: D.O. De Pree; U.S. Patent 3,725,259; April 3, 1973

Example 2: Regeneration of Amberlite CG-50 Resins Loaded with Fe^{++} or Fe^{+++} — The recovery of iron from CG-50 resin loaded with ferrous ion was determined with 1 N and 0.1 N solution of acetylacetone in methanol. The eluate stream was dark red in color indicating the iron to have been oxidized to the ferric state either on the resin or on contact with the ligand solution (ferrous acetylacetone is not red in solution). Because both the loaded resin and the chelate represent essentially neutral compounds, oxidation with air should occur readily. No further oxidation as indicated by color change occured on air blowing a selected sample.

The concentration of iron in the eluate samples was determined by colorimetry [$Fe(AA)_3$ per methanol was found to follow Beer's Law] at a wavelength of 440 mμ using a Beckman DK-2 spectrometer. The results obtained are shown in Figure 2. These data are roughly comparable to those obtained with ferric loaded resin and show that iron is readily removed from CG-50 resin using a β-diketone reagent in methanol.

Similar results were obtained with the ferric loaded resin, Figure 3.2c. The recovery curve for ferric iron with 0.1 N acetylacetone is essentially identical to that obtained in the ferrous experiment. The higher, narrower peak obtained with the 1 N regenerating solution indicates a difference in eluate fraction take-off time. The peak concentration of iron is 2,400 ppm, indicates a relatively high efficiency.

Example 3: Regeneration of manganese loaded CG-50 resin was determined with acetylacetone in methanol. A total of two and one-half grams of manganese loaded resin [3.0 milliequivalents of Mn^{++}/gram resin] charged to a ¼ inch chromatographic column was eluated with 100 ml of a 1 N solution of acetylacetone in methanol. Three milliliters fractions of eluate were collected and analyzed for Mn^{++} by flame photometry at a wavelength of 403 mμ using a Beckman DK-2 spectrometer. The analyses of the eluate fractions shown below demonstrate the efficiency of the regeneration of the resin.

Fraction	Concentration Mn^{++}, ppm
1	4
2	0
3	112

(continued)

Fraction	Concentration Mn^{++}, ppm
4	456
5	496
6	512
7	504
8	472
9	504
10	480
15	464
18	504
22	480
24	469
27	464
30	440
33	432

Example 4: Regeneration of Amberlite CG-50 Resin Loaded with Al^{+++} — Regeneration of aluminum loaded CG-50 resin with 0.1 N acetylacetone in methanol indicates successful regeneration but at a far lower efficiency, Figure 3.2d. It is possible that equilibrium conditions were not reached and it is obvious that modified acetylacetone or alternate chelating agents may be far more efficient. Aluminum is not present in most acid waste streams but does occur in many Western United States acid mine drainage streams.

Example 5: Preparation of Synthetic Acid Mine Water — One liter of a synthetic acid mine water (composition furnished by Dr. J.M. Shackelford, Physical Science Administrator, Federal Water Pollution Control Administration, Washington, D.C.) was prepared by dissolving the following salts in 0.01 N sulfuric acid and diluting with 0.01 N sulfuric acid to one liter.

Salt	Weight, grams
$CaSO_4 \cdot 2H_2O$	0.344
$MnSO_4 \cdot H_2O$	0.024
$Al_2(SO_4)_3 \cdot 18H_2O$	0.186
$FeSO_4 \cdot 7H_2O$	0.997
$MgSO_4 \cdot 7H_2O$	0.246

The pH of the solution was found to be 2.4. The metal ion composition, as determined by emission spectroscopy, was found to be as follows in ppm: Fe, 205; Al, 20; Mn, 10; Mg, 27; and Ca, 95.

The pH of 100 ml of the synthetic acid mine water was adjusted to 4.5 with finely divided calcium carbonate. The addition was made very slowly because CO_2 evolution causes a gradual rising of the pH. This solution was then passed through a ¼" by 9" chromatographic column containing 2.5 grams (dry weight) of water swollen Amberlite CG-50 resin (Mallinckrodt, H^+ form). Three milliliter samples of eluate were collected and found to have a pH of 3.1 and were analyzed for the various cations by emission spectroscopy. The data contained was as follows:

	Cation Concentration, ppm				
Fraction	Fe	Al	Mn	Mg	Ca
1	<15	<15	-	- -	- -
5 and 6	157	<1	9	27	450
9 and 10	155	<1	8	22	300
15	187	<1	8	16	285
20	187	<1	7	18	370
25	182	<1	8	21	327
30, 31 and 32	168	<1	8	10	215

These data indicate that all of the aluminum, none of the manganese and approximately 25% of the iron were removed by a single pass. The increase in acidity resulting from the exchange of H^+ ion from the column reduces the efficiency of the transition metal removal.

The eluate fractions were combined, the pH readjusted to 4.5 and the treatment repeated using the same resin. An equivalent reduction in transition metal ions was achieved by the recycle as indicated by the following analysis. Twenty-five ml fractions of eluate were collected.

	Cation Concentration, ppm				
Fraction	Fe	Mn	Al	Mg	Ca
1	145	7.5	<2	19	265
2	124	8.0	<2	25	400
3	111	7.5	<2	24	360
4	140	7.5	<2	20	360

The spent cation exchange resin from the treatment (two cycles) described above was treated in the chromatographic column with 100 ml of a 2 N solution of acetylacetonate in methanol. The solution was prepared by diluting 20 grams of acetylacetonate with methanol to 100 ml. Three milliliter fractions of eluate were collected and analyzed by emission spectroscopy for iron, aluminum, manganese, magnesium, and calcium.

The results, shown below, indicate a very effective regeneration of the column by removal of the iron and aluminum. As manganese and little calcium or magnesium were deposited on the column, these ions are not present in the eluate.

	Cation Concentration, ppm				
Fraction	Fe	Mn	Al	Mg	Ca
1	12	<1	<1	<1	<10
2	600	70	<1	<1	<10
3	59	45	<1	<1	<10
4	13	50	<1	<1	<10
5	9	24	<1	<1	<10
6	8	14	<1	<1	<10
7	17	14	<1	<1	<10
8	4	11	<1	<1	<10
15	8	8	<1	<1	<10
20	4	7	<1	<1	<10

The chelate solution (dark red in color) from fraction 2 (approximately 2 ml remained after removing analytical sample) was treated with concentrated NaOH in methanol containing a trace of water (2%). A dense precipitate of $Fe(OH)_3$ settled leaving a straw colored solution. The precipitate was readily removed by filtration.

The water eluates from the cation exchange resin treatment of the synthetic acidic mine effluent were treated with calcium carbonate and adjusted to a pH of 6.0.

Electrolysis Technique Using Surface Active Agent

M. Ichiki and M. Ishii; U.S. Patent 3,766,035; October 16, 1973; assigned to Mitsui Mining & Smelting Co., Ltd., Japan describe a method of electrolytic purification of the wastewater containing the ions of metals such as Cd, Zn, Pb, Cu, Ni, Fe, Cr, Mn, Hg, Al, etc., which is characterized by the process: conducting the electrolysis upon adding 1 to 50 grams of an organic solvent per one m^3 of the liquid (i.e., wastewater) to undergo the electrolysis, together with a nonionic surface active agent necessary for emulsifying the organic solvent; and floating the solid matter made by the electrolysis in order to be separated.

Example 1: The wastewater arising from Cd plating was subjected to cyan decomposition and Cr reduction to be followed by neutralization. Subsequently, to the thus obtained supernatant (which contained about 2.0 ppm of Cd and about 0.3 ppm of Cr) was added a mixture of the organic solvent and surface active agent as shown in the following table (result of tests). By supplying the resulting emulsion to a 15 liter electrolytic cell, provided with the anodes consisting of aluminum plates and having the total anode area of 14 dm^2 and the cathodes consisting of iron plates and having the total cathode area of 3 dm^2, from its upper end at the flux of 150 ml/min while discharging the purified wastewater through the lower end of the electrolytic cell, electrolysis was conducted continuously by applying DC 1 amp. The wastewater coming out of the electrolytic cell after purification was subjected to sampling of regular intervals to analyze its Cd content and Al content. The results are shown in the following table.

As is evident from this table, the purified wastewater obtained through electrolysis of the emulsion prepared by adding a mixture of the organic solvent and surface active agent to the raw wastewater contained but a very small quantity of metals and showed a high clearness.

	Additive (gr./m.³ of waste water)					
	Non-addition		Solvent naphtha, 10; polyoxyethylene lauryl ether, 5		Water-soluble cutting fluid (manufactured by Johnson Co.), 45	
Analytical value	Cd (p.p.m.)	Al (p.p.m.)	Cd (p.p.m.)	Al (p.p.m.)	Cd (p.p.m.)	Al (p.p.m.)
Number of days following the start of electrolysis:						
1 day	0.2	38	0.02	6.0	0.01	5.7
2 days	0.4	44	0.02	6.3	0.01	5.7
3 days	0.1	52	0.02	6.8	0.01	5.7
4 days	0.2	81	0.01	6.1	0.01	5.6
5 days	0.2	80	0.01	6.1	0.01	5.8
6 days	0.2	102	0.02	6.2	0.01	5.8
7 days	0.2	97	0.01	6.2	0.02	5.8
8 days	0.1	97	0.01	6.2	0.01	6.0
9 days	0.25	99	0.01	6.3	0.01	5.8
10 days	0.2	86	0.01	6.1	0.01	5.6

Example 2: When the same method as in Example 1 was applied to various wastewaters containing the ions of heavy metals such as Zn, Pb, Ni, Fe, Cr, Mn, Hg, etc. respectively, it was possible to obtain a purified wastewater containing but an extremely small quantity of heavy metal ions and having a high clearness, respectively.

Removal of Ionic Metallic Impurities

J.P. McKaveney; U.S. Patent 3,766,036; October 16, 1973; assigned to Occidental Petroleum Corporation describe a process for removing ionic metallic impurities from aqueous solutions which comprises contacting an aqueous solution containing ionic metallic impurities with a solid particulate silicon-metal alloy; the alloy containing up to about 64.0 weight percent of silicon based upon the total weight of the alloy and the remainder of the alloy being composed of one or more metals, the alloy having a higher electrochemical potential than the metallic ionic impurities sought to be removed as compared to impurity metals when in their free metallic state. In the process, less obnoxious ions are introduced to the aqueous phase from the alloy and certain of the ions may be hydrolyzed and redeposited on the surface of the alloy.

In a case, the metallic alloys which can become loaded with ionic metallic impurities and are then removed from the aqueous phase, are regenerated by oxidizing the ionic metallic impurities away from the surface of the alloy particles, permitting reuse of the alloy particles in the purification portion of the system. The oxidation may be carried out by chemical and/or electrochemical means. In general, it is preferred to utilize alloys containing from about 1.0 to 65 weight percent silicon based upon the total weight of alloy.

In the process, it must be borne in mind that the silicon present in the alloy modifies the overall electrochemical potential of the alloy and thus the alloys have a lower electrochemical potential than the pure nonsilicon metals which the alloy contains. The preferred net electrochemical potential of the alloys found useable in the process ranges between from 0.35 to 2.5 volts (values in volts referred to herein are to the hydrogen-hydrogen ion couple as zero and are for unit activities and temperature of 25°C). The following examples illustrate the process.

Examples 1 through 6: Two aqueous solutions were prepared containing 25 parts per million of each of the following elements: arsenic, cadmium, copper, mercury and lead as ionic metallic impurities dissolved therein. Thirty grams per liter of sodium chloride were contained in one of the solutions. Fifty milliliter samples of each of the solutions were placed into 250 ml flasks, which flasks contained 50 ml volumes of various silicon-metal alloy particles (the particles were sized so that they could pass through USS 8 mesh sieve). The flasks were then stoppered and vigorously shaken on a mechanical shaker. The treated solutions were then filtered to remove all solids and the filtrates of each sample were collected and analyzed by standard atomic absorption analysis methods for the amount of arsenic, cadmium, copper, mercury and lead contained therein. The results of the analysis are given in the following table.

The compositions of the six different alloys are given in Column A and the numbers following the elements show the weight percent of the element in the alloy based upon the total weight of the alloy. Some of the alloys used in the examples contained trace amounts of other elements, i.e., carbon, manganese, etc. These trace quantities of other elements did not impair the efficiency of the operation of the process.

Alloy of Example 2 was not used to remove the ions from the salt water solution due to the fact that the alloy overreacted with the aqueous salt water solution (pH of 2.0) as indicated by excessive dissolution of the alloy. The alloys used to extract the ionic metal impurities in this example were easily regenerated by washing the alloys with a dilute aqueous solution of hydrochloric acid containing a samll amount of hydrogen peroxide.

		Water base, all at 25 p.p.m. to start					Salt water base (30 g. NaCl per liter) 25 p.p.m. to start				
Example	Column A alloy composition	Arsenic p.p.m.	Copper p.p.m.	Cadmium p.p.m.	Lead p.p.m.	Mercury p.p.m.	Arsenic p.p.m.	Copper p.p.m.	Cadmium p.p.m.	Lead p.p.m.	Mercury p.p.m.
1	Calcium 32.4 Silicon 63.1 Barium 0.36 Iron 4.1	2	<1	<1	<1	<1	<1	<1	<1	<1	<1
2	Aluminum 20.5 Barium 12.4 Calcium 10.6 Silicon 39.0 Iron 17.5	<1	<1	<1	<1	<1	Not treated				
3	Iron 30.1 Titanium 69.8 Silicon 0.03	<1	<1	25	<1	<1	14	13.4	25	25	<1
4	Iron 43.0 Aluminum 8.0 Silicon 49.0	16	<1	25	23	<1	16	3	25	23	<1
5	Iron 46.0 Magnesium 9.0 Silicon 45.0	<1	<1	<1	<1	<1	<1	<1	<1	<1	<1
6	Iron 49.9 Aluminum 1.0 Silicon 49.0 Calcium 0.1	13	21	25	25	<1	20	21	25	22	20

NOTE: <1 means that the material contained less than one p.p.m. and as such was undetectable in the solution.

Examples 7 through 15: For the following examples there was treated a solution containing as impurities ionic chromium, zinc and mercury and having a pH of 5.6. For Examples 7 to 11, the alloy employed (Alloy A) 46.0% by weight iron, 9.0% magnesium by weight and 45% silicon by weight and had a mesh size of +60 mesh. The amount employed in each instance was 150 grams.

For Examples 12 to 15, the alloy employed (Alloy B) contained about 32 to 33 weight percent calcium, the balance being essentially silicon. The amount employed in each test was 142 grams and the particle size was +80 mesh. In all tests the columns employed had 30 mm ID and were filled to 150 mm in depth with the alloy. The results are shown in the following table in which there was a magnesium exchange for Alloy A and a calcium exchange for Alloy B. The parent solution contained 7.6 ppm Cr (VI), 9.9 ppm Zn (II), and 99,000 ppb Hg (II).

Example	Eluate Volume Cut (ml)	Flow Rate (ml/min)	Cr (VI) (ppm)	Cr^{++} Percent Removal	Zn (II) (ppm)	Zn^{++} Percent Removal	Hg (II) (ppb)	Hg^{++} Percent Removal
7	400-500	10	5.4	29	0.01	99	2.6	99.9
8	500-550	1 hr, static	2.3	70	0.01	99	2.4	99.9
9	900-1,000	2.5	5.3	30	0.01	99	0.8	99.9
10	1,000-1,050	Overnite, static	<0.5	>93	0.01	99	2.0	99.9
11	pH = 3.0 separates	14	2.6	66	0.01	99	4.8	99.9
12	400-500	10	6.3	17	0.01	99	10.8	99.9
13	500-550	1 hr, static	6.0	21	0.01	99	47.2	99.5
14	900-1,000	2.5	7.2	5	0.01	99	3.6	99.9
15	1,000-1,050	Overnite, static	3.5	54	0.01	99	75.6	99.3

Precipitation and Centrifugation

J.M. Haviland; U.S. Patent 3,725,266; April 3, 1973; assigned to Haviland Products Company describes a process which comprises adding one or more precipitating agents to the metal compound contaminated wastewater to precipitate the metal compound contaminant to form an aqueous metal compound sludge.

After settling, e.g., in settling tanks or mechanical clarifiers, the aqueous metal compound sludge is separated from the supernatant liquid and transferred to a centrifuge. The aqueous metal compound sludge is then concentrated by centrifugation. The metal compound contaminants are then reclaimed from the sludge. For example, the concentrated metal compound sludge can be dissolved and transferred to an electrolysis tank where the free-metals can be recovered by electrolysis. Alternatively, the concentrated metal compound sludge can be transferred to a dissolution tank where the metal compound contaminant can be dissolved and sold, or used in solution form, or recovered from the solution in a purified state by recrystallization and sold or used in the dry form.

The particular precipitating agents used will depend upon the particular metal compound contaminants involved. The preferred precipitating agents are hydroxides and carbonates, such as, sodium hydroxide, calcium hydroxide and sodium carbonate, which produce the insoluble metal hydroxides and carbonates, respectively. In addition coagulants can be added to improve settling of the precipitate. It has been found that polyelectrolytes, such as for example, Separan and Primafloc produce especially satisfactory coagulation.

Examples 1 through 10: In Examples 1 through 9, the aqueous metal sludges were produced by precipitating the metal hydroxides from metal compound contaminated wastewater with sodium hydroxide. After settling, the sludges were then pumped to a Barret Clarifuge, for centrifugation from 1,400 to 2,100 gravities. In Example 10, the same procedure was followed, except the metal carbonate was precipitated with sodium carbonate.

Example	Aqueous Metal Sludge	Solid Conc. Before Centrifugation (ppm)	Flow Rate to Centrifuge (gal/min)	Solid Conc. of Sludge After Centrifugation (ppm)	Solid Conc. of Water Effluent (ppm)
1	Aluminum hydroxide	8,000	20	129,000	1,750
2	Aluminum hydroxide	25,000	15	155,000	1,500
3	Copper hydroxide	10,500	10	110,000	1,600
4	Chromium hydroxide	29,000	15	195,000	1,150
5	Chromium hydroxide	12,500	20	220,000	1,200
6	Nickel hydroxide	9,000	20	150,000	1,800
7	Nickel, chromium, copper and zinc hydroxides	17,000	15	185,000	1,200
8	Chromium, copper and zinc hydroxides	11,600	15	143,000	1,350
9	Nickel, chromium, copper and zinc hydroxides	60,000	15	400,000	3,000
10	Copper carbonate	6,000	10	110,000	1,600

Example 11: Concentrated aluminum hydroxide sludge produced according to the procedure of Examples 1 to 9 having a solids concentration of 155,000 ppm was dissolved in concentrated sulfuric acid to produce an aluminum sulfate solution useful in the paper and leather tanning industries and in phosphate removal operations in waste treatment plants.

Example 12: Concentrated chromium hydroxide sludge produced according to the procedure of Examples 1 to 9 having a solids concentration of 220,000 ppm was dissolved in concentrated sulfuric acid to produce a chromium sulfate solution useful in the leather tanning industry.

Example 13: Concentrated nickel hydroxide sludge produced according to the procedure of Example 1 to 9 having a solids concentration of 210,000 ppm was dissolved in fluoroboric acid and subjected to electrolysis in a DC electrolysis tank, resulting in the deposition of nickel metal on the cathode.

Multiple Matrix Magnetic Separation

J.J. Nolan, P.G. Marston and L.M. Lontai; U.S. Patent 3,770,629; November 6, 1973; assigned to Magnetic Engineering Associates, Inc. describe a multiple matrix magnetic separation device having an electromagnetic coil means for providing a magnetic field volume in the space encompassed by an electromagnetic coil. There are a plurality of magnetic matrices stacked within the space encompassed by the electromagnetic coil. A plurality of flow control means are disposed between the magnetic matrices and one on each end of the stack of magnetic matrices.

Each of these flow control means disposed between the magnetic matrices includes a distribution network for distributing the slurry to undergo magnetic separation to its corresponding adjacent magnetic matrix and a collection network for collecting the slurry from its other corresponding adjacent magnetic matrix after the slurry has undergone separation. The flow control means at one end of the stack of magnetic matrices includes one of the distribution and collection networks and the flow control means at the other end of the stack includes the other network. There are inlet means for delivering slurry to undergo separation to the distribution networks and outlet means for receiving from the collection networks slurry that has undergone separation.

The magnetic matrix generally includes a ferromagnetic material such as steel wool, steel balls, or steel tacks enclosed in a canister which may be made of a stainless steel or other material of low magnetic permeability. The electromagnetic coil, multiple matrices and flow control members may be enclosed in a ferromagnetic return frame to increase the efficiency of the magnetic circuit. The technique of stacking magnetic matrices adjacent to each other to achieve a predetermined matrix area instead of using a single matrix of the same area may be employed to reduce the volume of the ferromagnetic return frame surrounding the coil.

Each additional layer used in the stack to accomplish a particular area requirement reduces the diameters of the top and bottom portions of the return frame, the intermediate cylindrical portion and the electromagnetic coil and increases the length of the coil and cylindrical portion of the frame in the direction of slurry flow. The number of matrices used to accomplish a particular process capacity should be optimized for maximum efficiency and in some cases it may even be determined that a single layer is the optimum configuration. But in those configurations where two or more matrices arranged in a stack provide greater efficiency than a single large matrix the multiple matrix separation device becomes a particularly useful and efficient device.

There is shown in Figure 3.3a a multiple matrix magnetic separation device **10** including a ferromagnetic return frame **12** consisting of disc-shaped pole pieces or pole sections, upper and lower portions **14** and **16**, and a cylindrical intermediate portion **18** within which is disposed electromagnetic coil **20** which is cylindrical in shape. The pole sections, upper and lower portions **14** and **16**, are ferromagnetic members of substantial mass located closely adjacent the matrices.

The pole portions by their size and positions provide a low reluctance path for the magnetic field whereby the major portion of the magnetomotive force (MMF) provided by the coil is presented by the pole portions across the matrices and MMF drops in the rest of the are minimized.

FIGURE 3.3: MULTIPLE MATRIX MAGNETIC SEPARATION DEVICE

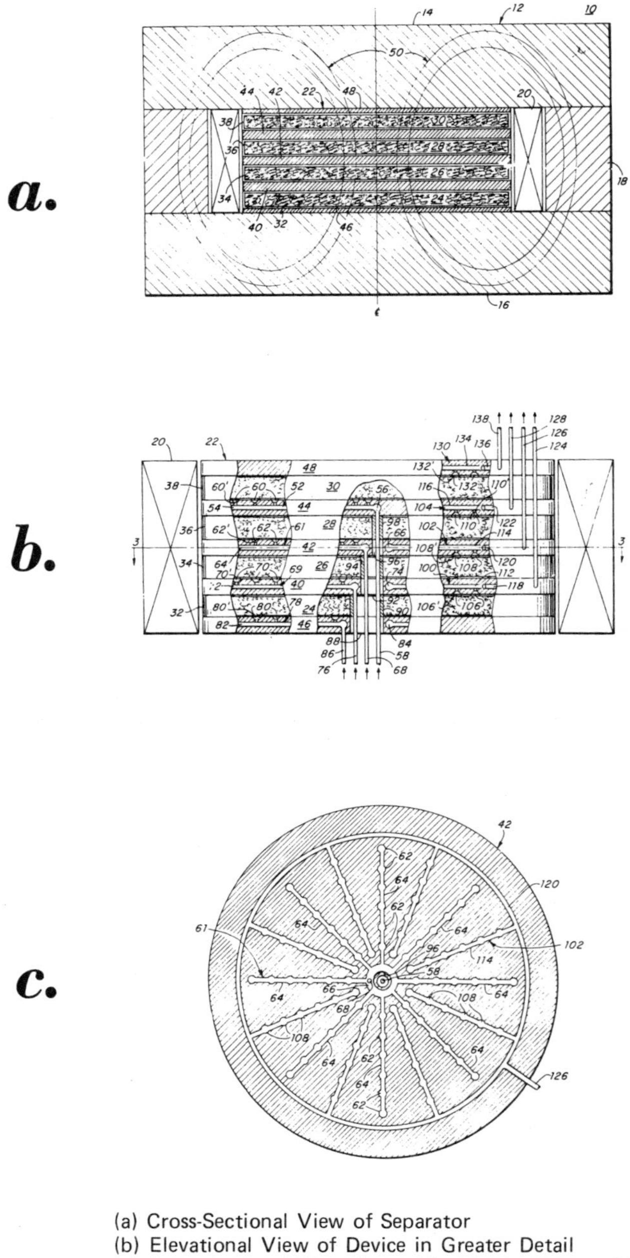

(a) Cross-Sectional View of Separator
(b) Elevational View of Device in Greater Detail
(c) Cross-Sectional View Along Line **3–3** of Figure 3.3b

Source: J.J. Nolan, P.G. Marston and L.M. Lontai; U.S. Patent 3,770,629; Nov. 6, 1973

The presence of the pole portions, particularly the parts directly above and below the matrices, serve to concentrate the field and direct it through the matrices parallel to the direction of slurry flow. Centrally located and encompassed by electromagnetic coil **20** is a multiple matrix array **22** which includes in this specific embodiment four steel wool magnetic matrices **24, 26, 28,** and **30,** each of which is contained in a stainless steel canister **32, 34, 36,** and **38,** respectively. Between each pair of these matrices are flow control members **40, 42,** and **44,** each of which contains a distribution network and a collection network, not shown in Figure 3.3a, but depicted in detail in Figure 3.3b.

At each end of the array **22** are two more flow control members **46** and **48**, each of which includes only one network, either a collection network or a distribution network, not shown here, but which appear in detail in Figure 3.3b. The determination of which type of network each of flow control members **46** and **48** includes depends upon the direction of flow of the slurry through the matrices: if the flow enters the matrices from the lower side and leaves them from the upper side, then flow control member **46** contains a distribution network and flow control member **48** contains a collection network.

If the slurry passes through in the other direction, the converse is true, as will be apparent from the description of Figure 3.3b. Inlet and outlet means for delivering and receiving slurry to and from the distribution and collection networks, respectively, are not shown in Figure 3.3a, but are shown in Figure 3.3b. The magnetic field **50** provided by electromagnetic coil **20** passes generally vertically through array **22** in a direction generally parallel to the direction of slurry flow through the matrices. The ferromagnetic return frame **12** which is useful to promote efficient use of the magnetic field **50** when the diameter of the coil is large and its length in the direction of the slurry flow is small, becomes less so as the multiple matrix device tends toward a smaller radius configuration: a coil longer in the direction of slurry flow but smaller in terms of radius. Such a reduction influences the magnetic design such that in some cases the return frame, particularly the portion **18**, may become of insubstantial value.

Flow control member **44**, Figure 3.3b, includes a distribution network **52** that feeds slurry to the matrix **30**. The distribution network includes a plurality of radial extending passages **54** which extend outward from the center of the member like spokes of a wheel and are fed by a central chamber **56** which is supplied by inlet pipe **58**. Each of passages **54** includes a plurality of ports **60**, each of which communicates with the lower edge of matrix **30**, through ports **60'** in canister **38**. Suitable sealing between the canisters and flow members is provided to prevent leakage. Sealing may be accomplished by cylinders surrounding the matrix and between the flow control members. In any event, suitable sealing is provided in array **22** to prevent leakage.

Similarly, flow control member **42** includes distribution network **61** having ports **62** disposed on radial passages **64** fed by annular channel **66** which in turn is supplied with slurry by inlet pipe **68**; flow control member **40** includes distribution network **69** having ports **70** in radial passages **72** fed by annular channel **74** supplied with slurry by inlet pipe **76.** Since the flow of slurry through the matrices of array **22** has been assumed as upward from bottom to top, flow control member **46** includes only a distribution network **78** whose ports **80** in radial passages **82** are fed by annular channel **84** supplied with slurry by inlet pipe **86**. Inlet pipes **58, 68** and **76** are accommodated in bore **88** of flow control member **46** and sealing sleeve **90** in matrix **24**. Inlet pipes **58** and **68** are accomodate as well by bore **92** in flow control member **40** and sealing sleeve **94** in matrix **26**; and pipe **58** is accommodated by bore **96** in flow control member **42** and sealing sleeve **98** in matrix **28**. Ports **62, 70** and **80** in flow control members correspond with ports **62', 70'** and **80'** in canisters **36, 34,** and **32,** respectively.

Each of flow control members **40, 42** and **44** also includes a collection network **100, 102, 104** having ports **106, 108, 110** connected to radial passages **112, 114, 116** which discharge into annular channels **118, 120, 122** which expel the slurry through outlet pipes **124, 126, 128,** all respectively. Since the direction of slurry flow through the array **22** is upward from bottom to top, flow control member **48** also includes a collection network **130** having ports **132** connected with radial passages **134** which discharge into annular chamber **136** that removes the slurry through outlet pipe **138.** Ports **106, 108, 110** and **132** in flow control members **40, 42, 44** and **48** correspond with ports **106', 108', 110'** and **132'** in canisters **32, 34, 36,** and **38,** respectively.

Although as shown in Figure 3.3b the slurry is fed into the matrices by inlet pipes which enter in the center of the array and is removed by outlet pipes which are connected to the periphery of the array this is not a limitation. The positions of the inlet and outlet pipes may as well be reversed, or some of the inlet pipes may be at the center and some at the periphery and likewise with the outlet pipes, or all pipes could be at the center or at the periphery. Various other delivery removal schemes may be used. Inlet pipes **58, 68, 76, 86** are connected in parallel to a source of raw feed and outlet pipes **124, 126, 128** and **138** are connected in parallel to an output line or other receptacle.

The distribution network **61** and collection network **102** in flow control member **42** is shown more completley in Figure 3.3c. Distribution network **61** includes eight passages **64**, each of which contains six ports **62** that feed slurry to magnetic matrix **28.** Each of passages **64** extends radially outwardly from annular channel **66** which is concentric with and located near the center of flow control member **42.** Annular channel **66** receives slurry from inlet pipe **68** disposed in bore 92. Collection network **102** includes a plurality of passages **114** extending radially inwardly from annular channel **120** which remove the slurry through outlet pipe **126.** Each passage **114** includes six ports **108.**

The distribution and collection networks shown in Figures 3.3b and 3.3c are but one configuration for such networks and many other designs and variations are possible. For example, ports may be added to annular channels **66** and **120.** The channels and passages may be progressively reduced in diameter as the distance from the inlet and outlet pipes increases: the diameter of the channel **66** and passages **64** may be reduced as the distance from the inlet pipe **68** increases. Similarly, the passages **64** and channel **66** may equally as well operate as a collection network and the passages **114** and channel **120** operate as the distribution network. In addition, the generally radial design pictured in Figure 3.3c is not a necessity to either the distribution or collection networks as any suitable rectilinear, curvilinear, or combinations of paths and shapes may be used.

V. Haensel; U.S. Patent 3,696,032; October 3, 1972 describes a process in which relatively large premagnetized iron particles are introduced into a fluid stream to scavenge smaller magnetic particles. The fluid stream is then passed through an annular space created by an open-ended conduit supported within a casing. A train of permanent magnets is moved through the conduit to attract the scavenged magnetic particles to the outer surface, and the particles are swept along the surface by the traveling magnetic field to a particle collecting chamber.

MERCURY

Sodium Sulfide

J.-M. Lalancette; U.S. Patent 3,790,370; February 5, 1974; assigned to Universite de Sherbrooke, Canada describes a process for removing metal ions such as mercury, copper, cadmium, zinc, lead, nickel or silver from polluted waters. The process involves causing the metal polluted water and a sulfur-containing precipitating agent to diffuse through peat whereby a water-insoluble sulfide of the metal is formed and is adsorbed on the peat and the metal-free water is recovered.

The process is particularly suitable for the treatment of mercury-containing wastewater and the removal of mercury from such water is described in detail by reference to Figure 3.4 which is a flow sheet illustrating the separation of mercury from wastewater and the recovery of mercury.

FIGURE 3.4: REMOVAL OF MERCURY FROM WASTEWATER USING SODIUM SULFIDE

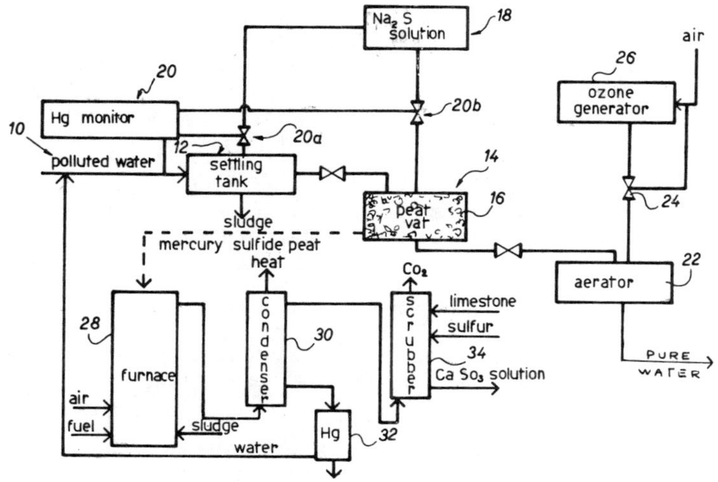

Source: J.-M. Lalancette; U.S. Patent 3,790,370; February 5, 1974

From the settling tank **12**, the water containing mercury either solution, emulsion or suspension in colloidal or in soluble salt from is transferred to the precipitating and filtering vat **14** containing commercial peat **16**. The peat **16** may be previously impregnated with a a precipitating solution **18 of** sodium sulfide or the precipitating solution **18** and the water from the settling tank are allowed to diffuse together through the peat.

Alternatively, the polluted water from the tank is transferred to the peat vat and then the sodium sulfide solution is allowed to diffuse through the polluted water impregnated peat vat. Upon contact of the mercury contaminated water and of the sodium sulfide there is formed a water-insoluble precipitate of mercury sulfide which is adsorbed on the peat.

The filtered water is then allowed to stand in aerator **22** and any excess sodium sulfide therein can readily be oxidized by aeration through adjustment of valve **24** or treatment with ozone from the ozone generator **26**. The concentration of the precipitating solution is adjusted according to the concentration of the mercury in the wastewater through the mercury monitor **20** which will also control the flow of the sodium sulfide solution **18** into the settling tank **12** through valve **20a** or through the peat vat **14** through valve **20b**. An amount slightly higher than the stoichiometric amount required to convert the mercury to insoluble mercury sulfide is used.

Alternatively, the mercury contaminated wastewater and sodium sulfide may be mixed and stored in the tank **12** and the mixture then is allowed to diffuse through vat **14**. Another possibility is that before allowing the mercury contaminated water to diffuse through vat **14**, the peat **16** is first saturated with the sodium sulfide solution or the polluted water.

In practice it has been found that the pH of the mercury contaminated water can vary from 3 to 12 but when the pH is outside of this range the pH of the water can be adjusted by acidification with an acid such as hydrochloric acid or alkalinization with an alkaline metal hydroxide such as lime or sodium hydroxide.

With regard to the peat the amount used is preferably in a ratio of at least 15 to 1 in weight of the precipitating agent, this ratio being preferred because it avoids and adverse effects of the sodium sulfide or other precipitating agents used on the structure of the peat moss.

In fact lower ratios of peat to precipitating solution will cause disintegration of the peat structure thus voiding its adsorption action. Also when using this ratio, the peat moss will easily retain from 5 to 7% of its weight of mercury.

The peat containing the mercury sulfide is removed from the peat vat, compressed and dried for a few days to bring the moisture content down to about 50%. It is then transferred to a furnace **28** and burnt at the lower portion of the furnace. The burning is carried out with fuel and air at a temperature of from 400° to 500°C, the air intake being adjusted to obtain a minimum of smoke and the flue gases formed as submitted to pyrolytic temperatures in the upper portion of the furnace **28**, for example 500° to 700°C before leaving the furnace.

The pyrolytic temperature will avoid the formation of dimethylmercury or other volatile mercury derivatives which are a most toxic gas. Also with the pyrolytic temperatures there is avoided the production of tar or other condensable gas while favoring the formation of water and mercury vapors and sulfur dioxide.

The gases are then passed through a water-cooled condensor **30** whereby metallic mercury is recovered from the condensed water **32**. Any uncondensed vapors of mercury and sulfur dioxide are then eliminated by passing in a scrubbing tower **34** containing limestone and elemental sulfur. There is formed some calcium sulfite which can be used in the pulp and paper industry.

Carbon dioxide gas is also formed in the scrubber **34** and is allowed to escape in the atmosphere. There may also be formed some mercury sulfide which can be reprocessed in the furnace with the next batch. The condensed water which represents about 1% of the treated water may be recycled with the polluted water **10**. The following examples illustrate the process.

Example 1: A solution (1 liter) of 2% of mercury (as $HgCl_2$) was stirred in the presence of metallic mercury. After allowing the solution to settle, it was flown over five pads of peat, 1 inch thick and 5 inches in diameter. Each pad weighed approximately 55 grams and had been impregnated with a 1% solution of Na_2S. The effluent water had the same volume as the starting solution, since the peat was saturated with moisture. The concentration of mercury was lower than 1 ppb as determined with a "dial Atom" apparatus (Jarrell Ash) with a Hg hollow cathode.

After this treatment, the peat was wrung and left to dry for a few days. The moisture content of the peat was then decreased from 85% to a value of the order of 50%, the weight being reduced accordingly. This dry peat was then burned at 400°C using 0.1 cc of air per gram of peat per minute. The effluent gases were heated at 600°C for approximately 30 seconds, before leaving the furnace. The condensate was collected and the mercury recovered by decantation. Starting with a solution containing 2.71 grams of $HgCl_2$, 2.0 grams of mercury was recoverd, a quantitative yield.

The cooled gases were passed through a 1" x 24" tube containing wet limestone chips about one-fourth inch in diameter mixed with 5% in weight of sulfur. The removal of SO_2 was quantitative and no mercury vapors could be found in the effluent gases.

Example 2: By proceeding in the same manner as in Example 1 and using an acid solution of mercury acidified to pH 3 with HCl, the results were identical to Example 1.

Example 3: The procedure of Example 1 was used using a basic solution of mercury made alkaline to pH 12 with sodium hydroxide. The results were identical to Example 1.

Example 4: The procedure of Example 1 was used except that the water solution of mercury was rendered "hard" by addition of 1% in weight of calcium ($CaCl_2$), magnesium ($MgSO_4$) and iron [$Fe_2(SO_4)_3$]. The results of the experiment indicated that the method was not affected by the hardness of water, the elimination of Hg being quantitative.

Example 5: A solution of 1.35 grams of cupric chloride in 1 liter of water was treated as in Example 1. The effluent water (approximately one liter) contained no copper (test by H_2S). In the ashes of the combustion of peat, a mixture of copper and copper oxide was found weighing 0.70 gram.

Inorganic Sulfides

W.E. Dean and C.M. Dorsett; U.S. Patent 3,674,428; July 4, 1972; assigned to PPG Industries, Inc. describe a process for removing mercury from an alkaline solution having a pH of at least 7. The process involves intermixing with the solution a sulfur compound in an amount sufficient to provide sulfide ion in an amount greater than the stoichiometric equivalent of mercury, oxidizing the soluble mercury polysulfide and recovering the insoluble mercury sulfide.

The sulfur compound employed can be any soluble or sparingly soluble sulfur compound which will provide sulfide ion in the solution. Exemplary of suitable compounds are the alkali metal sulfides such as sodium sulfide, potassium sulfide, and lithium sulfide; the alkaline earth metal sulfides such as magnesium sulfide, calcium sulfide, strontium sulfide, or beryllium sulfide.

W.E. Dean and C.M. Dorsett; U.S. Patent 3,749,761; July 31, 1973; assigned to PPG Industries, Inc. describe a process for removing mercury from solution and particularly from mercury electrolytic cell effluent. The process comprises intermixing with the solution a sulfur compound in an amount sufficient to provide sulfide ion to react with the mercury and other ions present therein which react with sulfide ion and treating the resultant solution with an adsorbent, preferably activated carbon, to remove the soluble mercury.

Water-Insoluble Organic Sulfides and Disulfides

R.P. Williams; U.S. Patent 3,769,205; October 30, 1973; assigned to Philips Petroleum Co. has found that mercury compounds soluble in water can be removed by extraction with one or more members of the group consisting of water-insoluble organic sulfides and disulfides. Using this process, all but trace amounts of the contaminant mercury can be removed from the water so treated. A particularly preferred alkyl sulfide is di-n-butyl sulfide. A particularly preferred alkyl disulfide is di-n-butyl disulfide. The following example illustrates the process.

Example: In order to demonstrate the effectiveness of this process, standardized solutions of mercuric chloride (14,000 ppm) and mercurous chloride (3.45 ppm) in deionized water were prepared and submitted to successive extractions with di-n-butyl sulfide, di-n-butyl disulfide and a mixture containing both. The mixture is obtained as a kettle bottom concentrate in the commercial synthesis of n-butyl mercaptan. That mixture, termed di-n-butyl sulfide concentrate, had the following analysis:

Components	Weight Percent
Lights	4.5
Di-n-butyl sulfide	87.9
Di-n-butyl disulfide	3.6
Unidentified heavies	3.2
Mercaptan sulfur	0.78
Specific gravity, 0.8481	

The extractions were carried out at 25°C by stirring the aqueous solution of the mercury salt with 20% by volume of the extracting agent for thirty minutes. After a quiescent period during which the layers separated, an aliquot of the aqueous phase was removed and analyzed by atomic absorption spectroscopy. Deionized water was used to minimize background absorption by other materials present in distilled water and thus simplify analysis. In order to obtain reproducible analyses all glassware was cleaned with sodium dichromate cleaning solution prior to use. The results of these extractions are summarized in the following table.

Residual Mercury Values in Aqueous Mercuric Chloride Solutions

Original mercury value (p.p.m.)	No. of extractions	Extracting agent	Residual mercury values (p.p.m.)
14,000	0	di-n-Butyl sulfide	14,000
	1	do	86.6
	2	do	2.52
	3	do	0.40
14,000	0	do.[1]	14,000
	1	do.[1]	62.2
	2	do.[1]	0.2
	3	do.[1]	0.13
14,000	0	di-n-Butyl disulfide	14,000
	1	do	11,200
	2	do	9,600
	3	do	8,600

[1] Concentrate.

The above data show that di-n-butyl sulfide concentrate was somewhat more effective than di-n-butyl sulfide in removing mercury from aqueous mercuric chloride solution. The di-n-butyl disulfide was least effective in removing mercury from aqueous mercuric chloride.

Residual Mercury Values in Aqueous Mercurous Chloride Solutions

Original mercury value (p.p.m.)	No. of extractions	Extracting agent	Residual mercury values (p.p.m.)
3.45	0	di-n-Butyl sulfide	3.45
	1	do	0.066
	2	do	0.0078
	3	do	0.0029
3.45	0	do.[1]	3.45
	1	do.[1]	0.15
	2	do.[1]	0.015
	3	do.[1]	0.011
3.45	0	di-n-Butyl disulfide	3.45
	1	do	0.11
	2	do	0.025
	3	do	0.026

[1] Concentrate.

The above data show that di-n-butyl sulfide was somewhat more effective than di-n-butyl sulfide concentrate in removing mercury from aqueous mercurous chloride solution. The di-n-butyl disulfide was least effective in removing mercury from aqueous mercurous chloride.

Sodium Borohydride

A process described by *E.L. Cadmus; U.S. Patent 3,764,528; October 9, 1973; assigned to Ventron Corporation* involves the removal of soluble inorganic mercury compounds from waste streams by mixing with the stream an amount of sodium borohydride about 100 to 150% in excess of the stoichiometric amount required to reduce its inorganic mercury content to metal as a precipitate. If organic mercurials are present in the stream, they are converted to soluble inorganic mercury salts by chlorination of the stream before mixing the stream with borohydride.

A sample of process waste from the manufacture of phenyl mercuric acetate contained 8.86 g/l of mercury, 4.34 g/l as inorganic mercury (mercuric acetate), and the balance, 4.52 g/l as PMA. The solution was made alkaline with NaOH to a pH of 10, and chlorine gas passed through. Samples were withdrawn periodically, boiled to expel chlorine, treated with $NaBH_4$ and filtered:

Sample	Chlorination Time (minutes)	--- Sodium Borohydride Recovery--- %Hg as Metal	%Hg not Recovered
1	0	49.8	6.8
2	15	68.3	4.0
3	30	95.7	0.2
4	60	98.8	0.0
5	150	99.6	0.0

An industrial application takes process waste from both inorganic and organic mercury production, and after treatment of the organic stream to remove organic mercury, a waste solution containing 30 to 300 ppm of mercury is treated at pH 7, with 3 gallons, of 12% $NaBH_4$ per 5,000 gallons of waste. After a reaction time of 1 hour, the slurry is pumped to a settling tank where the bulk of the mercury sludge settles. The solution decanted from the settling tank contains about 10 ppm of Hg suspended and then discharged as plant effluent. When the filters were of 5 micron size, Hg in the effluent was 3 to 4 ppm; when 1 micron filters were used, the concentration of Hg in the final effluent was 0.5 ppm. The Hg level can be reduced even further, to less than 0.1 ppm, if a final filtration through activated carbon is used.

Mercaptide in Carrier Material

N. Yokota, S. Tokuda, Y. Ito and H. Takatomi; U.S. Patent 3,755,161; August 28, 1973; assigned to Osaka Soda Co., Ltd., Japan describes a treatment process for removing metals from a gaseous phase or liquid phase metal-containing material using a solid treating agent consisting of a carrier material and a compound deposited thereon, the compound being selected from (a) an organic compound capable of forming a mercaptide compound of a metal, which compound has the —SH radical or an alkali salt in its molecule and may also have a member or members selected from the group consisting of the —N=, —S—, —NH—, —N=N— and —NH—NH— radicals in the same molecule; (b) an organic compound capable of forming a mercaptide compound of a metal, which compound has the >C=S radical in its molecule and also have a member or members selected from the group consisting of the —N=, —S—, —NH—, —N=N—, and —NH—NH— radicals in the same molecule; and (c) an organic compound capable of forming a chelate compound of a metal, which compound has the –OH radical in its molecule and also a member or members selected from the group consisting of the –N= and $-NH_2$ radicals in the same molecule.

Examples 1 and 2 and Control 1: Three hundred milliliters of carbon tetrachloride in which were dissolved 150 mg of dithizone were flowed down through a glass tube of 8 mm inside diameter packed with 5 grams of active carbon thereby causing the adsorption to and deposition on the active carbon layer of dithizone. The amount of dithizone adsorbed and deposited was 150 mg. This dithizone-supporting carrier was left standing for 24 hours to cause the carbon tetrachloride to volatilize. This was followed by flowing a saline solution prepared from mercuric chloride and sodium chloride (mercury content 1.4 mg/l) down through the tube at the rate of 130 cc per hour. This test was repeated ten times. In every case the mercury concentration of the effluent became less than 5 ppb.

Further, about 5 grams of the dithizone-supporting carrier prepared as in Example 1 were added to 200 ml of saline containing 1.4 mg/l of mercury, the contact being carried out for one hour with gentle stirring (Example 2), with the result that the mercury concentration of the aqueous solution became less than 5 ppb.

A control, the experiment was carried out exactly as in Example 1 except that the adsorption to and deposition on the active carbon of the dithizone was not carried out and 5 grams of active carbon were used (Control 1). The test was repeated 10 times and in every instance the mercury concentration of the effluent was above 420 ppb. The results obtained in the foregoing experiments are shown in the following table.

Experiment	Organic Compound	Organic Compound Adsorbed (wt. %)	Carrier	Carrier Weight (grams)	Metal Removed	Metal Concentration (ppb) Before Treatment	After Treatment
Example 1	Dithizone	3	Active carbon	5	Hg	1,400	<5
Example 2 (batch method)	Dithizone	3	Active carbon	5	Hg	1,400	<5
Control 1	- -	-	Active carbon	5	Hg	1,400	>420

Example 3 and Control 2: The experiments were carried out as in Example 1 and Control 1 except that 2-mercaptobenzothiazole and silica gel were used as the organic compound and carrier, respectively, and the carrier weight, amount of the organic compound desposited thereon by adsorption and the Hg concentration of the material treated were varied. The results obtained as shown in the following table.

Experiment	Organic Compound	Organic Compound Adsorbed (wt. %)	Carrier	Carrier Weight (grams)	Metal Removed	Metal Concentration (ppb) Before Treatment	After Treatment
Example 3	2-mercaptobenzothiazole	5	Silica gel	10	Hg	13,900	<5
Control 2	2-mercaptobenzothiazole	-	Silica gel	10	Hg	13,900	>13,000

Anthracite Coal

P. De Angelis, A.R. Morris and A.L. MacMillan; U.S. Patent 3,736,253; May 29, 1973; assigned to Sobin Chlor-Alkali, Inc. describe a method for removing mercury from water contaminated with mercury comprising contacting water-containing metallic mercury with finely divided anthracite coal. The mercury laden water can be first treated with sodium borohydride to reduce dissolved mercury to the metallic form. The water treated with reducing agent, is filtered through a pressure leaf filter coated with a filter aid, contacted with anthracite coal and contacted with chelating resin.

The anthracite coal used to remove the mercury is a standard finely divided anthracite coal used for filtering drinking water such as Anthrafilt or Filt-O-Cite. One commercial material has a density of approximately 52 pounds per cubic foot, a volume of about 39.2 cubic feet per ton, a specific gravity of about 1.52 and an angular shape of grain.

Generally, the finely divided anthracite coal particles range in size from about size 7, having a particle size of 2⁷⁄₁₆ inch by 1⅝ inch to about size ½ having an effective particle size of 0.10 to 0.20 mm. It is particularly preferred to use about size No. 1 having an effective particle size of from about 0.60 to 0.80 mm.

The preferred borohydride is that normally supplied as a stabilized water solution (SWS) which is 12% $NaBH_4$, 42% NaOH, with the balance being water. The amount added is calculated based on the weight of mercury in the wastewater. Theoretically, sodium borohydride has 8 electrons available and mercuric mercury can accept 2 electrons. Therefore, one molecule of sodium borohydride can reduce 8 atoms of mercuric mercury. Converting this to weight, 38 pounds of sodium borohydride can reduce 800 pounds of mercuric mercury which is a weight ratio of 1:21.

The preferred chelating resin used is a commercially available chelating resin known as Resinous Mercury Adsorbent. The particular chelating resin used, however, is not critical. In place of the preferred chelating resin one can use the chelating resin of U.S. Patent 3,083,070. The following examples illustrate the process.

Example 1: A sample of water containing finely dispersed metallic mercury and/or other mercury compounds in solution was passed through a Gooch crucible filled with 3 grams of No. 1 anthracite coal filter media to determine its effectiveness. The sample showed 0.760 ppm Hg before filtration and 0.0091 ppm after filtration. This represents a 98.8% removal of the Hg.

Example 2: About 5 gal/min of Hg laden water which had been treated with $NaBH_4$ was fed to a plant size filter bed of No. 1 anthracite coal. Runs I and II were each run continuously and tests taken when the run was initiated and after the time period was ended (see table).

Run	Test	ppm Mercury Before	After
1	Initial	0.66	0.075
	Next day	0.66	0.123
2	Initial	0.53	0.008
	Two days later	1.6	0.340

The anthracite coal causes a major reduction of the concentration of mercury in the water even when the initial mercury level is quite low. Moreover, the anthracite coal maintains its effectiveness for a relatively long period of time.

Example 3: Sodium borohydride was added to a mercury laden stream flowing at a rate of 3.0 gallons per minute. The mercury was reduced and the stream filtered in series through an Adams Filter containing porous carbon tubes, a Cuno porous cartridge filter having a pore size of 1 micron and a 50 gallon drum filled with finely divided anthracite coal. The following mercury levels in parts per million mercury were obtained.

Date	Inlet Adams filter	Inlet Cuno	Inlet drum filter	Outlet drum
4/12:				
12:30 p.m.	5.9	.62	.32	.06
8:45 p.m.	7.7	3.0	.50	.092
4/13:				
9:45 p.m.	20.0	.42	.24	.088
1:45 p.m.	18.4	3.5	3.2	1.4
7:00 p.m.	5.5	8.0	7.6	3.7
4/14	8.1	4.1	4.1	1.8

Nut Wastes

According to a process described by *A.C. Waiss, Jr., J.A. Kuhnle and M.E. Wiley; U.S. Patent 3,719,473; March 6, 1973; assigned to U.S. Secretary of Agriculture* water containing mercury is contacted with a nut waste, whereby the mercury is removed from the water. In a typical example, the mercury-containing water is passed through a bed of the nut waste contained in any suitable vessel. The effluent, no longer contaminated with mercury, can then be reused, for example, in industrial processes, or it can be deposited in a river, lake, or stream without contaminating such water course.

In the process, the nut waste acts as an absorbent in that mercury dissolved in the entering water as inorganic or organic compounds becomes bound to the waste. The action is thus not a mere matter of filtration, but one believed to involve chemical combination between the dissolved mercury and the waste.

Peanut skins are preferably employed as the absorbent. The process encompasses the use of nut wastes of all kinds, for example, hulls, skins, woody shells, expeller meal, and pellicles of walnuts, peanuts, pecans, almonds and chestnuts. All of these substances have the ability to absorb a substantial quantity of mercury per unit weight of the material. For Example, 100 parts of peanut skins will absorb 30 to 80 parts of mercury.

When the absorbent has taken up so much mercury that it loses its effectiveness, it is treated to recover its content of mercury. This is done by transferring the material to a combustion chamber provided with an arrangement for passing the evolved vapors through a condenser. In the combustion chamber the material is heated at a temperature which converts the organic matter to carbon. In the presence of this carbon, a known reducing agent, the absorbed mercury is converted to elemental mercury, which is collected in the condenser. In this way the mercury is recovered and can be recycled in industrial processes, e.g., electrolysis of brine, preparation of fungicides, etc. The process is demonstrated by the following examples.

Example 1: Nut Waste Absorption of Mercuric Ion from a Concentrated Solution — Peanut skins were ground and sieved to pass 1 mm openings. Then, 5 grams of these skins were packed in a column 30 cm in height and 12 mm in inside diameter. These skins were washed with twenty times their volume of water.

The column was loaded with 5 ml of an aqueous solution of mercuric acetate containing approximately 100 mg of mercuric ion per ml. The column was then washed three times with 100 ml of distilled water. These washings were analyzed for mercuric ion content by atomic absorption, which is sensitive to 1 ppm of mercury ion. Walnut expeller meal was prepared in a similar manner to that of peanut skins, and applied in the same manner.

Nut Waste Absorption of Mercuric Ion from a Concentrated Solution

Absorbent	Mercuric ion applied, mg.	Murcuric ion concentration wash— 1	2	3	Total mercuric ion absorbed, mg.
Peanut skins	495	0	0	0	495
Walnut expeller meal	495	0	0	0	495

Example 2: Nut Waste Absorption of Mercuric Ion from a Dilute Solution — The procedure described in Example 1 was followed. However, in this case the columns were loaded with a mercuric acetate solution containing approximately 1 mg of mercuric ion per ml. The results are summarized in the following table.

Nut Wash Absorption of Mercuric Ion from a Dilute Solution

Absorbent	Volume of solution applied, ml.	Mercuric ion applied, mg.	Mercuric ion concentration, wash— 1	2	3	Total mercuric ion absorbed, mg.
Peanut skins	575	575	0	0	0	575
Walnut expeller meal	495	495	0	0	0	495

Keratin Derivatives

According to a process described by *M. Friedman; U.S. Patent 3,725,261; April 3, 1973; assigned to U.S. Secretary of Agriculture* mercury is effectively removed from water or other liquids by contact with a keratin derivative, e.g., reaction products of reduced wool with N-vinylimidazole, N-vinylpyrrolidone, or 2-vinylpyridine. The following examples illustrate the process.

Example 1: Preparation of Keratin Derivatives — Reduced wool was prepared by treating 10 grams of wool with 4 ml of tri-n-butylphosphine in a mixture of 350 ml of pH 7.8 tris buffer and 350 ml of n-propanol, and shaking the mixture for 2 days at room temperature under nitrogen. After this period, 4 ml of 2-vinylpyridine was added to the mixture containing reduced wool. The system was flushed with nitrogen and shaken at room temperature for an additional 2 days.

The modified wool was removed from the mixture, washed with water, and air-dried. Reduced wool modified with other vinyl monomers were prepared as described above, replacing the 2-vinylpyridine by N-vinylimidazole and N-vinylpyrrolidone, respectively.

Example 2: Mercury Absorption Tests — Samples of the keratin derivatives prepared in Example 1 and a sample of untreated wool were tested for their ability to absorb mercury, using the following procedure in each case.

One gram of the keratin material was treated with 100 ml of 0.2 M aqueous mercuric chloride solution for 24 hours at room temperature. The keratin material was then removed from the solution, washed thoroughly, and analyzed for mercury by atomic absorption technique. The results obtained are tabulated below:

Sample	Modifying agent	Mercury absorbed grams of Hg per gram of keratin
1	N-vinylimidazole	0.6
2	N-vinylpyrrolidone	0.7
3	2-vinylpyridine	0.8
Control	None (untreated wool)	0.5

Example 3: Mercury Absorption Tests — In this example the most effective keratin studies (the 2-vinylpyridine derivative) was compared with native and reduced wool. The use of native and reduced wool in this example is by way of comparison only; they form no part of this process.

In runs a, b, and c, 1 gram of the keratin material was treated with 50 ml of 5×10^{-4} M (100 ppm) aqueous mercuric chloride solution for 24 hours at room temperature. The keratin material was then removed from the solution, washed thoroughly, and analyzed for mercury by an atomic absorption technique. Runs d, e and f were the same as above except that a 5×10^{-3} M aqueous mercuric chloride solution was used. The results are summarized below.

Run	Absorbent	Concentration of mercury in solution, p.p.m. Initial	Final	Amount of mercury absorbed by keratin material, p.p.m.
a	Native wool	100	1.8	98.2
b	Reduced wool	100	0.1	99.8
c	Reaction product of reduced wool and 2-vinylpyridine.	100	0.3	99.7
d	Native wool	1,000	225	775
e	Reduced wool	1,000	1	999
f	Reaction product of reduced wool and 2-vinylpyridine.	1,000	16	984

Amalgam Formation

P.F. Waltrich; U.S. Patent 3,704,875; December 5, 1972; assigned to Pennwalt Corporation describes an apparatus for the treatment of effluent streams containing very low concentrations and even trace amounts of mercury in both elemental form and as mercuric ion. The stream is passed through a closed vessel having a bed of reactive metal particles, such as zinc, magnesium, aluminum, and iron, so that the mercury forms an amalgam on the metal surface. The bed is regenerated and the mercury recovered by heating the bed while the vessel is under vacuum and condensing the vaporized mercury. The lower temperature permitted by the reduced pressure and the absence of air precludes oxidation and clumping of the metal granules.

Recovery from Sediment

V.A. Stenger; U.S. Patent 3,679,396; July 25, 1972; assigned to The Dow Chemical Co. describes a method for recovering mercury from sediment at the bottom of a body of water by overlaying the sediment with a layer of metal containing at least one of aluminum, magnesium and zinc. Preferably, the metal is in sheet form for more efficient interception of mercury vapor rising from the sediment. After removal of the metal from the water body, the mercury, which forms an amalgam on the surface of the metal, may be recovered by distillation from the metal surface.

Example 1: A sludge containing 600 ppm by weight of mercury was obtained from the bottom of a mercury contaminated pond. Seven pellets of particulate zinc weighing about 0.8 gram each were washed with dilute hydrochloric acid (1:10), rinsed and air dried. Six of the pellets were then mixed in a flask with 10.2 grams of the sludge and 30 ml of water. This quantity of sludge contained 6,100 μg of mercury. The flask was allowed to stand at ambient temperature for an extended period. At certain intervals, one pellet was removed, rinsed with water, and analyzed for associated mercury.

	Exposure, days	Mercury found, micrograms
Pellet No.:		
1	[1] 0	0
2	2	11
3	7	39
4	12	62
5	20	73
6	40	119
7	165	520

[1] Blank.

From these data, it is apparent that the presence of the zinc pellets resulted in significant separation of metallic mercury from the sludge. Recovery of the mercury from the zinc-mercury amalgam is achieved by heating the metal at reduced pressure and above about 100°C but below that temperature at which the zinc itself begins to distill in significant amounts. Preferably such a distillation is conducted from about 150° to 200°C and a pressure below about 1 mm of mercury. After recovery of the mercury by distillation, the zinc is cleaned as by washing with a dilute mineral acid and reused.

Example 2: River mud containing a trace amount of mercury was placed in a petri dish. Four coupons of galvanized steel, 22 gauge sheet cut into 3 x 3 cm squares, were placed on top of the mud and sufficient water was added to completely cover the mud and the metal coupons. After a certain interval of exposure, one coupon was removed and replaced with an identical coupon. The one removed was analyzed to determine the mercury absorbed on its surface.

The results of such tests on two muds containing about 25 and 131 parts per million by weight mercury, respectively, are reported in the following table.

Exposure interval (days)	Micrograms Hg per square	
	Mud No. 1, 25 p.p.m. Hg	Mud No.2, 131 p.p.m. Hg
0	0.4	0.4
3	2.2	1.5
6	1.2	10.7
27	1.1	42
128	0.8	73
160	0.9	125

The above data were obtained by dissolving the entire sample in aqua regia, making the solution to volume, and analyzing aliquots for mercury by the so-called cold atomic absorbance method. Results are significant to within approximately one microgram in the 0 to 10 microgram range. For amounts of more than 10 micrograms, the results are significant within about 10% of the reported values. Thus the data indicate that practically no mercury was recovered from the sediment containing 25 ppm. This sediment is thought to contain mercury in the form of oxide or sulfide rather than as metallic mercury. The sample containing 131 ppm mercury appeared to be capable of releasing a significant fraction of its mercury to the galvanized iron.

OTHER SPECIFIC PROCESSES

Nuclear Fuel Wastes

B.J. Newby and D.W. Rhodes; U.S. Patent 3,781,217; December 25, 1973; assigned to U.S. Atomic Energy Commission describe a method of preventing or retarding the formation of excessive amounts of a calcium fluorozirconate precipitate from fluoride containing waste solutions generated during the reprocessing of zirconium alloy nuclear fuels.

In the process, boric acid or a borate salt is added to an aqueous solution containing calcium, zirconium, and fluoride values in order to prevent the formation of a calcium fluorozirconate gel and repress the precipitation of calcium fluorozirconate. In particular, boric acid or a borate salt is added to such an aqueous solution of fluoride containing radioactive wastes generated during the reprocessing of zirconium alloy nuclear fuels which solution is to be converted to solid form by calcining in a fluidized bed and to which solution calcium nitrate is to be added to prevent fluoride volatility during calcining.

The boron containing species which are present in aqueous solutions of boric acid and borate salts and which are introduced into the wastes solution by the addition of boric acid or a borate salt prevent the formation of a calcium fluorozirconate gel and retards the formation of a precipitate, both of which would pose problems in the transporting of the aqueous wastes to the fluidized bed calciner and in the injecting of the wastes into the fluidized bed through the wastes feed nozzles. The repressive effect on the precipitation of calcium fluorozirconate obtained by increasing the amount of borates added to the solution can be appreciated from a consideration of the following illustrative examples.

Example 1: An appropriate amount of calcium nitrate was added to 30 ml of a waste solution which was 3.2 M in fluoride and to which no borates had been added, this amount being an amount sufficient to give a ratio of calcium concentration to fluoride concentration of 0.55 M calcium to 1.0 M fluoride which ratio is also the standard ratio obtained prior to calcining by the addition of calcium nitrate in the subject process. The precipitate which resulted from the addition of the calcium nitrate was separated and compacted to yield 10 cc of compacted calcium fluorozirconate.

Example 2: A sufficient amount of a solution of boric acid to yield 2 g/l of boron was added to 30 ml of a similar waste solution which was 3.5 M in fluoride. Again the appropriate amount of calcium nitrate was added to the resulting solution. The granular precipitate which formed following the addition of the calcium nitrate was separated and compacted as in Example 1 to yield 6.3 cc of compacted calcium fluorozirconate.

Example 3: Boric acid was added to 30 ml of waste solution which was 3.5 M in fluoride and very similar to the waste solution of Example 2 in an amount sufficient to give 13 g/l of boron of which amount it is estimated, based on the solubility characteristics of the boric acid under the test conditions and the presumption that a saturated solution was formed, 6 g/l of boron were actually in solution. An appropriate amount of calcium nitrate was added to the resulting solution and the granular precipitate which formed following the addition was again separated and compacted as in the previous examples. The yield of calcium fluorozirconate from this solution was 3.4 cc of the compacted precipitate.

Metal-Working Lubricants

A process described by *L.G. Treat and P. Hart; U.S. Patent 3,687,845; August 29, 1972; assigned to The Dow Chemical Company* concerns a method for purifying oil-in-water emulsions, such as are employed in the cleaning, lubricating and/or cooling of metal surfaces.

For instance, it is commonplace to utilize oil-in-water emulsions to lubricate and cool metals which are being fabricated as by rolling, drawing, cutting, milling, drilling and the like shaping operations. The emulsion is usually sprayed onto the surface being worked and sometimes the equipment utilized. It then drains off the workpiece or equipment and is recovered for reuse.

Frequently, in the use of such emulsions, foreign oils and greases accumulate from hydraulic oils, bearing lubricants and protective coatings to introduce a substantial quantity of a foreign oil phase, otherwise referred to herein as tramp oil. Usually, but not necessarily, such oils are of a substantially greater viscosity than the oil of the emulsion phase. If allowed to remain in the emulsion, the foreign oils result in the loss of desirable lubricity and the discoloration or grease staining of workpieces.

This process involves subjecting an oil-in-water emulsion to a processing sequence which comprises the following steps. To an oil-in-water emulsion containing, as basic ingredients, a water-immiscible oil, an anionic or nonionic surfactant and water and containing, as a foreign ingredient, a tramp oil phase is added a small amount of a high molecular weight, water-soluble, anionic polymer with sufficient agitation to disperse the polymer throughout the emulsion. The emulsion is energized, such as by heating at an elevated temperature sufficient to induce coagulation of the disperse tramp oil phase which then floats to the surface of the emulsion. Separation of the coalesced tramp oils is readily accomplished as by skimming or withdrawing the underflow emulsion.

The polymer utilized in the process is characterized by having along the polymer chain a plurality of carboxylate and/or sulfonate groups having the formulas $-COOM$ and $-SO_3M$, respectively. In these formulas, M is a water-soluble cation, such as for example, an alkali metal, ammonium, water-soluble amine or hydrogen. By high molecular weight is meant the polymers are of flocculant grade, i.e., they are capable of inducing the formation of flocs when added in small concentrations to aqueous suspensions of an oppositely charged disperse phase. A polymer has sufficient molecular weight for the purposes of this process when a 0.5% by weight solution thereof in water at pH 4 and 25°C exhibits a Brookfield viscosity of at least about 10 centipoises, preferably at least about 100 centipoises.

In a specific example, oil-in-water emulsion containing 5% of a mineral oil emulsified with a petroleum sulfonate was treated with a copolymer of acrylamide and sodium acrylate. The polymer had been prepared by polymerizing acrylamide and hydrolyzing about thirty weight percent of the initially available carboxamide groups to sodium carboxylate groups. The polymer had a 0.5% water solution Brookfield viscosity of about 2,000 centipoises, which corresponded to a molecular weight of approximately 2 million. To each 1 liter of emulsion was added 40 ml of a 0.1% by weight solution of the polymer. Dispersion of the polymer throughout the emulsion was promoted by about 5 minutes of mild agitation with a laboratory paddle stirrer.

The polymer addition increased the charge condition of the system from about 50 to 72 microamperes, as measured with a Leeds and Northrup Hydroscan meter. The latter instrument yields a measurement which is a function of the zeta potential of the emulsion and is generally proportional to the magnitude and polarity of the charge condition of the emulsion particles. The meaning of the measurement is described in U.S. Patent 3,399,133.

The treated emulsion was then heated to about 60°C and maintained under quiescent conditions for about 10 minutes. The tramp oils including, i.e., carrying, some dispersed dirt, coalesced and rose to the top of the emulsion phase. The resulting film of oil was easily separated by skimming. In subsequent operations, the treated emulsion was readily filtered through a diatomaceous earth coated filter to remove further impurities.

In a manner similar to the foregoing, the removal of tramp oils from oil-in-water emulsions is promoted by the addition of flocculant grade sodium polystyrene sulfonate, potassium polyvinyltoluene sulfonate, sodium styrene-maleate copolymers, ammonium methacrylamide-acrylate copolymers, acrylamide-oxazolidone copolymers and sodium acrylate-pyrrolidone copolymers.

Steel-Making

G.C. Almasi and W. Slusarczyk; U.S. Patent 3,714,037; January 30, 1973; assigned to Crucible Inc. describe an apparatus and method for treating industrial wastewater containing particles of magnetic material to make the particles readily removable from the water by conventional techniques, such as the use of a settling tank or filter.

In the process, water passes through conventional cylindrical piping to conventional particle-removal means such as a filter or settling tank. The piping is intersected at a location up stream from the particle-removal means by a restricted passage having a large width to height ratio and preferably being of generally rectangular cross section with the cross-sectional area being substantially equal to the cross-sectional area of the piping. Along the width and exterior of the restricted passage is positioned a plurality of opposed electromagnets which when energized produce within the passage a magnetic field the lines of force of which are substantially at right angles to the lengthwise, fluid-flow direction through the passage. In this manner the magnetic particles within the liquid are magnetized and agglomerated during flow through the restricted passage.

Consequently, the particles are more easily removable by the conventional filter or settling tank than would have been the case in their finely divided, unagglomerated form. During agglomeration of the magnetic particles nonmagnetic material, which either has adhered to the particles or is trapped between particles during agglomeration, is also removed from the liquid. It may be seen therefore, that with this process although only magnetic particles will agglomerate this incidentally results in removal of nonmagnetic contaminants from the water.

Typically the electromagnets are energized to produce a magnetic field having an intensity of about 900 to 1,200 gauss at the center of the restricted passage. The magnetic field should be varied in intensity depending upon the size of the particles to be removed, the ratio of particles to liquid volume, liquid flow rate, and the shape of the particles. By varying the intensity of the magnetic field it is possible to magnetize the particles without attracting them toward the electromagnets and thus causing them to adhere to the interior surface of the passage. If, of course, the intensity of the field is such as to cause adherence of the particles this could result in blockage of the passage and shutdown of the system.

Forging Compounds

W.G. Mattice and D.A. Sanford; U.S. Patent 3,725,262; April 3, 1973; assigned to Waste Research and Reclamation Co., Inc. describe a process for reclaiming a graphite-containing forging compound where the removal of fine metallic contaminants (e.g., metallic shavings of less than 200 mesh size, 0.074 mm) is accomplished utilizing magnetic separation. The forging compound is initially preheated to increase the viscosity and, after undergoing filtering to remove solid contaminants larger than the graphite, the fine magnetically attractive contaminants are magnetically separated out of the forging compound. The compound is then sequentially heated to cause dehydration and cooled to ambient temperature while undergoing blending to provide a homogeneous product.

Typically, a forging compound is composed of a heavy viscosity mineral oil (e.g., 90 wt), finely divided graphite (e.g., 5 to 45% by weight of oil), various animal fats (e.g., 1 to 15% by weight of oil), and a small quantity of calcium soap suspended in the mineral oil. The graphite is generally amorphous and commonly passable through a 200 mesh screen (i.e., the graphite particles are commonly about 0.074 mm and smaller in size). For some applications, clay, or additional calcium soap is added to the mineral oil to increase its viscosity. Such forging compounds are commercially available.

In use, a forging compound often times becomes contaminated with such contaminants as metal scales, metal filings, and floor sweepings. For the most part, these contaminants are magnetically attractive. The size of the contaminants will range from very small particles (e.g., 50 microns) to relatively large objects (e.g., nuts and bolts). To obtain high quality reclamation, all such solid contaminants (both large and small) must be removed from the final product.

As shown in Figure 3.5a a contaminated forging compound undergoes preheating **10** for increasing the viscosity of the compound. This facilitates flow and further operation on the compound (e.g., filtering) within the reclamation system. The preheated compound then undergoes a separation operation **11** including (1) centrifugal filtering to remove the solid contaminants larger than the graphite (e.g., larger than about 0.07 mm) and (2) removal of the magnetically attractive contaminants by magnetic separation. The forging compound then undergoes a heating operation **12** to remove volatile liquid contaminants.

Since water is the primary liquid contaminant, heating the forging compound to a temperature above 212°F (e.g., 250°F) substantially removes all the liquid contaminants. The dehydrated compound then undergoes a further or second separation operation **13** similar to operation **11**. It should be understood, however, that operation **13** is unnecessary in the case where operation **11** removes the solid contaminants to such an extent that the forging compound is of sufficiently high quality for reuse. Finally, the forging compound undergoes a cooling and blending operation **16**. In this operation, the heated forging compound is cooled to ambient temperature while undergoing a blending operation for providing a homogeneous product.

In Figure 3.5b, a system generally designated **20** is illustrated for achieving the reclamation process. A description of the apparatus comprising system **20** follows.

FIGURE 3.5: RECLAIMING FORGING COMPOUNDS

a.

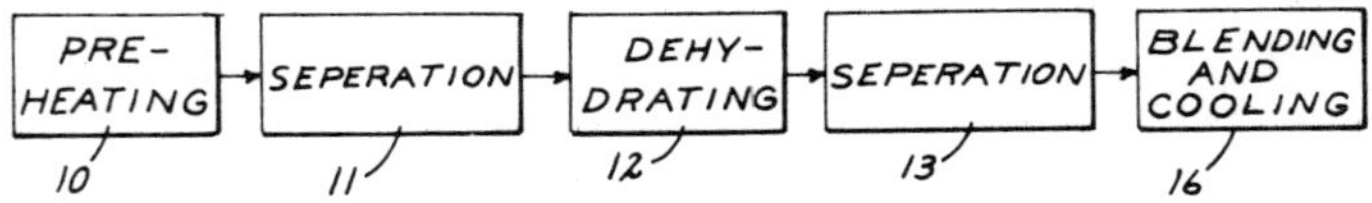

Flow Diagram

b.

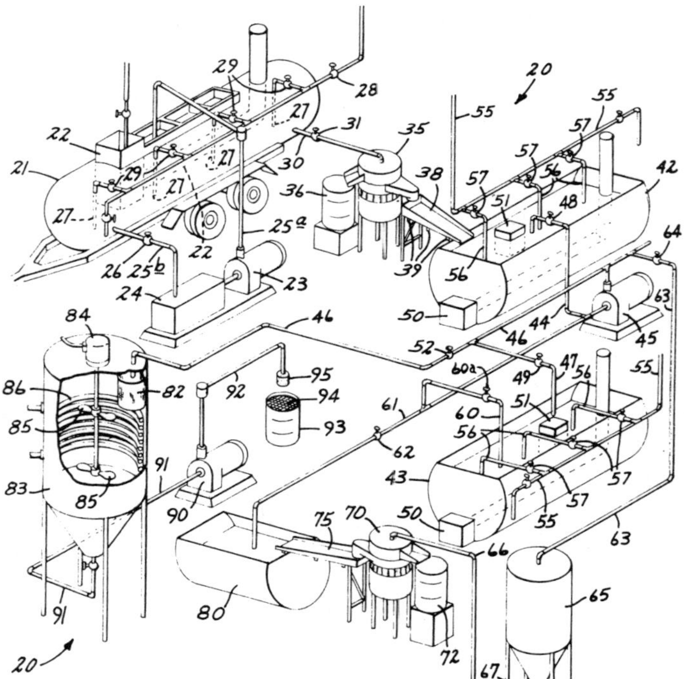

Detail of Apparatus

c.

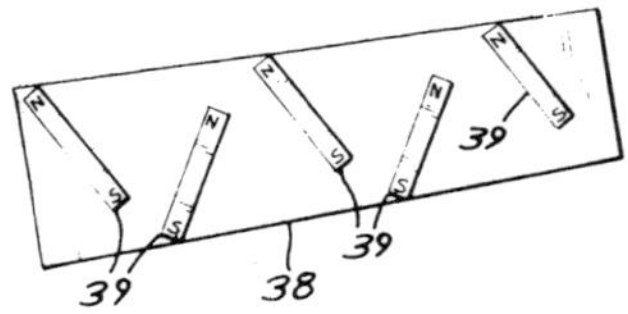

Enlarged Bottom View of the
Magnetic Separation Apparatus

Source: W.G. Mattice and D.A. Sanford; U.S. Patent 3,725,262; April 3, 1973

The Preheating Apparatus: As shown, a contaminated forging compound is brought to the reclamation apparatus in a mobile tank **21**. A conventional propane heater **22** extends along the bottom of tank **21** for preheating the forging compound decreasing the viscosity thereof. The tank **21** is connected to a conventional liquid pump **23** and a circulating tank **24** through conduits **25a** and **25b**. A valve **26** in conduit **25b** allows termination of the flow of forging compound through the pump **23** and circulating tank **24**.

Preferably, tank **21** is provided with a plurality of stationary steam nozzles **27** which are positioned so as to inject jets of steam into tank **21** along the bottom thereof. The steam nozzles are connected to a source of steam and valves **28** and **29** are provided for controlling injection of steam into tank **21**. It should be understood, however, that a movable steam lance insertable into tank **21** through an opening in the top can be utilized instead of the stationary nozzles **27**.

The forging compound is exhausted from tank **21** to the next stage of operation through a conduit **30** in liquid communication with tank **21**. A valve **31** in conduit **30** controls the flow of forging compound.

The Initial Separation Apparatus: The forging compound flowing through conduit **30** is exhausted into an electrically energized filter or separator **35**. The separator is preferably a screening device that vibrates about its center of mass. One such separator is the commercially available Sweco Vibro-Energy Separator Model LS30C66. The screen (not shown) utilized in the separator preferably has a mesh size in the range of 20 to 50 mesh. A stainless steel screen having a 35 mesh with 0.0215 inch openings has been found particularly desirable.

The forging compound is exhausted from the separator onto a magnetic separator in the form of a downwardly inclined, magnetic trough **38**. As can be seen particularly well in Figure 3.3c, the magnetic trough has a plurality of bar magnets **39** fastened to the underside thereof in an alternating diagonal pattern. This pattern has been found to facilitate maximum flow rate of the forging compound along the trough. It should be understood that numerous alternative magnetic separators can be utilized. For instance, a separator having rotating magnets and a wiper for continuously removing the contaminants collected by the magnets can also be utilized.

The Dehydration Apparatus: The forging compound which flows the entire length of magnetic trough **38** is exhausted into an open dehydration tank **42**. As shown, an overflow dehydration tank **43** is connected to tank **42** through a conduit **44**, a pump **45**, and conduits **46** and **47**. The tank **43** receives forging compound introduced into tank **42** subsequent to the compound reaching a given level within tank **42**. If necessary, additional tanks can be utilized to receive the excess compound dependent on the size of mobile tank **21** and the speed at which the reclamation process is to proceed. Conduits **44** and **47** include conventional valves **48** and **49**, respectively, for controlling the flow of forging compound. A valve **52** is positioned in conduit **46** downstream form the connection of conduit **47** thereto.

A conventional propane heater **50** extends along the bottom of each of the tanks **42** and **43**. As explained subsequently, heaters **50** raise the temperature of the forging compound within tanks **42** and **43** to a temperature above the vaporization point of water (the primary liquid contaminant). Apparatus for automatically terminating the heating operation when the forging compound reaches a given temperature (e.g., 250°F) is preferably provided in tanks **42** and **43**.

A conventional circulating pump **51** is positioned within each of the dehydration tanks **42** and **43** for circulating or agitating the forging compound within the respective tanks during the dehydration operation. A conventional gear driven pump has been found particularly desirable. Steam injection means are associated with each of the dehydration tanks **42** and **43** for causing additional agitation of the forging compound. As shown, the steam injection means includes a conduit **55** connected to a steam source (not shown) and a plurality of branch conduits **56** connected to conduit **55**. Conduits **56** extend into tanks **42** and **43** to a position adjacent the bottom. A valve **57** in each of the conduits **56** controls the flow of steam therethrough.

The Second Separation Apparatus: An outlet conduit **60** extends into tank **43** to a position adjacent the bottom thereof. Conduit **60** is connected to a conduit **61** having a valve **62** therein and which communicates with pump **45**, and a conduit **63**. Conduit **63** has a valve **64** and empties into a holding tank **65** having a conduit in communication therewith connecting tank **65** to a second Sweco separator **70**. A valve **67** positioned with conduit **66** controls the flow of forging compound therethrough to separator **70**. Separator **70** is preferably a Sweco Model S18 separator which is similar to separator **35** except for the utilization therein of a smaller mesh screen to further filter the forging compound. A 80 mesh stainless steel screen is preferred. A barrel **72** is provided to receive the discharge waste from separator **70**. The forging compound filtered within separator **70** is exhausted onto a magnetic trough **75** similar to the magnetic trough **38** previously described.

A holding tank **80** receives the forging compound exhausted from the magnetic trough **75**. A screen **82** mounted within a holding tank **83** is connected to holding tank **80** through conduit **46**, pump **45** and conduit **61**. Typically, the screen **82** is of a mesh size in the range 80 to 200 mesh. The forging compound filtered by screen **82** is exhaused into tank **83**.

The Blending and Cooling Apparatus: A conventional electrical blender or mixer **84** is mounted in holding tank **83**. Blender **84** includes blades **85** within tank **83** for agitating the forging compound therein. One suitable blender is the Lightnin Mixer Model LDG-150. Numerous alternative blenders are commercially available. A cooling coil **86** is positioned within tank **83** extending circumferentially about the walls thereof.

A pump **90** is connected to tank **83** through conduit **91**. A storage container **93** having a screen **94** mounted in the opening therein receives the reclaimed compound exhausted from a conduit **92** connected to pump **90**. Preferably, conduit **92** is pivotally mounted to pump **90** so as to be movable to a number of storage containers which may be positioned adjacent to container **93**. A conventional constant flow magnetic filter **95** is mounted at the outlet of conduit **92** above screen **94**.

The Reclaiming Operation: The operation of the reclamation system can be described as follows. The contaminated or waste forging compound is delivered to the reclamation system in tank **21**. The propane heater **22** is activated to heat the forging compound to a temperature in the range 160° to 210°F (e.g., 180°F). When this temperature is reached, valve **26** is opened and pump **23** is activated to continuously recirculate the forging compound within tank **21**. Concurrently, steam is injected into the tank **21** through steam nozzles **27** to further agitate the compound. Such recirculation and agitation replaces in suspension any material which has settled to the bottom of tank **21**.

This operation is allowed to continue until the contaminated compound is generally homogeneous. At this phase of the process, a test (or series of tests) is ordinarily conducted to determine the viscosity of the forging compound. Where necessary, oil is added to decrease the viscosity of the compound to the desired level or until it is at least flowable for movement through the reclamation equipment.

When the preheating operation is completed, valve **31** is opened and preheated and homogeneous compound is exhausted through conduit **30** into separator **35**. Separator **35** removes solid contaminants in the forging compound having a size greater than the graphite suspension (e.g., larger than 0.07 mm). On exhaustion from separator **35**, the forging compound flows through the magnetic trough **38**. Here, the magnetically attractive solid contaminants are removed from the forging compound by magnets **39**. The size of the contaminants removed will range from very small (e.g., 50 microns) to relatively large objects.

After the first separation, the forging compound flows continuously into dehydration tank **42**. Where tank **42** is too small to contain the entire quantity of forging compound undergoing reclamation, excess compound is pumped into tank **43**. Propane heaters **50** then heat the compound to a temperature in the range 212° to 275°F, preferably 250°F, so as to remove all the water contaminant in the compound. The operation of propane heaters **50** is terminated by conventional temperature control apparatus after the compound temperature reaches 250°F. During the heating operation, circulating pumps **51** continuously circulate the forging compound within tanks **42** and **43**. This greatly reduces the amount of graphite in the heated compound which settles to the bottom of the tanks.

At the end of the dehydration operation, a jet of steam is injected into tanks **42** and **43** through conduits **56**. This causes agitation of the forging compound assuring that any graphite which may have settled out of the compound is replaced in suspension. It is highly desirable that this steam be as dry as possible so as not to add moisture to the dehydrated compound. For example, a steam jet at 300°F and 120 psi is desirable.

With valve **62** in its closed position and valves **60a** and **64** in their open position, the dehydrated forging compound is pumped out of tanks **42** and **43** through conduits **44, 46, 47, 60, 61** and **63** into the holding tank **65**. During the pumping operation, steam is continuously injected into the tanks **42** and **43** until the entire quantity of compound is pumped therefrom.

On opening of valve **67**, the compound is exhausted from tank **65** and flows through the separator **70** and magnetic trough **75**. After this second separation operation, valves **48, 60a** and **64** are closed and valves **52** and **62** are opened. The forging compound is then pumped out of holding tank **80** through conduits **61** and **46** and through screen **82** for further filtration.

The forging compound is blended and cooled by blender **84** and cooling coils **86**. The blender **84** agitates the forging compound with tank **84** until a good homogeneous product is obtained. The cooling coils **86** bring the temperature of the forging compound to approximately 140°F. The concurrent blending and cooling provides a homogeneous product with a minimum amount of stratification. During this operation, graphite, oil, or other solid additives or agents are added to the forging compound to bring the product within the tolerances desired. The amounts of agents needed to be added are determined by conventional viscosity and percent graphite tests.

The reclaimed forging compound is pumped out of holding tank **83** through conduits **91** and **92**. Conduit **92** exhausts the reclaimed forging compound into a storage container **93**. The conventional constant flow magnetic filter **95** positioned at the outlet of conduit **92** magnetically attracts and removes from the compound any remaining magnetically attractive contaminants. The screen **94** positioned over barrel **93** prevents contaminants from accidentally (or otherwise) entering the barrel.

Lead

C. Lores and R.B. Moore; U.S. Patent 3,770,423; November 6, 1973; assigned to E.I. du Pont de Nemours and Company describe a process for treating an aqueous medium containing dissolved organic and inorganic lead compounds, for example, aqueous effluent from the manufacture of tetraalkyl lead, with an alkali metal borohydride to substantially reduce the level of dissolved lead in the aqueous medium.

In the following examples, parts given are on a weight basis unless otherwise stated; ppm signifies parts per million of lead by weight based on the method described by S.R. Henderson and L.J. Snyder in *Analytical Chemistry* 33, 1172 (1961).

Example 1: Two samples of tetraalkyllead process aqueous effluent (settling basin water) containing 8.2 ppm of total dissolved lead and having a pH of 9.0 were treated with sodium borohydride. To the first sample (A) was added 0.5 ppm of $NaBH_4$ and to the second sample (B), 2.5 ppm of $NaBH_4$. The samples were stirred for 16 hours at 20°C, filtered and analyzed for lead. Sample A showed 5.6 ppm of lead and Sample B, 0.4 ppm of lead.

Example 2: Example 1 was repeated except that Sample A contained 1.0 instead of 0.5 ppm of $NaBH_4$ and samples of effluent were removed at intervals and analyzed for lead. The results are shown below.

Stirring Time, Hours	Sample A	Sample B
0	8.2	8.2
1	5.8	5.2
2	5.1	4.6
4	3.2	2.6
16	2.4	0.45

Example 3: To each of four samples of the same settling basin water used for Example 1 were added 2.5 ppm of $NaBH_4$. The pH of each sample was adjusted to the pH shown below with the aqueous sodium hydroxide or hydrochloric acid as needed. The samples were stirred 6.5 hours at 20°C, filtered and analyzed for lead. The results are shown below.

pH	Ppm of Lead
6.7	6.4
9.3	0.65
11.2	3.4
12.0	4.1

Example 4: Four samples, (A), (B), (C) and (D), of settling basin water obtained as in Example 1 but containing 6 ppm of total dissolved lead were treated as follows:

(A) Added 1 ppm of a commercial flocculating agent (Atlasep 1A1), stirred 8 minutes, and let settle for 16 hours;

(B) Added 2 ppm of $NaBH_4$, stirred 8 minutes, added 1 ppm of the commercial flocculating agent, stirred 2 minutes and let settle for 16 hours.

(C) Added 1 ppm of the commercial flocculating agent, stirred 8 minutes, added 2 ppm of $NaBH_4$, stirred 2 minutes and let settle for 16 hours.

(D) Added 2 ppm of $NaBH_4$, stirred 2 minutes and let settle for 16 hours.

At the end of the settling time, the supernatant liquid was decanted and analyzed for lead. The results are shown below.

Sample	Ppm of Lead
A	6
B	3.4
C	1.0
D	1.9

Example 5: A sample of aqueous effluent containing 5.75 ppm of dissolved lead (about 90% in trialkyllead ions) and having a pH of 8.6 was stirred for 15 minutes at 20°C in the presence of 2 ppm of dissolved sodium borohydride and filtered within 5 minutes thereafter. The total lead concentration of the filtrate was 0.5 ppm.

Organic Lead from Alkyllead Process

E.R. Taylor, Jr.; U.S. Patent 3,697,567; October 10, 1972; assigned to E.I. du Pont de Nemours and Company describes a process for removing dissolved alkyllead compounds from aqueous effluent produced in the manufacture of alkyllead compounds.

The aqueous effluent from the manufacture of alkyllead compounds containing dissolved organic lead is reduced in soluble organic lead content by the process which comprises: (A) contacting the effluent with a metal more electropositive than lead but essentially nonreactive with water until the dissolved organic lead is converted into an insoluble lead-containing product and (B) separating the insoluble lead-containing product from the aqueous effluent. Thus when aqueous effluents from the manufacture of alkyllead compounds containing dissolved organic lead are contacted with a metal selected from the group consisting of iron, zinc, aluminum and magnesium until the dissolved organic lead is converted into an insoluble lead containing compound and the insoluble lead containing compound is separated from the aqueous effluent the dissolved organic lead content is reduced.

This process is based on the discovery that elemental metals more electropositive than lead but essentially nonreactive with water are highly effective in removing water-soluble organic lead products from wastewater of the manufacture of alkyllead compounds. The organic lead products removed from such waste streams constitute valuable recovery products.

This process is particularly applicable to the treatment of waste effluent from the manufacture of (1) tetraethyllead by the reaction of sodium lead alloy with ethyl chloride, (2) tetramethyllead by the reaction of sodium lead alloy with methyl chloride in the presence of an aluminum halide and (3) mixed methyl and ethyl alkyllead compounds by the redistribution of tetramethyl- and tetraethyllead mixtures with Lewis acid catalysts.

All of these processes tend to produce in the aqueous effluent water-soluble organic lead compounds not precipitatable by adjustment of pH to alkaline range, e.g., 8 to 9.5, such as those of the type $R_{4-n}PbX_n$ where R stands for methyl or ethyl, X for halide, such as chloride, or hydroxide, depending on pH of the solution, and n equals 1 or 2.

Ordinarily the aqueous wastewater to be treated will contain from 10,000 to 30,000 ppm of lead in the form of water-soluble organic lead compounds not precipitatable by pH adjustment. Levels higher than 30,000 ppm of soluble organic lead compounds, when encountered, can also be successfully reduced.

The elemental metal used in this process should be more electropositive than lead but essentially nonreactive with water. Such a metal is typified as being in the range from lead to magnesium in the electromotive series.

Examples of metals useful in this process which are more readily available and inexpensive in comparison to others within the above limits include iron, zinc, aluminum and magnesium. The most preferred metal is aluminum because its derivatives are relatively nontoxic and it is effective in alkaline media. The metal used need not be pure, the metal used may have impurities which do not interfere with its effectiveness in this process. The form of the metal should be such that a reactive surface is present. This may be achieved by having a high surface to volume ratio, such as in chopped sheet, foil, shavings, chips, wire or powder form.

In the following examples which illustrate the process, the parts given are on a weight basis unless otherwise stated.

Example 1: To an agitated equimolar mixture of tetraethyllead and tetramethyllead (the tetramethyllead is introduced as 80% solution in toluene), 0.3 weight percent of boron trifluoride diethyl etherate, based on the weight of the alkyllead, is added. The reaction is is carried out at room temperature. After five minutes, the reaction mixture is washed with alkaline solution (1% aqueous sodium hydroxide) to remove the catalyst. The washed mixture of alkylleads is then blended in the usual manner with a requisite amount of ethylenedibromide and/or ethylenedichloride, antioxidants and dye. Analysis of the redistributed mixture was found to be as follows:

Tetraalkyllead	Mol Percent
Tetramethyl	4.2
Trimethylethyl	24.6
Dimethyldiethyl	42.4
Methyltriethyl	24.4
Tetraethyl	4.4

Redistribution may be carried out in the ethylenedibromide and/or the ethylenedichloride normally present in the commercial antiknock blends.

Example 2: 6,700 parts aqueous effluent from Example 1 produced, by washing the reaction mixture with 1% aqueous sodium hydroxide was charged in a stream still. This effluent contained 0.4% residual sodium hydroxide, the reaction products of the catalyst with sodium hydroxide and 27,700 ppm dissolved lead, 89% of which was organic lead essentially as a mixture of $(CH_3)_n(C_2H_5)_{3-n}Pb^+$ ions in which n is 0, 1, 2 or 3 and 11% of which was divalent inorganic lead. Ten parts sodium carbonate were added to anticipate a final reacted pH of about 7.45 parts chopped aluminum foil were added.

The charge was heated to the boil by steam injection and then steam-distilled until no more water-immiscible condensate formed. About 1½ hours' steam distillation was required. 213 parts water-immiscible condensate was recovered consisting essentially of tetra (mixed ethyl and methyl) compounds as $(CH_3)_n(C_2H_5)_{4-n}Pb$. The undistilled effluent from steam distillation contained 190 ppm dissolved lead, 10% of which was organic lead and 90% of which was inorganic lead. The organic lead was therefore reduced from 24,600 ppm to 19 ppm.

Rare Earth Ions

H.L. Recht and M. Ghassemi; U.S. Patent 3,692,671; September 19, 1972; assigned to North American Rockwell Corporation describe a method for chemically removing trivalent rare earth ions from wastewater by treating the trivalent rare earth ion-containing water with a carbonate salt, such as an alkali metal carbonate, to form an insoluble rare earth carbonate precipitate and then separating the precipitate from the water. The rare earth ions may be regenerated in the form of a soluble rare earth salt, for example, by treating the separated rare earth carbonate precipitate with acid, such as hydrochloric or sulfuric acid.

Example 1: A series of 6 tests was run in which solutions of lanthanum nitrate [$La(NO_3)_3$] were added to stirred phosphate solutions which contained Na_2HPO_4 at a concentration of 12 milligrams P/liter (3.87×10^{-4} M). Solutions of pH 5, 5.5, 6, 7, 8 and 9 were prepared by adjustment with additions of either hydrochloric acid or sodium hydroxide solutions. To these solutions were added 210 milligrams per liter of sodium bicarbonate (2.5×10^{-3} M). The lanthanum additions were at a 1:1 lanthanum-to-phosphate ratio. The precipitates were allowed to settle and the solutions then filtered. The precipitates and filtrates were analyzed.

The phosphate-phosphorus concentrations were reduced from 12 milligrams per liter to less than 0.01 milligrams per liter at pH 5 and pH 5.5. Residual phosphate-phosphorus concentrations of about 0.07, 0.15, 0.36 and 0.86 milligram per liter were obtained at pH 6, 7, 8 and 9, respectively. Similar experiments conducted in the absence of sodium bicarbonate using lanthanum additions at a 1:1 lanthanum-to-phosphate molar ratio and a concentration of 12 milligrams P/liter yielded residual phosphate-phosphorus concentrations of less than 0.01 milligrams per liter over the pH range 5 to 9.

Example 2: A 2:1 lanthanum-to-phosphate molar ratio (7.72×10^{-4} M La) was used at a pH of 7.2 following the procedure of Example 1 and resulted in a residual phosphate-phosphorus concentration of less than 0.01 milligrams per liter.

Example 3: To stirred solutions of lanthanum nitrate which were 7.72×10^{-4} M in $La(NO_3)_3$ were added 210 milligrams per liter of sodium bicarbonate (2.5×10^{-3} M) in a series of three tests. Solutions of pH 6, 7 and 8 were prepared by addition of either hydrochloric acid or sodium hydroxide solution. The pH 6 solution developed fine precipitates which agglomerated into larger particles on standing. The pH 7 and 8 solutions immediately developed large flocs which settled readily. The precipitates were allowed to settle and the pH 7 and 8 solutions then filtered and the filtrates acidified with hydrochloric acid to a pH of less than 4 and analyzed for lanthanum by an extraction-flame photometric technique capable of detecting lanthanum concentrations as low as 0.5 milligram La/liter (3.6×10^{-6}M). No lanthanum was detected in either the pH 7 or 8 filtrates.

Example 4: A sample of acid bleach waste from a draft pulping mill was diluted 10 to 1 (10 parts diluent to 1 part acid bleach waste) with distilled water. The color of the diluted sample at pH 7 was 1,000 platinum-cobalt (Pt-Co) units (see *Standard Methods for the Examination of Water and Wastewater,* 12th edition, page 127 et seq). The sample was placed in a stirring apparatus and stirred rapidly at 90 rpm for 2 minutes. The stirring rate was then reduced to 20 rpm and the stirring was continued at this rate for an additional 10 minutes. After the stirring was completed, the sample was allowed to settle for 20 min.

After the period of quiescent settling, sodium bicarbonate was added to the solution until a pH greater than about 6 was reached to precipitate unreacted lanthanum as lanthanum carbonate. Following the addition of sodium bicarbonate, the supernatant was filtered through Whatman No. 40 filter paper and the color of the filtrate was determined at a wavelength of 420 mμ using a Lumetron Colorimeter (Model 402E) which had been calibrated against platinum-cobalt color standards. The color of the supernatant after filtration was measured at an adjusted pH of 7 to be 400 platinum-cobalt units which represented a 60% reduction in color.

Recovery of Vanadium and Chromium from Vicalloy Etching Process

M.S. Burg, L.M. Carbonaro and E.M. Songco; U.S. Patent 3,752,759; August 14, 1973; assigned to Western Electric Company, Incorporated, have described a process involving the depuration of ammonium persulfate solutions which have been spent in etching vicalloy. Specifically, the process relates to removal of cobalt, vanadium and iron, which are in a highly oxidized state, from an acid solution.

Vicalloy is generic (page 405, *Ferromagnetism* by Bozarth, 1st ed, pub 1951 by Van Nostrand) for alloys of the type disclosed in U.S. Patent 2,190,667, U.S. Patent 2,298,225 and U.S. Patent 2,519,227. A typical vicalloy composition comprises 52% cobalt, 37.25% iron, 10% vanadium, 0.050% manganese and 0.026% silicon, and is laminated on an aluminum substrate for manufacturing magnetic memory cards.

In the manufacturing process, a resist pattern which outlines magnets, is applied by known techniques on the exposed vicalloy surface. During ensuring processing, which is accelerated by heat, 92% of the vicalloy is dissolved in an aqueous etchant solution comprised of ammonium persulfate (a powerful oxidizing agent) highly acidified with sulfuric acid, for leaving a magnet patern on the substrate. Except that etched silicon will precipitate from the etchant in the form of silicon dioxide, the spent etchant will contain dissolved salts of all of the other foregoing metals of vicalloy, the metals being oxidized to their highest valence states by reason of the etching reactions.

The levels of the concentrations of the dissolved metals render the spent etchant unfit for disposal as sewer outfall, existing governmental regulations prohibiting such disposal of dissolved metals exceeding a concentration of 5 parts per million.

Controlled addition of a base to the spent ammonium persulfate solution will remove all the dissolved metals except the cobalt, whose trivalent state is complexed with ammonia to form the highly stable cobaltic hexammine ion $[Co(NH_3)_6]^{+++}$. Boiling the resulting solution to remove ammonia will liberate the cobalt ions from their complexed ammine state and cause them to precipitate out of solution as insoluble cobalt hydroxide. However, for commercial operation, an uneconomical expenditure could be entailed for an ammonia gas scrubbing installation to prevent pollution of the air with the exhaust of reaction. According to this process, highly oxidized cobalt and vanadium are removed from an aqueous solution by adding a soluble reducing agent to the solution. Then, the pH of the solution is adjusted to form insoluble compounds of cobalt and vanadium.

In the process, spent etchant having a pH in the range of about between 0.8 and 2 is collected in a storage tank from which it is removed for batchwise treatment. Trivalent cobalt (Co^{+++}), particularly hexammine cobaltic complexed cations and pentavalent vanadium (Va^{+++++}) which are originally present in the spent etchant in each batch are reduced, respectively, to bivalent cobalt cations, and lower valent vanadium. The trivalent hexammine cobaltic cation is a stable, highly soluble ion, which will not react with a readily soluble hydroxide to form an insoluble cobaltic hydroxide. However, the cobalt can be reduced to bivalent cobalt, which will react with a hydroxide forming an insoluble cobalt hydroxide.

Pentavalent vanadium (its highest oxidized state) forms salts as vanadates and oxides which are highly soluble in both acid and alkaline solutions. On the other hand, ions containing tetravalent and trivalent vanadium will react chemically with a base to form compounds which are highly insoluble at a pH in the range of between 6 and 8 and less insoluble at higher pH levels. The following examples illustrate the process.

Example 1: Add 21 pounds of finely divided or powdered iron to a 500 gallon batch of spent ammonium persulfate in a reaction tank equipped with a heating coil and an agitator. Heat and maintain the temperature of the contents preferably within the range between 120° and 150°F for about 12 minutes while constantly agitating. Thereafter, pump 33½ lb of sodium hydroxide in the form of a 30% solution into the reaction tank and continue agitation of the contents for an additional 12 minutes while maintaining the heat at the level aforesaid.

Transfer the resulting slurry by removing the entire contents of the reaction tank to a centrifuge. Then spin down the particulate matter in a centrifuge until solids are packed. The resulting almost dry solids are comprised of vanadium salts, iron hydroxides and a dark pink clear liquid containing cobalt. Thereafter, pump the liquid back to the reaction tank.

Next, pump 24½ pounds of sodium hydroxide in the form of a 30% solution into the reaction tank holding the dark pink liquid. Again, heat and maintain the contents of the reaction tank at a temperature preferably within the range between 120° and 150°F for about 12 minutes while constantly agitating. Pump the resultant slurry to a centrifuge; and spin the slurry until the solids are separated from the liquid. A second batch of almost dry solids comprised of insoluble cobalt containing materials will separate from the second supernatant liquid. The latter will be sufficiently free of metal contaminants for permitting its discharge into the public sewer.

However, before discharge, transfer the supernatant liquid to a reaction tank and add about 16 pounds of sulfuric acid in the form of a 30% solution. The end point of this last addition will be reached in about ten minutes and is determined with a stable pH reading of between 6 and 8 is obtained.

Example 2: Add 106.4 pounds of ferrous sulfate in the form of a saturated solution (a concentration of 1.33 pounds per gallon of water) to 500 gallons of spent ammonium persulfate etchant in lieu of the divided iron of Example 1. Thereafter, follow the remainder of the steps of Example 1.

Example 3: Add 21 pounds of finely divided or powdered iron to a reaction tank containing a 500 gallon batch of spent ammonium persulfate. Then heat to and maintain the solution at a temperature preferably in a range between 120° and 150°F for about 12 minutes while constantly stirring. Thereafter, from a reagent tank, introduce to the spent etchant 58 pounds of sodium hydroxide in the form of a 30% solution. Simultaneously, continuously agitate the contents of the reaction tank for an additional 12 minutes while maintaining the temperature in the aforesaid range. Pump the slurry which develops into a centrifuge by transferring thereto all of the contents of the reaction tank; and spin the contents of the centrifuge for a period sufficient to separate the solids from the liquid. Almost dry solids comprised of materials containing cobalt, vanadium and iron compact in the bottom of a centrifuge and the supernatant liquid will be substantially water clear.

The supernatant liquid thus formed will have metallic impurities at a level which falls within the range permitted for discharge as outflow to the public sewer. However, because the supernatant liquid then has a pH between 11 and 13, transfer the liquid again to a reaction tank. There, neutralize it with sulfuric acid in a manner provided in the final step of Example 1. This last neutralization step may produce a light precipitate of some insoluble vanadium salts. In each event, these may be separated from the liquid by transferring the neutralized mass from the reaction tank to a centrifuge and spinning. Thereafter, the resultant liquid can be discharged to the public sewer.

In Examples 1 and 2 the spent etchant, following treatment with the reducing agent, is alkalinized in a stepwise fashion. In the first of these steps the pH increased to a level in the range of between about 6 and 8. In the second step the pH increased to a level in the range of between 11 and 13. On the other hand, in Example 3, in a single step, the pH of the spent persulfate solution elevated from an acid condition to a level of between 11 and 13. The alkalinization procedures of Examples 1 and 2 are indicated, while the procedure of Example 3 is contraindicated in the event it is desired to recover the cobalt and vanadium separately for subsequent exploitation commercially.

METAL FINISHING

CYANIDES

Removal of Anionic Complexes Using Anion Exchange Resin

W. Fries; U.S. Patent 3,788,983; January 29, 1974; assigned to Rohm and Haas Company describes the use of ferrous ion as the complexing agent for cyanide, with subsequent adsorption employing a weakly basic ion exchange resin. The advantages of ferrous ion are two-fold:

(1) The ferrous ion incorporates all the cyanide into an anion complex ($Fe(CN)_6^{-4}$). This complex is not subject to further reaction of excess ferrous ion, because it is a very stable species. Also, the excess ferrous ion is precipitated quantitatively as ferrous hydroxide at the pH's of complex formation, and is therefore not available for further reaction. All the cyanide remains in solution as the ferrocyanide anion. It has been further discovered that the presence of small quantities of ammonia aid in the formation of the ferrocyanide complex.

(2) The ferrocyanide complex is extremely stable at all pH's in water. Therefore, it is able to be adsorbed by weakly basic resins operating in the acid salt form. No other practical cyanide complex has yet been demonstrated as stable enough for this.

The selectivity of these resins for the ferrocyanide complex is just as high as for the quaternary resins; however, the weakly basic resins may be simply regenerated by ion exchange techniques in which the ferrocyanide is removed as an entity. The complexes adsorbed by quaternary resins must be destroyed within the resin matrix before regeneration can take place. The overall process consists of adding excess ferrous ion in the form of ferrous sulfate, or ferrous ammonium sulfate (used if no indigenous ammonia is present in the stream) to the waste stream containing cyanide. The pH of the stream is adjusted above 9 to form the ferrocyanide complex. If free ammonia needs to be added, then it should be in an amount sufficient to maintain the complete complexing of the cyanide through the adsorption step. This may range as high as 500 to 1,000 ppm.

At this point, it is necessary to eliminate the solid phase of the waste stream which consists of indigenous solid matter as well as the excess ferrous ion as ferrous hydroxide. By using one ion exchange bed of a weakly basic resin in the salt form to adsorb the preformed ferrocyanide ion, complete removal of cyanide to less than 1 ppm is achieved. Also, regeneration may be accomplished simply with sodium hydroxide solution (1 to 10% is recommended), giving a highly concentrated effluent of $Na_4Fe(CN)_6$, which is relatively safer to handle than CN^-, nontoxic, and marketable. The resin and solution pH must be adjusted into the range of 4 to 7 prior to the adsorption step. The resin pH may be adjusted in two ways. For relatively strongly basic nonquaternary resins ($pk_a > 9$), a small quantity of a very weakly basic buffer such as $NaHCO_3$ may be passed through the bed to neutralize the small quantities of acid remaining from the conversion. Or for the ordinary weakly basic resins, a small quantity of the resin (approximately 10%) can be left unconverted by the acid step. Upon mixing the resin bed with air, again the small excesses of acid in local areas of the resin will be neutralized. If the adsorption process is not kept above pH 4, the formation of a blue cyanide complex precipitate, $Fe_3(CN)_6$, occurs and it is passed right through the resin bed. The overall chemistry of the process is as follows.

(1) Complex formation reaction:

$$2FeSO_4 + 6NaCN + 2NaOH \xrightarrow{pH>9} Na_4Fe(CN)_6 + Fe(OH)_2 + 2Na_2SO_4$$

(2) Settling of solids
(3) pH adjustment of solution (4 to 7)
(4) Adsorption of complex:

$$4Resin—N\cdot\tfrac{1}{2}H_2SO_4 + Na_4Fe(CN)_6 \xrightarrow{pH\ 4\ to\ 7} (Resin—N)_4\cdot H_4Fe(CN)_6 + 2Na_2SO_4$$

(5) Regeneration of resin:

$$(Resin—N)_4\cdot H_4Fe(CN)_6 + 4NaOH \longrightarrow 4Resin—N + Na_4Fe(CN)_6 + 4H_2O$$

(6) Resin preparation of adsorption step:

$$Resin—N + \tfrac{1}{2}H_2SO_4 \longrightarrow Resin—N\cdot\tfrac{1}{2}H_2SO_4$$

(7) Rinse with water
(8) pH adjustment of acidified resin

Step (8) may be accomplished by, for example, (a) use of small quantity (approximately 10% of resin stoichiometry) of a weak base, or (b) postmixing of resin bed which has been treated with 90% or less of H_2SO_4 required in step (6). Multiple cycle studies that have demonstrated that streams of up to 2,000 ppm cyanide (as free cyanide, or mixtures of free, complexed, and cyanide precursors) have been reduced to levels substantially below 1 ppm cyanide by this process.

The resins recommended for this process are weakly basic in nature, that is with average functionality of pK_a between 6 to 10. Their cross-linked polymeric structure may comprise: (a) A polymerizable ethylenically unsaturated monomer having about 2 to 100 weight percent of at least one polyvinylbenzene monomer selected from the group consisting of divinylbenzene, trivinylbenzene, alkyl divinylbenzenes having from 1 to 3 alkyl groups of 1 to 2 carbon atoms substituted in the benzene nucleus and any other bifunctional vinyl moiety; and (b) From 0.1 to 99% by weight of one of acrylic acid, methacrylic acid and maleic acid, which are cross-linked with one of the above mentioned polyvinylbenzenes; and (c) From 0.1 to 99% epichlorohydrin cross-linked internally; and (d) From 0.1 to 99% ethylene dichloride and polyamine condensate in which internal cross-linking is used also.

Example 1: The cyanide in a waste stream containing 500 ppm CN^- + 5,000 ppm of Na_2SO_4 was adjusted to pH 11.0. Excess ferrous ammonium sulfate was added to give a CN^-/Fe ratio in excess of 6. After allowing the excess iron precipitate to settle for 1 hour, the clear solution was decanted, filtered, adjusted to pH 7.0, and passed through a weakly basic resin at a flow rate of 2 gal/ft^3/min. Resin A is a relatively strong weakly basic resin (pK_a = 9.5). A capacity of 30 mg CN/ml resin was obtained with an average leakage of 0.2 ppm CN^- and 1 ppm Fe. The Na_2SO_4 and other dissolved solution passed to the effluent untouched.

Regeneration of the resin was accomplished using one normal (1 N) NaOH. Finally, the resin was converted with H_2SO_4, rinsed and adjusted in pH to 4 to 7 by passing through a small quantity of 1 N $NaHCO_3$. 100% regeneration of the iron and cyanide in the form of pure $Na_4Fe(CN)_6$ was achieved. A total of 16 complete cycles based upon this process were successfully performed.

Example 2: A wash stream containing 500 ppm CN + 5,000 ppm Na_2SO_4 was adjusted to pH 11.0 with NaOH. 300 ppm Fe^{++} in the form Fe_2SO_4 and 500 ppm $(NH_4)_2SO_4$ were added. After allowing the excess iron precipitate to settle for 1 hour, the clear solution was decanted, filtered, adjusted in pH to 5.5 and passed through a weakly basic resin (sulfate form) at a flow rate of 2 gal/ft³/min.

A capacity of 8 mg CN/ml resin was obtained with an average leakage of 0.5 ppm CN. The other dissolved solids passed to the effluent untouched. Regeneration was accomplished using 1 N NaOH. After rinsing, the resin was treated with 1 normal H_2SO_4 equal to 90% of the resin's functional group. This left 10% of the functionality in the basic form. The resin was then agitated under water by air lancing. The resin was then prepared for a second cycle of adsorbing the ferrocyanide influent.

Conversion to Nontoxic Biodegradable Compounds

A.C. Lauria and J.L. Owens; U.S. Patent 3,697,421; October 10, 1972; assigned to Miles Laboratories, Inc. have found that undesirable cyanide compounds in waste effluents can be converted to nontoxic biodegradable materials by treating such waste effluents with a starch conversion syrup. Preferably the waste effluent is also treated with a metal chelating composition when such effluent contains heavy metal cyanides.

In carrying out the process the waste effluent containing undesirable cyanide compounds is mixed with a starch conversion syrup at a temperature from about 18°C (64°F) to about 100°C (212°F) for from about 15 minutes to about 4 hours. The higher the reaction temperature, the shorter can be the reaction time. A reaction temperature of 18°C will require a reaction time of about 4 hours, while a reaction temperature of 100°C can employ a reaction time of about 15 to 30 minutes. During this reaction the toxic cyanide portions of such compounds are converted to nontoxic materials, such as glucoheptonates. Such materials are biodegradable and thus can be further broken down by natural processes to also reduce the overall pollution which would otherwise be created by adding the original untreated waste effluent to various natural bodies of water, rivers or streams. The following examples illustrate the process.

Example 1: A starch conversion syrup having a Dextrose Equivalent of 70 and containing 82 weight percent dissolved carbohydrate solids was added to a waste effluent containing sodium cyanide in an amount of 5 to 6 lb dissolved carbohydrate solids per pound of sodium cyanide in the effluent. The resulting mixture was heated at 100°C for 30 minutes. The resulting treated effluent was free of detectable cyanide.

Example 2: A starch conversion syrup having a Dextrose Equivalent of 36 and containing 80 weight percent dissolved carbohydrate solids was added to a waste effluent containing sodium cyanide in an amount of 20 lb dissolved carbohydrate solids per pound of sodium cyanide in the effluent. The resulting mixture was heated at 100°C for 30 minutes. The resulting treated effluent was free of detectable cyanide.

Example 3: A starch conversion syrup having a DE Equivalent of 82 containing 85 wt % dissolved carbohydrate solids was added to a metal plating waste effluent containing heavy metal cyanide compounds, such as cadmium cyanide and zinc cyanide, in an amount of 8 lb dissolved carbohydrate solids per pound of cyanide compounds in the effluent. Disodium salt of ethylenediaminetetraacetic acid was also added in an amount of 3 lb per pound of cyanide compounds in the effluent. The resulting mixture was heated at about 100°C for 1 hour and then allowed to cool for 1 hour. The resulting treated effluent was free of detectable cyanide.

Formaldehyde

According to a process described by *Z. Csürös, I. Petró, Z. Dusza, Z. Simó, I. Turcsán, and J. Vaczulin; U.S. Patent 3,729,413; April 24, 1973; assigned to NIKEX Nehezipari Kulkereskedelmi Vallalat, Hungary* sewage is decyanized with formaldehyde and the product recovered. The essence of the method is that a formaldehyde solution of at least pH 8 alkalinity is added under intensive agitation to the sewage, the solution allowed to stand at temperatures ranging from room temperature to 100°C for 1 to 50 hours and heated or boiled. To improve the economy of the method in given cases the ammonia is absorbed and the formed glycolic acid recovered from the solution. The method can be applied to the detoxication of both dilute and concentrated cyanide-containing sewage. The following examples illustrate the process.

Example: To 1 kg of sewage containing 10 weight percent of sodium cyanide (sodium cyanide content 100 grams equal about 2 gram mols of NaCN), 250 grams of a 37 weight percent formaldehyde solution (about 3.1 gram mols of formaldehyde) which has been first made alkaline (pH 8) with sodium hydroxide is added under intensive agitation. The solution is boiled for 2 hours under a reflux condenser and the ammonia formed in the reaction absorbed in water. At the end of the 2 hours boiling the cyanide ion concentration of the solution is about 0.5 mg per liter.

When 1 mol of cyanide decomposes 1 mol of glycolic acid is formed. Thus according to calculation the above solution will contain 2 gram mols (152 g) of glycolic acid. Glycolic acid is conveniently recovered by way of its calcium salt. The hot solution is acidified with a 20 weight percent hydrochloric acid (about 500 ml) and 102 grams of calcium carbonate is added under constant agitation.

After one day standing the precipitated calcium glycolate is filtered and dried at 50°C under 30 mm Hg vacuum. The weight of the dry calcium glycolate is 342 grams, which, calculated for sodium cyanide, corresponds to a 90% yield. The glycolic acid can be liberated from its calcium salt by adding the salt to 400 grams of 30 weight percent sulfuric acid. The precipitated calcium sulfate is separated by filtration and the solution evaporated under vacuum to the desired concentration. In this way 127 grams of glycolic acid can be obtained. Calculated for sodium cyanide the glycolic acid yield is about 84%.

Copper Salts and Oxygen

D.G. Hager and J.L. Rizzo; U.S. Patent 3,650,949; March 21, 1972; assigned to Calgon Corporation describe a method of removing cyanide from water in which copper salts and oxygen are added to the water upstream of a bed of activated carbon. The copper and cyanide form a complex of uncertain composition. The copper complexed cyanide thus formed is adsorbed on the activated carbon, which catalyzes the oxidation of the cyanide to carbon dioxide and nitrogen.

Control of Floor Spillage

L.E. Lancy and I.E. Wittmann; U.S. Patent 3,714,039; January 30, 1973; assigned to Dart Industries Inc. describe a process for treating spillage and waste solutions from a line of metal finishing processing solution tanks to prevent the forming of toxic hydrogen cyanide gas. The continuous process involves flowing a waste solution receiving and conditioning liquid along a floor catch basin. The conditioning liquid with its liquid waste content from a lower end of the catch basin is collected in an end-positioned sump having sufficient capacity to allow safe treatment of the probable worst condition with regard to the spillage of acid and cyanide solution containing tanks of the processing line. A pH indicator is provided at the sump for immediately indicating the approach of the solution to or the attainment of an acid pH due, for example, to a major spillage or a complete loss of solution from an acid containing tank of the line. This indicator is connected to cause the opening of a normally closed valve for immediately supplying a neutralizing alkaline chemical, such as sodium hydroxide, to the solution in the sump and bring it up to and maintain it within an alkaline pH range to inhibit the formation of toxic hydrogen cyanide gas. The solution from the sump is continuously flowed into a conditioning tank, at which time, a soluble alkaline chemical, such as an earth metal hydroxide, is added to raise its pH, and a cyanide reacting chemical, such as sodium or calcium hypochlorite or chlorine, may be added to react with any cyanide content to destroy cyanide chemicals through oxidation by chlorine in the alkaline range. The solution is moved from the conditioning tank into a reservoir where it is held to precipitate out and settle the precipitated metal salts and provide a fully reconditioned aqueous liquid or solution for reutilization.

CHROMIUM

Reducing Sugar

C.H. Elges III, P.R. Haskett, D.J. Bauer and R.E. Lindstrom; U.S. Patent 3,784,669; January 8, 1974; assigned to U.S. Secretary of the Interior describe the recovery of metal values from chrome etching solutions. Chrome etching solutions are extensively used in processing copper or copper alloys, such as brass. These solutions, the active ingredient of which is chromic acid, usually consist essentially of an aqueous solution of sodium dichromate and sulfuric acid, and generally have a pH of about 0 to 1.

Spent chrome etching solutions, from treatment of copper or copper alloys, contain Cr(VI) (hexavalent chromium), Cr(III) (trivalent chromium) and Cu(II) (cupric copper), typically in concentrations of about 50 to 70 g/l of chromium and about 20 to 40 g/l of copper. These spent solutions are usually discarded by injection into disused oil wells or by pumping into municipal dumping areas, and are in an active form for severe ground water pollution. Small amounts of the solution are used in tanning leather, but this utilization is negligible compared to the supply.

It has been found that chromium and copper may be recovered, in the form of oxides, from the spent etching solutions by means of a process in which the spent solution is initially treated with a reducing sugar to reduce Cr(VI) to Cr(III), to complex Cr(III) and copper in solution, and to provide an excess of reducing agent. Copper is then precipitated as Cu_2O by addition of sufficient base to raise the pH of the solution to about 10 to 12. After separation of the precipitated Cu_2O, the solution is heated to eliminate the reducing agent, and chromium is then precipitated as the hydrous oxide by adjustment of the pH to about 6 to 9. By means of this procedure, both chromium and copper values are salvaged from the spent etching solution, which can then be disposed of without presenting a pollution problem. In addition, the process provides an efficient separation of the chromium and copper values from each other.

The sugar may be any reducing sugar that is soluble in the spent etching solution, such as sucrose, glucose, maltose, etc. As discussed above, the function of the sugar is to effect initial reduction of the Cr(VI) to Cr(III), and, in a subsequent step, reduction of cupric copper to cuprous. In addition, it is believed that the process involves formation of a complex with the Cr(III) and the cupric copper, thus permitting separate precipitation of the metals when the pH is adjusted to the appropriate value. The complexing agent is believed to be a degradation product of the reducing sugar and is formed by oxidation of the sugar by the Cr(VI). This degradation product-complexing agent serves to keep the Cr(III) in solution when the copper is precipitated as Cu_2O.

The reducing sugar is added in amount sufficient for initial reduction of the Cr(VI) to Cr(III) and for subsequent reduction of cupric copper to cuprous copper. These reactions are essentially stoichiometric and the amount of sugar will, therefore, depend on the amounts of Cr(VI), Cr(III) and copper in the spent etching solution. For most efficient results, however, it is usually desirable to use about 2 to 4 times the stoichiometric amount required for the total metal values present. This will usually require the use of about 3 to

5 mols of sugar per liter of spent etching solution. Temperature, pressure, and reaction time are not critical in the treatment of the spent etching solution with the sugar, ambient conditions of temperature and pressure usually being satisfactory. A reaction time of about 20 to 30 minutes is generally sufficient. Base is then added to raise the pH of the solution to a value of about 10 to 12, preferably 11.5. Sodium hydroxide is the preferred base, but other strong bases may also be used. Again, ambient conditions of temperature and pressure are usually satisfactory. This results in precipitation of the copper as Cu_2O, which is then removed from the solution by any conventional means such as filtration.

The remaining solution is then heated to a temperature of about 50° to 80°C, and is held at this temperature for a period of about 5 to 10 minutes in order to destroy the complexing agent holding the chromium in solution. The solution is then cooled to ambient temperature and the pH adjusted to about 6 to 9, preferably about 7, by addition of acid. The preferred acid is sulfuric, but other acids such as hydrochloric or nitric may also be used. This results in precipitation of the chromium as the hydrous oxide. Again, the precipitate is removed by filtration or other conventional means.

Recovery of chromium and copper from the respective oxides may be readily accomplished by means of conventional processes such as hydrogen reduction of the Cu_2O and conversion of chromium oxide to ferrochrome. Or, the metals may be utilized directly in the form of their oxides for a variety of conventional uses such as pigments, ceramics, catalysts, glasses, electroplating, fungicides, etc. The process is illustrated by means of the following examples.

Example 1: 25 grams of a spent etching solution containing copper (30 g/l), chromium(VI) (30 g/l), and chromium(III) (30 g/l) was treated with 5 grams of sucrose and stirred at ambient temperature for 24 hours. The pH was then adjusted to 11.5 (from about 1.2) by the addition of NaOH. Cuprous oxide precipitated and was filtered from the solution. The copper oxide product analyzed less than 0.1% chromium, and the filtrate contained only 5 ppm copper. The chromium in solution was then precipitated by heating the solution to a temperature of 60°C for a period of 10 minutes to destroy the complexing agent, and adjusting the pH to 7 with H_2SO_4. The precipitate analyzed greater than 99.9% hydrous chromium oxide and the solution contained about 100 ppm chromium(III).

Examples 2 through 8: The procedure of Example 1 was repeated except that the amount of sucrose was varied. Results are shown in the following table.

	Sucrose added, g.	Cr(VI) in solution, g/l.	Cr(III) in solution, g./l	Cr in Cu ppt., percent
Example:				
2	0.0	30.0	30	-----------
3	.4	.10.0	50	-----------
4	.6	0.1	60	-----------
5	.8	<0.1	60	32.0
6	2.0	<0.1	60	0.6
7	4.0	<0.1	60	<0.2
8	5.0	<0.1	60	<0.1

Although 0.6 gram of sucrose was sufficient to reduce all of the chromium(VI) to chromium(III), this amount was not sufficient to insure separation of chromium from copper. A greater amount of sucrose (2 to 5 grams) was required for a 99+% copper oxide product.

Example 9: The procedure of Example 1 was repeated except that the amount of base added to precipitate the Cu_2O was varied. Adding 5 grams of NaOH precipitated all but 280 ppm of copper from the solution, whereas 7 grams of NaOH precipitated all but 60 ppm. The optimum amount of base, therefore, depnds on the degree of copper precipitation required as well as the initial acid content of the etching solution.

Biochemical Treatment

S.V. Yakovlev, J.V. Voronov, V.N. Korenkov, A.B. Nevsky, V.A. Dobrikova, T.A. Karjukhina, I.N. Churbanova and J.M. Laskov; U.S. Patent 3,755,156; August 28, 1973; describe a method for biochemical treatment of industrial wastewater containing combined oxygen in the form of inorganic compounds of hexavalent chromium, chlorine or mixtures of the compounds. In the process, the industrial wastewater is mixed with municipal sewage containing organic matter and the municipal sewage is subject to mechanical purification from contaminants present therein. The mixed water is then fed to an unaerated tank, where biochemical reduction of the inorganic oxygen-containing compounds through the use of activated sludge as well as biochemical oxidation of organic matter are effected simultaneously.

Biochemical reduction of inorganic oxygen-containing compounds have been effected, the mixed water together with the activated sludge is delivered from the unaerated tank to a settler and the precipitated sludge adapted to the inorganic oxygen-containing compounds is returned to the unaerated tank. The method provides for treatment of industrial wastewater featuring higher concentration of inorganic oxygen-containing compounds of chromium and chlorine as well as for a more intensified process of reduction of the compounds.

Lead Compound on Vermiculite

According to a process described by *G.J. Nieuwenhuls; U.S. Patent 3,791,520; February 12, 1974;* water containing chromium anions in the parts per million range are removed by contacting the water-laden chromium solution with a water-insoluble lead compound, such as lead oxide, lead carbonate or lead hydroxide, adsorbed in the pores of an essentially inert, porous, particulate matrix such as vermiculite. The lead compound reacts with the chromium in the water to form insoluble lead chromate. The effluent may be passed through an ion exchange column containing an acidic cation exchanger in the hydrogen form for removing any lead ion solubilized during passage through the bed of adsorbed lead compound.

The weight ratio of lead compound to carrier material may range from 1:1 to 1:4 and preferably 1:2. Best results were obtained using one gram lead carbonate to 9.93 grams ground vermiculite (Zonolite Company), the vermiculite having a particle size ranging from $\frac{1}{16}$" to $\frac{1}{32}$". In this proportion the lead carbonate/vermiculite has a specific gravity of about 1.0 and provides an uninterrupted flow of water solution with no back pressure.

Acidification and Reduction Techniques

R.G. Robertson; U.S. Patent 3,716,485; February 13, 1973; assigned to Ayteks International Corporation describes an apparatus and method for precipitating and/or destroying a chemical material comprising a source of material connected in series with an acidification unit, a reduction unit, and a pH adjustment unit and with detention chambers between each unit and a container to allow settling of the newly formed material. This apparatus provides a circuit through which the reaction takes place within engineered volumes of piping surrounding the unit.

Figure 4.1a and 4.1b show that the unit is made up of the three detention chambers **17**, **25** and **32**, which are connected in series with the sight-glasses **14**, **18**, **26** and **33**. These are, in turn, connected in series with the sump **11** and the holding tank **38**. A pump **13** circulates liquid from the sump through the line **12** and through the system. Acid is added by means of a proportional acid injection pump **41** through line **16** and check valve **V** to the acid injection point **15** in line **12**. The detection chamber **17** is sufficiently large in cross section and is internally baffled in such a manner to provide approximately one minute of turbulent detention in a typical example.

A pH cell **19** senses the pH of the liquid after it passes through sight-glass **18** and controls a valve **20** so that when the pH is above a predetermined value, the liquid is recirculated through line **21** back to the sump **11**. If the pH is at or below the proper value, it proceeds into line **22**. pH cell **19** passes an electrical signal to proportional controller **44** which, in turn, transmits a signal to proportional acid injection pump **41**. The controller **44** will balance the pump output rate such that the rate will stabilize at the desired setpoint to obtain correct pH at the pH sensing cell. The liquid thence passes through line **22** to the reducing agent addition point **23** where a reducing agent injection pump **42** feeds reducing agent through check valve **V** and valve **24** and line **24'**. The liquid then continues through the detention chamber **25** where it is held and mixed for a period of five minutes in a typical installation, thence through sight-glass **26** to the oxidation reduction potential detector **27**. If the ORP reading is above or below the proper oxidation reduction potential range, the fluid is diverted through valve **28** into line **21** which recycles to sump **11**.

FIGURE 4.1: PROCESS FOR DESTROYING HEXAVALENT CHROMIUM

a.

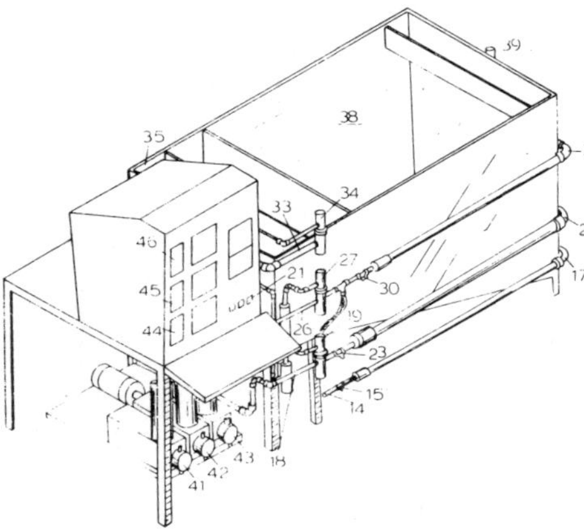

Isometric View of Machine

b.

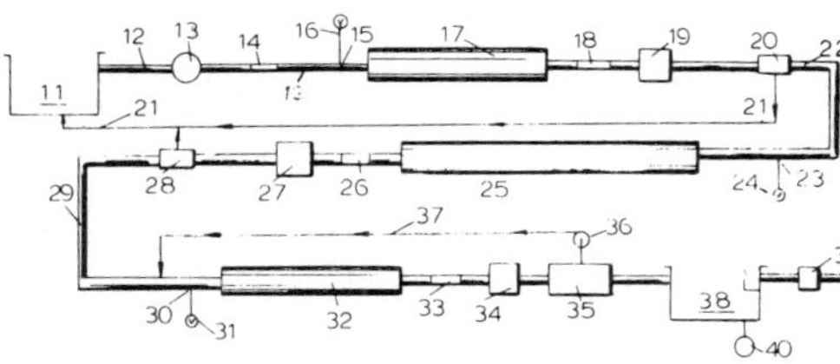

Schematic of Process

Source: R.G. Robertson; U.S. Patent 3,716,485; February 13, 1973

If the oxidation reduction potential falls within the acceptable range, the fluid will continue into line **29**. ORP cell **27** passes an electrical signal to proportional controller **45** which, in turn, transmits a signal to reducing agent injection proportional pump **42**. The controller **45** will balance the pump **42** output rate such that the rate will stabilize at the desired setpoint to obtain correct ORP at sensing cell **27**. The liquid thence passes through line **29** to the caustic addition point **30** where caustic is added by way of a proportional caustic injection pump **43** through check valve **V** and through line **31** and valve **31'**. The liquid then continues through the detention unit **32** where it is held and mixed for a period of one minute in a typical installation, thence through sight-glass **33** and pH sensor **34** and into detention tank **38** primary chamber **35**.

If the pH is below a predetermined level, the liquid is recirculated through pump **36** and line **37** to a point of injection in line **29** prior to caustic injection point **30**. If the liquid is within the desired limits of pH, the liquid passes undisturbed from primary chamber **35** into sump **38**. pH cell **34** passes an electrical signal to proportional controller **46** which, in turn, transmits a signal to proportional caustic injection pump **43**. The controller **46** will balance the pump **43** output rate such that the rate will stabilize at the desired setpoint to obtain correct pH at pH sensing cell **34**. The newly formed solids in the fluid are allowed to settle, for example, for 2 hours thence clear water is discharged through a pH cell and recorder **39** which will indicate the pH of the water discharged from the unit.

Example: A typical example of operation will be hexavalent chrome of 1,200 ppm Cr^{+6} at tank **11** which will be pumped by pump **13** at a rate of 5 gal/min through sight-glass **14**, chrome liquid will be yellow in color, and will have 60 cc/min of 66° Bé sulfuric acid injected at **15** and it will be detained for a period of one minute at **17** where its color will become a reddish-orange, which may be observed at sight-glass **18**. If the pH is below 2.5 at pH cell **19**, the liquid will proceed past point **23** where 12.6 lb/hr of 98% purity sodium bisulfite will be injected and shall be detained and mixed for a period of five minutes at **25** thus changing the color of the fluid to emerald green which can be observed at sight-glass **26**.

If the oxidation reduction potential at ORP cell **27** is less than 210 mv and greater than 190 mv, the liquid shall then proceed past point **30** where 6.8 lb/hr of 98% sodium hydroxide will be injected and shall be detained and mixed for a period of one minute at **32** thus changing the color of the fluid to powder blue which can be observed at sight-glass **33**. If the pH of the liquid at pH cell **34** is greater than 8.0, it shall pass into tank **38** where it shall be held for two hours to allow the chrome hydroxide to settle and allow the clear supernate to pass through pH recorder **39**. Periodic removal of settled sludge will be accomplished through blowdown port **40**. Typical chemical reactions that will occur within the above circuitry are as follows:

Circuit 1:

$$2Na_2CrO_4 + 2H_2SO_4 \rightleftharpoons H_2Cr_2O_7 + H_2O + 2Na_2SO_4 \rightleftharpoons 2H_2CrO_4 + 2Na_2SO_4$$

Circuit 2:

$$6NaHSO_3 + 3H_2SO_4 + 4H_2CrO_4 \longrightarrow 2Cr_2(SO_4)_3 + 3Na_2SO_4 + 10H_2O$$

Circuit 3:

$$Cr_2(SO_4)_3 + 6NaOH \longrightarrow 2Cr(OH)_3\downarrow + 3Na_2SO_4$$

Ion Exchange Column

A. Yagishita; U.S. Patent 3,761,381; September 25, 1973 describes a plant waste disposal unit so made that rinse water, containing a minor quantity of plating solution etc. adhering to the ware being transferred to the wash water tubs, is passed through an ion exchange resin layer for adsorption and removal of ions and the purified water thus obtained through ion exchange is circulated to one of the wash water tubs for reuse as wash water, while the hot water from a condenser connected to the heating and concentrating tower is admitted to another of the wash water tubs.

Referring to Figure 4.2, by way of giving a detailed explanation of the unit according to the process, a plurality of wash water tubs **2, 3, 4, 5** are connected in series with the plating tank **1**, the fourth tub **2** among the wash water tubs serving as the hot wash water tub. The wash water tubs **2, 3, 4, 5** are connected one with the other by means of siphons **15, 16** and **17**, and the fourth tub **2** is fitted with an overflow pipe **14**. To the first wash water tub **5** is connected a heating and concentrating tower **7** via feed pipe **6**. Heating of the liquid to be concentrated is effected by means of steam sent forth into a heating unit **12**, enclosed inside the tower **7**, via pipe **18**; and waste stream is exhausted via pipe **19**. Furthermore, a condenser unit **10** is annexed to the heating and concentrating tower **7** to condense the vapor evaporated from the tub water supplied into the tower heated by the heating steam. Cooling water is passed through a helically wound cooling pipe **11** inserted into the unit **10**; and this water is heated in the condenser unit **10** and passes through the pipe **11** to the fourth wash water tub or hot wash water tub **2**. Tub water from the first wash water tub **5** is pumped into the heating and concentrating tower **7** via feed pipe **6** and is heated there and concentrated to the level equivalent to that of the plating solution contained in the plating tank **1** before it is restored to the plating tank via pipe **8**. Water is supplied to the third wash water tub **3** via pipe **13**. The supply of the water is regulated by a balltap **13'** to keep the level of the tub water constant.

Supposing that a plating operation comprising moving the plated ware from the plating tank **1** and immersing the same consecutively in a series of wash water tubs is conducted in a chromium plating plant employing 10 kg of chromic acid per eight hours a day, and that the capacity of the first wash water tub **5** is 200 liters, the metallic articles plated in the plating tank **1** are washed consecutively in the wash water tubs **5, 4** and **3** and finally in the hot wash water tub **2**. If the above mentioned recovery unit is not utilized, approximately 85% of the chromic acid contained in the plating tank **1** would go onto the plated ware and be carried off from the plating tank **1**; and the chromic acid concentration levels of the wash water tubs **5, 4, 3** when the operation is finished would be approximately 45 g/l, 6.5 g/l and 0.6 g/l respectively.

On the other hand, if the recovery unit is utilized, supposing that the evaporative capacity of the heating and concentrating tower **7** is 100 kg/hr and that the capacity of the first wash water tub **5** is 200 liters and other conditions are also the same as above, the tub water contained in the tub **5** is diminished by ½ per hour, and the tub **5** is supplied with wash water from the second, third and fourth wash water tubs by means of siphons **15, 16, 17**. The wash water contained in the wash water tubs **5, 4, 3** is diluted; and the chromic acid concentration levels of the tubs **5, 4, 3** are maintained respectively at 15 g/l, 1 g/l and 0.08 g/l, or thereabout.

As will be seen from the above, the concentration level is apparently reduced to about 1/7 to 1/8 by using of the recovery unit as compared with that obtained when such unit is not used. However, such reduced concentration is not yet sufficient in view of the standards for preventing the environmental pollution. An ion exchange column **21** as shown in Figure 4.2 is annexed for instance to the second water tub **4** and to the third water tub 3. In this embodiment, tub water contained in the wash water tub **4** is pumped therefrom by pump **30** via pipe **20**, and is passed through the exchange column **21**, where the tub water is in contact with an ion exchange resin layer **22** provided in the column, whereby chromic acid etc. are adsorbed by the resin layer **22** and removed from the water. The purified water from the column **21** is returned by pipe **23** to the wash tub **3** for reuse as wash water.

FIGURE 4.2: PLATING WASTE RECOVERY UNIT

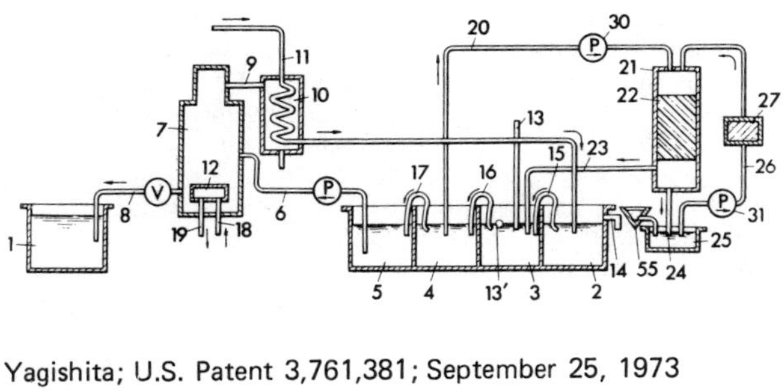

Source: A. Yagishita; U.S. Patent 3,761,381; September 25, 1973

The gauges of the pipes **20** and **23** are so designed that a prescribed flow rate of wash water circulating through pipes **20, 23** may be established to effect prompt removal of chromic acid etc., which are carried into the wash water tub **4** along with the plated articles. Normally, such capacity may be ten times the capacity of the wash water tub **4.** In treating tub water containing chromic acid, the reaction of the cation exchange resin contained in the cation exchange resin column **21** is shown as follows:

$$R(-OH)_2 + H_2CrO_4 \longrightarrow R-CrO_4 + 2H_2O$$

where R is a resin base. In regenerating the cation exchange resin saturated with chromic acid, the tank **25** is filled with a regenerant such as 4% caustic soda solution and the tank **25** is supplied with barium hydroxide in a suitable amount from the feeder **55.** In the first place, caustic soda solution is pumped by pump **31** into the ion exchange column **21** via pipe **26**, which is connected at one end through pump **31** with tank **25**, and at its other end through a filter **27** with ion exchange column **21**, for regeneration of the cation exchanger resin layer whereby a reaction takes place as shown in the following formula:

$$R-CrO_4 + 2NaOH \longrightarrow R-(OH)_2 + Na_2CrO_4$$

The chromic acid adsorbed to the cation exchange resin layer is discharged into the regenerant tank **25** via drain pipe **24** as sodium chromate solution together with excess caustic soda solution, where the sodium chromate solution is reacted with barium hydroxide, which is almost insoluble in the caustic soda solution and is converted to barium chromate, as shown in the following formula:

$$Na_2CrO_4 + Ba(OH)_2 \longrightarrow BaCrO_4 + 2NaOH$$

The barium chromate is then filtered off by the filter **27**, and caustic soda solution alone is supplied to the exchange column **21** to repeat the above mentioned reaction. Referring to the unit shown in Figure 4.2, the tub water contained in the first wash water tub **5** has the utmost chromic acid concentration, and is supplied to the heating and concentrating tower **7** for heating and concentration, and for subsequent return to the plating tub **1**.

The tub water in the wash water tub **4** situated ahead of the hot wash water tub **2**, and containing a minor quantity of plating solution carried over from the wash water tub **5** with the plated articles, is pumped into the cation exchange column **21** to remove plating chemicals through ion exchange. Purified water from the cation exchange column **21** is returned by pipe **23** to the wash water tub for reuse as wash water. On the other hand, cooling water is passed through helical pipe **11** in the condenser unit **10** of the heating and concentrating tower **7** for cooling of the vapor, and is there heated, and then supplied to the hot wash water tub **2** for use as hot wash water which rinses the plated metal articles and consequently dries them.

Electrolytic Treatment

M. Ishii and M. Ichiki; U.S. Patent 3,783,114; January 1, 1974; assigned to Mitsui Mining & Smelting Co., Ltd., Japan describe a method of treating the wastewater discharged in great quantities from the metal processing and plating shops by electrolysis. The process involves the following steps: conducting the electrolysis by introducing the industrial wastewater as the electrolyte into an electrolytic cell equipped with vertically suspended anodes and the horizontal cathodes disposed close to the bottom of the cell; removing the scum arising from the electrolysis (that is, the polluting substances coming up through the electrolyte to float to the surface) from the upper part of the electrolytic cell; and discharging the purified wastewater, that is, the electrolytically purified liquid, from the lower part of the cell. According to the method, the feed of wastewater and the discharge of electrolytically purified liquid are usually performed successively. The following example illustrates the process.

Example: An electrolytic cell **1** of 300 liters in capacity divided into 5 rooms by 4 partitions **10** as shown in Figure 4.3 and provided with the cathodes **3** made of iron plate (the cathodes measuring 600 mm x 800 mm x 2 mm in total) disposed horizontally along the bottom of cell **4**, the anodes **2** consisting of a pair of aluminum plates (measuring 400 mm x 300 mm x 10 mm each) disposed vertically within each room and a raking device **8** disposed above the cell so as to rake continuously the scum **7** generated therein, was filled with 240 liters of a wastewater **5** coming from a plating shop and comprising 0.5 ppm of zinc, 0.1 ppm of nickel, 3.0 ppm of chromium, 2.6 ppm of copper and a pH value of 7.8.

FIGURE 4.3: ELECTROLYTIC TREATMENT OF WASTEWATER

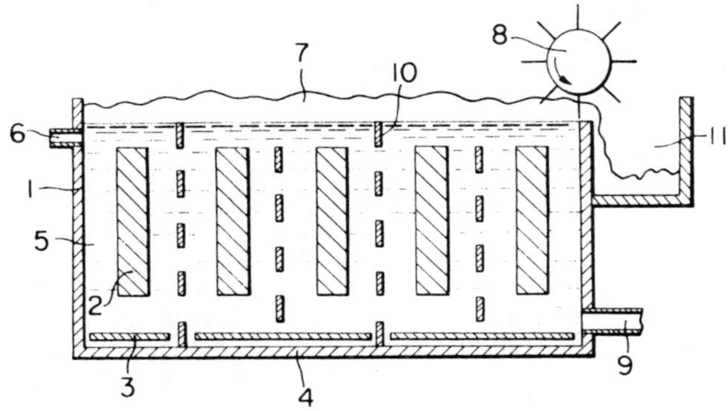

Source: M. Ishii and M. Ichiki; U.S. Patent 3,783,114; January 1, 1974

Subsequently, electrolysis was conducted by applying direct current of 6 amperes in toto. After one hour following the start of electrolysis, the wastewater coming from the plating shop was continuously supplied at the rate of 240 l/hr from the inlet **6** for wastewater to the upper part of the leftmost room of the electrolytic cell as shown in Figure 4.3, while the resulting purified liquid was continuusly taken out through the outlet **9** for purified liquid installed in the vicinity of the bottom of the rightmost room.

The scum **7** which had come to the upper part of the cell was raked out as occasion arose. Samples of the purified liquid were picked directly from the outlet **9** for purified liquid during the period of one hour following the application of electric current, and thenceforth were picked from the purified liquid discharged. When these samples were analyzed and further the cell voltage was measured to examine the rising rate thereof, the results were as shown on the following table which gives results of treatment of wastewater from the plating shop. The cell voltage at the start of electrolysis was taken as 1.00.

Time following the start of electrolysis	Concentration of purified liquid (p.p.m.)		cell voltage
	Chromium	Copper	
At the start of electrolysis	3.0	2.6	1.00
20 min. after	0.7	0.6	1.00
40 min. after	0.2	0.1	1.00
1 hr. after	0.2	0.0	1.00
24 hrs. after	0.3	0.1	1.00
48 hrs. after	0.2	0.0	1.01
96 hrs. after	0.1	0.0	1.02
144 hrs. after	0.2	0.0	1.01
240 hrs. after	0.2	0.1	1.08

As may be noted from this table, almost all of the heavy metals can be efficiently eliminated by the method, and there is scarcely observed any rise in the cell voltage even after 10 days' consecutive electrolysis. It is to be noted in this connection that the same tendency was observed even when the materials for the anode as well as the cathode were modified variously.

Mixed Plating Wastes

J.F. Zievers and C.J. Novotny, U.S. Patent 3,681,210; August 1, 1972; assigned to Industrial Filter & Pump Mfg. Co. describe a method and system for treating mixed rinse waters used to rinse workpieces removed from chromium and other metal treating baths. The process provides for moving the workpieces from a chromium plating solution directly into a reducing bath prior to rinsing, thereby to convert the hexavalent chromium ions carried by the workpieces to trivalent chromium ions. The workpieces are then moved through water rinse baths or sprays wherein the liquid carried by the workpieces is transferred to the water. Inasmuch as the rinse water does not contain any hexavalent chromium, it may then be mixed or combined directly with the rinse water from other metal plating stations such, for example, as nickel plating, copper plating, or zinc plating stations and the mixed waters subjected to the usual deionization treatment by passing them sequentially through cation and anion exchange resin columns.

ALUMINUM

Increased Sludge Yields

A process described by *L.E. Lancy; U.S. Patent 3,738,868; June 12, 1973; assigned to Lancy Laboratories Inc.* relates to the treatment of aqueous waste solutions containing dissolved aluminum and particularly, to minimizing the amount of flocculent material and to maximizing the amount of dry solid material content settled out or precipitated from the solutions.

After wet processing of aluminum, soluble aluminum compounds have to be precipitated out of the aqueous solution and separated from the clear supernatant. Aluminum hydroxide is recognized as an easily settleable and flocculent precipitate and has been used in water and sewage treatment applications by reason of such properties. However, for clarification of wastewaters in metal finishing processes, a definite problem has arisen by reason of the existence of floc in the form of aluminum hydroxide. In aluminum processing or treating line, a significant quantity of the metal is dissolved from the surface and the precipitated hydroxide does not compress in settling, but remains in a flocculent condition and thus, occupies a great volume of space and creates a significant weight of waste sludge for disposal.

The molecular weight of aluminum is 26.97 while the molecular weight of aluminum hydroxide is 77.99, representing a ratio of about 1 to 2.89. One gram of aluminum precipitated as the metal hydroxide in one liter of waste solution, either from aluminum sulfate or from sodium aluminate, will occupy a volume space of about 600 ml even after three hours of standing. Since the dry weight of the precipitated aluminum hydroxide is about 3 grams and the total weight of the 600 ml of sludge is about 600 grams, the dry weight in the total sludge is only 0.5%. From a practical standpoint, this means that for each unit of weight for example, for each pound of aluminum, 600 lb of sludge have to be dealt with before it is dewatered or dried.

The problem is further complicated from the standpoint that in metal finishing of alumnum, for example in anodizing, it can be assumed that an average of 1 to 2% of the aluminum is dissolved away in the process. Thus, employing the same unit figures, a plant that is processing 1,000 lb of aluminum per day, per week or per month has to dispose of significant quantities of sludgey waste. 2% of 1,000 lb of aluminum processed provides 20 lb of aluminum which, according to the above calculations, will create 12,000 lb of sludge.

Also, a great quantity of aluminum is dissolved in metal etching systems such as used in so-called chemical milling. Airplane frames are shaped by a chemical milling process and the metal removed may easily reach a figure of 1,000 lb/per day. The two most common chemical processing solutions that attain high concentrations of aluminum in aluminum finishing systems are described on the following page.

(A) A so-called basic aluminum etch containing free caustic soda that is continuously and slowly being enriched by aluminum which is dissolved from the surface of the metal workpiece and which is maintained until the aluminum concentration reaches a relatively high level at which time the solution may be dumped and a new one made up. Such a solution will contain aluminum in a range of 50 to 120 g/l at the time it is being dumped, with a free caustic soda content of about 30 to 50 g/l (NaOH). The aluminum is in the form of sodium aluminate, the basic hydroxide of aluminum.

(B) The second most common solution is an acid utilizing anodizing solution made up of about 250 to 300 g/l of sulfuric acid. After the aluminum content has increased, the solution has to either be dumped and renewed or some portion dumped and a new acid solution added to hold the aluminum concentration within a range of about 30 to 45 g/l.

Wash waters applied to a workpiece following either of the processes A or B contain the film that is dragged out from the process solution on surfaces of the workpiece which has been washed off by the operation. The wash water usually contains considerably less and never more than about 1% concentration of the process solution. On neutralization of such used dissolved aluminum containing wash water, the aluminum hydroxide precipitates as a flocculent precipitate which is not easily compacted, compressed or thickened in any known manner.

By way of example, a 1% solution of an aluminum etch solution of which the original etch solution contained 120 g/l of aluminum and 30 g/l of free NaOH, has about 1.2 g/l of aluminum. Neutralizing this solution to a pH of about 7 to 7.5 with a mineral acid, such as sulfuric produced a measured precipitated volume of sludge in a one liter graduated cylinder that, after one hour, represents a settled sludge of 94% or 940 ml of sludgey liquid. This sludge further settled after 24 hours to 40% volume of the initial one liter of neutralized waste solution. This indicates that the neutralized wash water, after an aluminum etching processing, will create a sludge volume of 940 grams of wet sludge containing 1.2 grams of original aluminum content. Even after 24 hours of holding for each 1.2 grams of aluminum in the waste water, there are 400 grams of sludge waste.

Since the original 1.2 grams of aluminum has formed about 3.5 grams of aluminum hydroxide, on a dry solid basis, the 940 grams of sludge contains 0.37% dry solids, and sludge even after 24 hours of settling, will contain less than 1% (0.86%) dry solids. A 24 hour settling period is economically not attainable, and the sludge cannot be removed in settling without pumping some additional water from the supernatant liquid when attempting separation.

It has been found, however, that in comparison with the large sludge volume attained by neutralization of wash waters when working with a dilute concentration of aluminum solution, the situation is entirely different when neutralizing the concentrate, itself. That is, it was discovered that by only slightly diluting an original etch solution to provide one containing two parts of concentrated etch and one part of water, with the new solution having a concentration of about 90 g/l of aluminum and neutralizing such waste with sulfuric acid to a pH value of 7 to 7.5, produced a total sludge of 98% of the one liter solution. This indicates a content of about 270 grams of aluminum hydroxide in one liter volume of sludge or 27% dry weight in the treated wet sludge. This proved to be an important key to a solution of the problem.

Thus, in carrying out the process, an important discovery was made from the standpoint of an aqueous caustic or alkaline etch waste solution of the A type and from the standpoint of an aqueous acid anodizing waste solution of the B type that the ratio of dry sludge content to wet sludge volume could be materially increased when neutralizing either of the solutions by increasing the precipitated aluminum concentration. This is important since aluminum processing lines very often contain both an aluminum etching bath or station, and a subsequent aluminum anodizing bath or station, with the drag-out and thus the wash solution of each having a dissolved aluminum content. Precipitation occurs from the standpoint of aqueous waste solution take off upon the addition of a neutralizing chemical, which in the case of a caustic carryover wash solution will be an inorganic acid and, in the case of an acid carryover wash solution, will be an earth metal alkali, such as calcium oxide or hydroxide. Another advantage was found to result from the standpoint that by combining two opposite types of waste solutions, they can in a sense react with each other in effecting neutralization, thus minimizing requirements for the additional chemicals. Also, importantly, combining such opposite types of wash water solutions enables an increase in concentration of the precipitated aluminum content to thereby lower the ratio of sludge volume to dry solid content resulting from neutralization.

Further, as to increasing the dry solids content, an important discovery was made that it is practical and highly advantageous to reconstitute and reuse the aqueous waste wash solutions and, that it could be accomplished to increase the aluminum concentration of the reused solution which is returned for neutralization, while at the same time, conserving the chemicals and wash water.

Importantly, it was determined that an aqueous solution for washing off drag-out from surfaces of a workpiece, as reconstituted by neutralizing a previously used aqueous waste wash solution, may be applied after neutralization without adversely affecting the finish of the workpiece in process, and that any residual precipitated aluminum hydroxide that may be carried back by the reconstituted solution will not adhere and can easily be removed after the workpiece has been rinsed. The process can be operated on a continuous basis with a continuous taking off of waste wash water, neutralizing and reconstituting it, and returning and reusing it in a washing bath or as a washing spray on the workpiece for removing both caustic and acid types of drag-out.

By way of example, doubling the aluminum etch solution addition to one liter of total rinse water by slowly accumulating precipitated aluminum with constant neutralization to a pH 7 with sulfuric acid, resulted in a sludge volume after one hour of settling of only 90%. This is significantly less than the sludge volume noted when neutralizing only ½ as much aluminum in the manner previously outlined. After 24 hours, the waste had further compacted to 63% sludge which indicates a 25% reduction in total sludge volume and a comparable increase in dry solids in the sludge. The used wash water as reused and continuously neutralized and reused will not damage or adversely adhere to the workpiece. It serves as a means of enriching the solution with fresh aluminum solution waste and of providing an increase in dry solid weight content of precipitated and settled out sludge.

The principles of the process were found to be applicable to a processing cycle involving an aluminum anodizing operation, although the waste wash water in the latter situation is acid in nature rather than alkaline. It was determined that when neutralization occurs as to the aluminum content from either an A or B type of solution, the precipitate is aluminum hydroxide which will increase slowly in sludge volume and much more in dry solid content as the rinse solution is used over and over again. By adding caustic soda or lime to a type B waste solution to attain neutrality for the wash water reuse, the same desirable results are attained from the standpoint of total sludge volume per gram of aluminum waste and at the same time, the dry solid content of the sludge is increased to a practical content of at least 1.5% or higher. Vacuum filtration or centrifugation of the dry solids in a sludge becomes economical and suitable where the dry solid content reaches about 1.5%.

In an exemplary utilization of 20 liters of chemical rinse solution, designated as example C, slow and periodic additions of caustic etch solution and acid anodizing solution were added in increments. The caustic etch solution contained 120 g/l of aluminum and 30 g/l of free NaOH, while the anodizing solution contained 30 g/l of aluminum and 300 g/l of sulfuric acid. The purpose was to test the procedure from the standpoint of the combined utilization with solutions of the two types represented by A and B. After each addition of the caustic etch solution, the pH was adjusted to neutrality using a sulfuric acid solution, and after each addition of the acid acidizing solution, the pH was adjusted by the addition of hydrated lime. Each increment was made to simulate the quantity of drag-in that has been determined could occur in a 20-liter chemical washing. Each wash water was used and recirculated from the two process steps to remove the drag-over process solution film remaining on the workpiece. The combined waste solution thus produced was recirculated at the rate of 300 ml/min as reconstituted wash water for workpieces, combined respectively from a caustic etching solution treatment bath and a sulfuric acid anodizing treatment bath. A neutral pH of about 7 was maintained with the chemical additions.

In example C, the total aluminum added to the 20-liter recirculated chemical rinse solution was 60 grams or 3 g/l. After the total additions were made from the standpoint of 3 g/l of aluminum from the etching waste solution and also from the anodizing waste solution, with periodic neutralization, the sludge volume was measured and found to be 85% by volume. The chemical wash solution thus constituted was kept in recirculation overnight. It was determined that even though the recirculation acted to stir precipitated flocculent aluminum hydroxide that the sludge was compacted to 55% by total volume, indicating about 9 grams of dry weight of aluminum hydroxide in each 550 ml of sludge volume, or a dry weight concentration of 1.98% (better than 1.5%). The 3 g/l of aluminum simulates the drag-out from an etching solution, such as occurs when a workpiece is processed carrying an adherent film.

Drag-out was also simulated by adding increments of an anodizing process solution to thus provide conditions prevailing where the workpiece is passed through both an alkaline etch solution and an acid anodizing solution and carries a wet drag-out film from each which cannot be discharged into the sewer system. Further improvement was attained in reducing the sludge volume and thereby increasing the dry weight of aluminum hydroxide in the sludge by heating the chemical rinsing solution above an approximate room temperature of 75° (preferably, to a minimum of about 100°F) through a range of up to about 180°F and maintaining a temperature of about 135° to 145°F while recirculating the solution. Sludge volume was reduced from 55 to 34.4% of the total volume, with an indicated 2.71% weight of dry solid content.

Further work along these lines indicated that sludge volume that is generated by neutralization of an aluminum etch (alkaline) solution was always greater than the sludge volume generated from neutralization of an anodizing (acid) solution, possibly due to the fact that the aluminum hydroxide precipitates in a different form or because the aluminum hydroxide, as precipitated occludes some sodium aluminate and is not pure aluminum hydroxide. A further advance was made by first acidifying an aluminum caustic etch wash solution addition for the reconstituted chemical wash solution with a mineral acid, such as sulfuric to a pH of below 6 to about 4, before neutralizing it with lime, $Ca(OH)_2$, followed by then neutralizing the solution to a final neutral pH of about 6 to 7, as an optimum, but tested also to a higher final pH of about 8 to 9.

It was discovered that 3 g/l of aluminum added to a waste wash water solution, as made acid before it was neutralized, created considerably less sludge than was attainable by simple neutralization. This is important not only from the standpoint of further increasing the proportionate dry solid content of the sludge in the caustic etch waste wash solution, but also importantly from the standpoint of making such a solution as reconstituted approach in its final treated characteristics, a reconstituted waste wash solution obtained from an acid anodizing operation. This thus enables obtaining a reconstituted solution from either type of washing operation that, as reconstituted, will have the same character and that may be freely used as an aqueous wash solution for workpieces that have been subjected to either or both a caustic etch or an acid anodizing treatment.

Thus, employing a two step treatment for a caustic type of wash water solution, as compared to the single type of treatment in which the alkaline content is directly neutralized and where only 1.2 g/l of aluminum was precipitated and the sludge volume after one hour was 94%, the resultant sludge volume utilizing 3 g/l aluminum, a preliminary acid pH of about 4 and neutralization above a pH of about 6 (about 6 to 7), was only 95%. Settling tests with this type of treatment and neutralization of the chemical wash solution were conducted at various temperatures with the results as indicated in Table 1. Table 1 illustrates the percentage of sludge or floc present after the solution was held for a specified time and at a specified temperature.

TABLE 1

	Percentage after—						
	15 min.	30 min.	60 min.	90 min.	120 min.	3 hrs.	24 hrs.
(1) 75° F	98.5	97.5	95	92	88.5	82.5	44
(2) 100° F		92	82.5	68	62	56.5	42
(3) 120° F	97	87.5	73.5	61	58	53.5	39
(4) 140° F	86	71	56	53	50	46.5	35
(5) 175° F		56	40.5	37			26

Another set of sample solutions was processed with the assumption that the chemical wash water solution would receive drag-out from both an aluminum etch treatment and from an anodizing treatment. Equal volumes of drag-out solution in 50 ml increments from the etch wash solution and subsequently from the anodizing solution were used. After each addition, the etching wash water solution was neutralized with sulfuric acid and the anadizing wash water solution with calcium hydroxide to produce a final pH of 9. Determinations have indicated that raising the pH to about 6 to 7 gives the same results. The total addition to a 3-liter chemical treatment wash solution was 7.5 grams of aluminum or 2.5 g/l of aluminum. The following Table 2 shows the sludge percentage for various test conditions:

TABLE 2

	Percentage after—						
	15 min.	30 min.	60 min.	90 min.	120 min.	3 hrs.	24 hrs.
(1) 75° F	97.5	95.5	91	87	83		42
(2) 100° F. (10 min.)	94.5	85	75.5	69.5	64		37.5
(3) 120° F. (10 min.)	84	67.5	58	53	49		36
(4) 140° F. (10 min.)	71	49	42.5	40	38		32
(5) 160° F. (10 min.)	67	49	42.5	39.5			34
(6) 180° F. (10 min.)	64.5	49	41.5	39			32.5

Recognizing that the chemical treatment wash water works best if the drag-out from a caustic etching solution is first acidified before it is lime neutralized, and that the drag-out from the acid anodizing solution need only be lime neutralized to produce comparable improved results, an additional test series was conducted. Equal volumes of etching and anodizing solutions in 50 ml increments each were added to a 3-liter chemical wash water solution of the previously mentioned type. Then the pH of the combined solution was reduced to about 2.5 after each addition of the aluminum etch wash water solution followed by thereafter neutralizing with lime, both the etching solution and the anodizing solution, all to a pH of 9. Determinations have indicated that no better results are obtained than where the pH is reduced to about 4 rather than below it. Considering that the total aluminum in three liters of chemical rinse solution was 7.5 grams of aluminum or 2.5 g/l of aluminum, tests indicated that the sludge volumes resulting were as follows:

TABLE 3

	Percentage after—						
	15 min.	30 min.	60 min.	90 min.	120 min.	3 hrs.	24 hrs.
(1) 75° F	45	27	26	26	26		26
(2) 100° F. (10 min.)	22	18	17.5	17.5	17		17
(3) 120° F	25	18	17.5	17	17		16.5
(4) 140° F	26	18.5	17	17	17		16
(5) 160° F	32	22	19.5	19	19		17.5
(6) 180° F	48	29	24.5	23	22		20

On further experimenting with a chemical wash to compare the advantages of preliminarily reducing the pH of the caustic etch wash water to about 4 before being neutralized, and in an endeavor to determine if neutralization to a pH of about 6 to 7 would be equally suitable, as compared to a pH of 8 or 9, additional experiments were conducted and have been labeled as shown in tables 4, 5 and 6.

TABLE 4

	Percentage after—						
	15 min.	30 min.	60 min.	90 min.	120 min.	3 hrs.	24 hrs.
(1) 75° F	99	98	97	94	91.5	87	62
(2) 120° F	98	96	92	89	86.5	79.5	60
(3) 160° F	97	94	89	85.5	83.0	76.5	59

Recirculated treatment rinse, 3 liters, containing 2.5 g/l aluminum; neutralized to final pH 6.8

TABLE 5

	Percentage after—						
	15 min.	30 min.	60 min.	90 min.	120 min.	3 hrs.	24 hrs.
(1) 75° F	63	41	38	38	36.5	35	35
(2) 120° F	51	35	34	34	34	34	33
(3) 160° F	42	34	33	33	33	33	32

Recirculated treatment rinse, 3 liters, containing 2.5 g/l aluminum. Fresh waste rinse water pH reduced to 4 before returning to the treatment reservoir and neutralized with lime to pH 6

TABLE 6

	Percentage after—						
	15 min.	30 min.	60 min.	90 min.	120 min.	3 hrs.	24 hrs.
(1) 75° F	58	39	36	35.5	35.5	35	35
(2) 120° F	40.5	34	33	33	32	32	32
(3) 160° F	46	45.5	44	43.5	43	43	42

Recirculated treatment rinse, 3 liters, containing 2.5 g/l aluminum; neutralized with lime to pH 8, after previously having reduced pH to 4 before neutralization

Aluminum neutralization has been found to be completed at a pH of about 4.7 and nearly all the aluminum is precipitated. Going to a pH of higher than 6 is wasteful of chemicals, but may be required from the standpoint of waste effluent discharge. From the standpoint of effective treatment of aluminum, a pH of 4.7 or higher is suitable, with 4.7 to 6 being a good range. Maintaining a pH of not more than 6 to 7 has the advantage that no free lime occurs in the solution that would tend to precipitate in the form of calcium carbonate on the workpiece when it is being rinsed or washed by the reconstituted solution. The last mentioned experiments of tables 4, 5 and 6 were carried out as before by adding to a three liter chemical treatment rinse solution equal volumes of etching and anodizing solutions in 50 ml increments each, reducing the pH in systems 5 and 6 to 4, but with no acidification being conducted in system 4. The total aluminum in the 3 liters was 7.5 grams or 2.5 g/l of aluminum, with the sludge volumes the same as indicated earlier.

In the system illustrated in Figure 4.4, a workpiece **9** is shown adapted to move in line from left to right through a caustic (or acid) etching solution bath, tank or zone **10** and thence into and through and out of a treatment wash solution bath, tank or zone **11**. It will be noted that overflow from the zone **11** is carried by return line **12** into a servicing or treatment tank or bath **13** of a reconditioning zone at which an inorganic acid, such as sulfuric acid, may be supplied through line **14** and control valve **15** to first convert the overflow representing the used aqueous waste wash solution from the zone **11** to an acid pH of less than 6, for better results of less than 4.7, and for optimum results, to a pH of about 4. An electric mixer **16** is shown extending into the tank or bath **13** and a pH probe is also shown for checking its attained pH.

FIGURE 4.4: WASTE TREATMENT OF ALUMINUM-CONTAINING SOLUTIONS

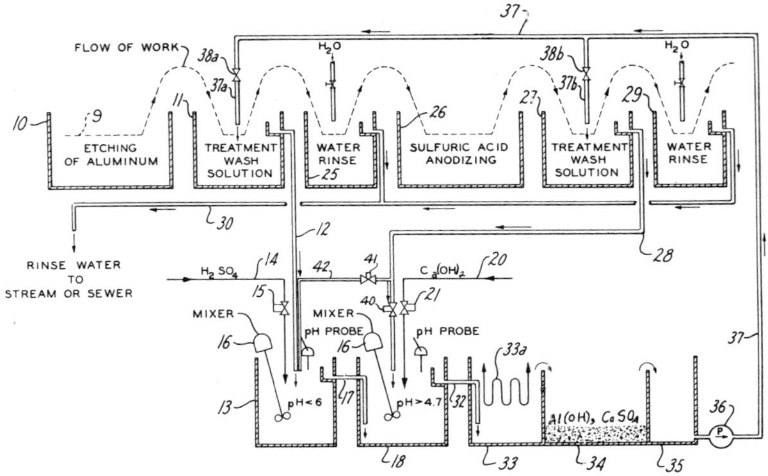

Source: L.E. Lancy; U.S. Patent 3,738,868; June 12, 1973

Solution overflow from servicing bath or tank **13** moves through line **17** into a second servicing or treatment tank or bath **18** at which complete neutralization may be accomplished, although from a waste treatment standpoint, a pH of 4.7 allowing a full precipitation of aluminum may be sufficient. Since the second tank **18** is receiving acid wash water from wash zone **27** through return line **28**, an alkaline earth metal hydroxide, such as calcium hydroxide, may be supplied through line **20** and control valve **21**. The tank **18** also has a mixer **16** and a pH probe and is shown adapted to receive and mix overflow from the servicing tank **13** and from the wash treatment zone **27**.

Following washing treatment in the tank, bath or zone **11**, the workpiece **9** then moves through water rinse tank, bath or zone **25**, and then into and through a sulfuric acid anodizing tank, bath or zone **26**. On leaving the zone **26**, the workpiece **9** is moved into and through treatment wash tank, bath or zone **27** which, like the zone **11** may be continuously supplied with reconstituted used aqueous washing solution, as moved along main return flow line **37** by pump **36**. Branch line **38a** supplies the used aqueous wash solution through control valve **39a** to zone **11**, and branch line **38b** and control valve **39b** supply the used aqueous solution to the zone **27**. The workpiece may then be moved from the zone **27** into a water rinse tank, bath or zone **29**. Since the water from the rinse zones **25** and **29** is relatively low in contaminants, it may as shown, be discharged from overflows through pipeline **30** to a stream or sewer.

Treated solution from the second tank **18** moves through overflow pipe **32** into a heating tank **33** which is shown provided with a steam coil **33a** to heat it to a temperature of at least 100°F. Thence, the heated solution, after a period of about 10 minutes heating in tank or bath **33**, passes over an overflow into settling tank or bath **34** where it may be permitted to cool and where it may be fully reconstituted from the standpoint of the settling and removal of the aluminum containing sludge. The reconstituted used wash solution then moves into reservoir or holding tank **35** from which it may be supplied to the processing line. Figure 4.4 illustrates a typical recirculating system which desirably combines, reconstitutes and reuses the aqueous wash solutions from both alkaline and acid treatment stations of the processing line, and enables supplying replenished solution continuously or without fluctuations, either from the standpoint of supply or pH. pH fluctuations may be eliminated by supplying sufficient chemicals to overcome chemical fluctuations of the returning waste wash solutions. Preferably, the separation will not only be accomplished in a manner to effect a reuse of the wash solutions containing dissolved aluminum, but also to first or preliminarily acidify the solution as it comes from a caustic or alkaline station before neutralization, in order to provide a maximum content of dry solids or ratio of dry solids to floc.

Polyacrylamide Resin

A process described by *C.J. Huffman; U.S. Patent 3,798,160; March 19, 1974; assigned to American Metal Climax, Inc.* relates to the treatment of aluminum waste liquors, and particularly to waste acid and alkaline liquors from aluminum anodizing processes. In the anodizing of aluminum, the metal is cleaned with a detergent solution, rinsed, etched with a dilute (about 3 to 10%) caustic soda solution, rinsed, anodized in a dilute (about 18 to 25%) sulfuric acid anodizing solution, and rinsed. The rinse liquors from both the alkali etching and the acid anodizing process constitute waste liquors, containing about 1% aluminum or less, which must be treated before they can be discharged into surface waterways or municipal sewage systems. They must be treated to adjust the pH and to remove settleable solids, particularly aluminum hydroxide. The problem of treatment of aluminum waste liquors has been growing in importance over the past several years to a point where it is a major issue in the industry. Lack of adequate waste treatment can lead to the shutting down of a plant

Acid and alkaline liquors containing dissolved aluminum values can be neutralized by addition of acid or alkali. The pH is usually adjusted to a value in the range of 6 to 8.5, usually about 7.5 to 7.8. This causes the precipitation of aluminum hydroxide which has its minimum solubility at a pH of 7.8. Aluminum hydroxide forms a milky precipitate which is slow to coagulate and difficult to remove from aqueous solutions in which it is suspended.

In this process, an anionic resin is added in relatively small quantities (one or four parts per million) to the aluminum waste liquor following the adjustment of pH of the liquor to the range of 6 to 8.5. The anionic resin aids and assists the rapid coagulation and precipitation of aluminum hydroxide which is formed in a suspension. By coagulation with the aid of the anionic resin, it is possible to clarify or thicken the aluminum hydroxide suspension to form a precipitate containing approximately 2 to 4% solids. This can be dewatered by a vacuum or pressure filtration to produce a thickened suspension containing 16 to 50% solids. The effluent or filtrate from this operation contains sodium sulfate which can be discharged to a sewage system or waterway without creating problems.

One advantage in concentrating the aluminum hydroxide suspension to 16 to 50% solids in accordance with this procedure is that there is produced an aluminum hydroxide paste which can be treated with concentrated sulfuric acid to produce an aluminum sulfate solution containing 28% aluminum sulfate which is commercially useful as a flocculent for other types of wastewater treatment.

The process is described in more detail in conjunction with Figure 4.5 which is a flow sheet for the treatment of aluminum anodizing waste liquors, having pH values between 4 and 12. These liquors are passed through a neutralizing tank **1** where the pH is adjusted to 7.5 to 8.5 with acid or alkali, as required. If the waste liquors are acidic (e.g., from the anodizing rinse) they are preferably treated with alkaline liquors from the alkali etching rinse, and if they are alkaline, an acid liquor is preferably used. Conventional automatic control equipment controls the addition of spent acidic or alkaline liquors to adjust the pH to the desired range. In either procedure the pH is brought to a value near 7.8, at which point aluminum hydroxide precipitates, forming a suspension ordinarily containing about 800 parts of solids per million parts of water. The aluminum hydroxide suspension is pumped to a circular thickener or clarifier **2** where a solution of high molecular weight anionic resin is added to facilitate settling of solids. The quantity of resin added (usually as 0.020 to 0.050% aqueous solution) is selected to produce a final resin concentration of 1 to 4 parts per million part of liquor. The solids settle to the bottom of the thickener **2** and the clear effluent overflows to the sewage system, control of this effluent being conducted by means of a turbidimeter **3** and a pH recorder **4**. The underflow from the thickener consists of 10 to 15% of the feed volume and is a soupy suspension containing 2 to 4% solids. It is fed to a drum-type vacuum or pressure filter **5**. A paste consisting of 16 to 20% solids builds up on the filter drum and is continuously removed by a knife as the drum rotates. The paste (or filter cake) is collected and removed. The filtrate is returned to the center well of the thickener **2** for reprocessing.

FIGURE 4.5: TREATMENT OF ALUMINUM WASTE LIQUORS

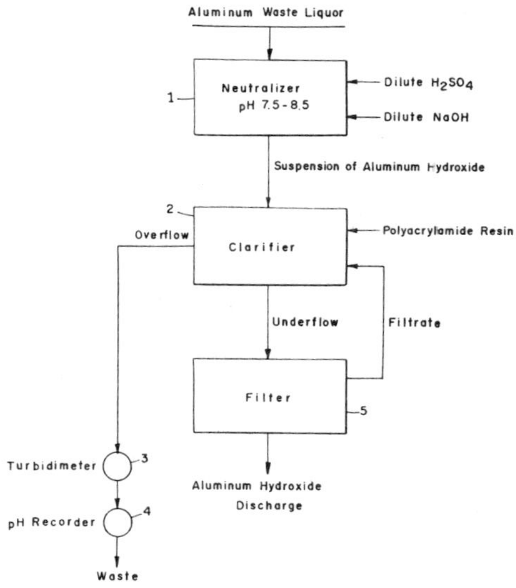

Source: C.J. Huffman; U.S. Patent 3,798,160; March 19, 1974

In a typical operation of the type described above, 350 gal/min of aluminum rinse liquor enters a 10,000-gal neutralizing tank where it is brought to pH 7.5 to 8.5 with spent acidic or alkaline waste liquor, forming a suspension of aluminum hydroxide (800 ppm). It is then pumped to the thickenener where a 0.025% solution of high molecular weight polyacrylamide resin is added. Flocculation occurs and the suspension is rapidly thickened and settles as a sludge in the thickener from which it is removed as an underflow at the rate of 40 gal/min. The overflow of 310 gal/min is discharged to sewer or river at pH 6.5 to 8.5 and maximum suspended solids of 20 ppm. The underflow (2 to 4% solids) is fed to a drum-type vacuum filter coated with a filter precoat of diatomaceous earth. A filter cake consisting of 16 to 25% solids builds up on the drum and is removed with a knife, collected in a hopper and disposed of as waste, or treated with sulfuric acid to produce a 28% aluminum sulfate solution suitable as a flocculent for other water treatment.

The diatomaceous earth is insoluble in the aluminum sulfate solution and can be removed by filtration. A roll discharge vacuum filter is also feasible. If a pressure filter is used instead of a vacuum filter, it is possible to produce a filter cake with 20 to 50% solids. With either filter, about 3 tons per day of a cake is produced which is substantially all aluminum hydroxide and which can be used as a commercial source of aluminum values.

In these operations the water which is used is ordinary municipal water of varying hardness and magnesium and calcium content. During the early operations, there is some precipitation of magnesium and calcium compounds in the thickener. With recirculation of the water in the system, however, magnesium and calcium values decrease and precipitation of compounds of these metals with aluminum hydroxide decreases. The temperature of the operation is usually ambient temperature (50° to 100°F). It has been found that higher temperatures (100° to 120°F) may be preferable and more rapid settling obtainable thereby. The anionic flocculating agent in this process is preferably a polyacrylamide resin of molecular weight from about 500,000 to 3,000,000. This is a polymer of acrylamide, which has the formula $CH_2{=}CH{-}CONH_2$. The polymer can be represented as

$$-(CH_2-\underset{\displaystyle CONH_2}{\underset{|}{CH}})_n-$$

where n is a large number between about 7,000 and 42,500. Related resins, such as those of homologs of acrylamide having the formula $R{-}CH{=}CH{-}CONH_2$ wherein R is methyl, ethyl or propyl, can also be used. The resins are soluble in water and are usually used as solutions of 0.05% to 0.1% concentration. In aqueous solution part of the amide groups are hydrolyzed to carboxyl groups, which are anionic in character. The hydrolysis is hastened by alkaline pH ranges. The resins are added to the aluminum hydroxide suspensions in quantities sufficient to make concentrations of 1 to 3 ppm.

The preferred form of polyacrylamide resin is that in which part (15 to 50%) of the amide groups are converted to carboxylic acid (or salt) radicals. They can be represented by the following general formula (for the sodium salt)

$$-[(CH_2-\underset{\displaystyle CONH_2}{\underset{|}{CH}})_x-CH_2-\underset{\displaystyle COONa}{\underset{|}{CH}}]_n-$$

where x is a number from about 2 to 4, inclusive, and n is a number from about 1,400 to about 14,000. These polymers have long straight chains with numerous anionic carboxyl groups spaced along the chains at intervals. The anionic groups form bonds between the polymers and the particles of aluminum hydroxide. Each polymer molecule thus agglomerates a number of aluminum hydroxide molecules and causes flocculation. The bonds are formed quickly and irreversibly and the settling of the aluminum hydroxide suspension occurs rapidly.

OTHER PROCESSES

Continuous Process

A.N. Sulfaro and A.H. Stein; U.S. Patent 3,715,308; February 6, 1973; assigned to Oxy Metal Finishing Corporation describe an apparatus and method for continuously treating and destroying toxic constituents in effluent waste streams from metal finishing processes such as aqueous rinse solutions containing dissolved hexavalent chromium and cyanide compounds present in concentrations which are lethal to marine ecology. The apparatus and process provide for a two-stage treatment of such waste streams which are continuously recirculated in each stage and their chemical characteristics are continuously monitored and adjusted so as to assure optimum conditions for promoting the desired chemical reactions for effecting a destruction of the toxic constituents and assuring that the effluent continuously discharged from the treating apparatus is devoid of any harmful constituents. Referring to Figures 4.6a and 4.6, the apparatus consists of a framework defining a first tank **10** and a second tank **12**, through which the aqueous waste solution sequentially flows while reagents are added to effect the desired chemical reaction with the toxic constituents. The first tank **10** is provided with an upright transverse partition **14** disposed approximately midway therealong defining a first compartment **16** and a second compartment **18** which are interconnected by means of a trough or weir **20** mounted along one side of the upper edge of the partition **14**.

FIGURE 4.6: CONTINUOUS TREATING OF WASTE STREAMS FROM METAL FINISHING PROCESSES

a.

b.

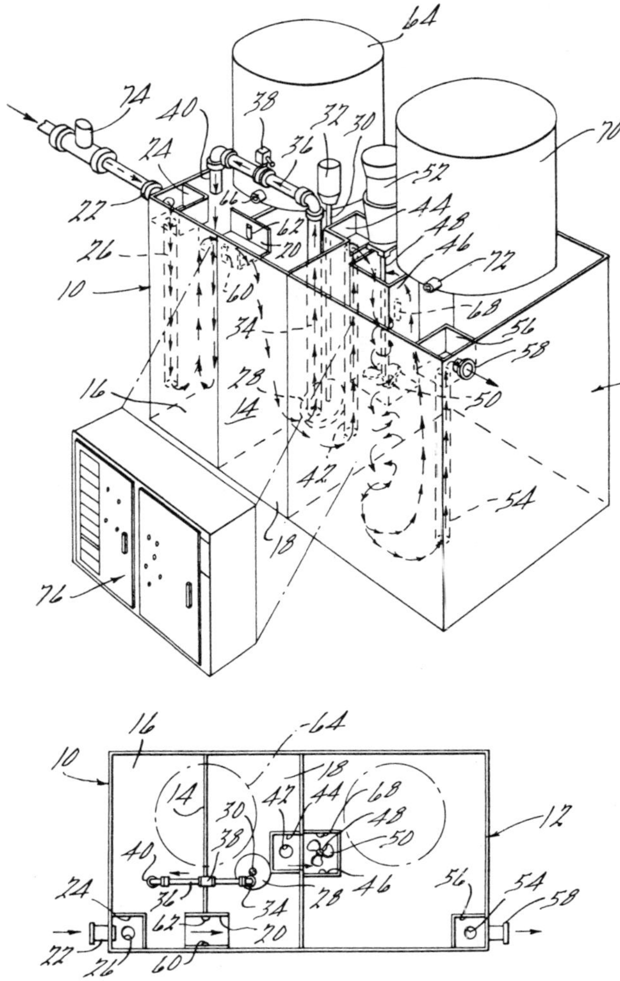

Source: A.N. Sulfaro and A.H. Stein; U.S. Patent 3,715,308; February 6, 1973

The waste solution, such as a water rinse solution containing dilute concentrations of dissolved hexavalent chromium or cyanide compounds derived from cyanide copper, cadmium and zinc rinses or other cyanide bearing rinses, enters the first compartment of the first tank **10** through a flanged inlet **22** which is disposed in communication with an inlet compartment **24** provided with an upright downcomer or standpipe **26** mounted in the base. Accordingly, the waste solution enters the inlet and passes downwardly in the direction of the arrows through the downcomer and is discharged from the lower end into admixture with the solution in the first compartment. The consolidated and mixed solution passes upwardly in the direction of the arrows and a portion continuously overflows through the action of gravity through the trough **20** into the second compartment **18** of the tank **10**.

A portion of the solution in the second compartment **18** is continuously withdrawn adjacent to the bottom and remote from the discharge side of the trough **20** by means of submersible pump **28** driven by a shaft **30** connected to a motor **32** mounted above the tank **10**. The discharge side of the pump **28** is connected to a riser conduit **34**, to the upper end of which a horizontal branch **36** is connected and is provided with a remotely actuatable injector **38** for continuously or intermittently introducing chemical reagent to the solution passing therethrough. The solution is discharged through the end of a downwardly directed discharge line **40** which is connected to the output end of the horizontal branch **36**. The recycled solution with make-up reagent is uniformly mixed with the solution in the first compartment and after sweeping downwardly and across the base of the tank, again flows upwardly and a portion is transferred to the second compartment through the overflow trough **20**.

A portion of the solution from the second compartment is continuously withdrawn through a normal overflow action at a rate corresponding to the preselected influx rate of waste solution into the system by means of a riser pipe **42** which extends with its inlet end spaced from the bottom of the second compartment upwardly and in communication with a liquid-tight chamber **44** mounted on the upper end portion of a partition separating tanks **10** and **12**. The liquid entering the chamber **44** overflows and is discharged into the upper end of an open-ended rectangular duct **46** through which a shaft **48** of a propeller-type mixer extends having a propeller **50** secured to the lower end and which is driven by a motor **52** drivingly coupled to the upper end.

The direction of rotation and orientation of the propeller **50** is such that the solution in the tank **12** and disposed within the duct **46** is caused to move downwardly, whereby a continuous mixing of the solution overflowing into the tank **12** and the solution is accomplished and the consolidated solution is recirculated in accordance with the direction of flow depicted by the arrows in Figure 4.6a.

A portion of the solution in treating receptacle **12** is continuously withdrawn at the preselected waste solution influx rate through a riser pipe **54** having its inlet end disposed adjacent to the bottom of tank **12** and remote from the point of entry of the treated solution from tank **10**. The withdrawn solution passes upwardly through the riser **54** into a discharge chamber **56** through a normal overflow action from which it is discharged through a flanged outlet **58**. The fully treated solution flowing through the flanged outlet can be readily discharged into a sanitary sewer or can be subjected to further settling if desired or required to remove the harmless but entrained solids therein.

The chemical characteristics of the solution during its progressive transfer from the flanged inlet **22** out through the flanged outlet **58** is continuously monitored at strategic locations to assure the addition of appropriate reagents including acids, bases and reactive constituents to effect a chemical destruction of the toxic constituents therein. In the specific arrangement shown, a first stage pH electrode **60** and an oxidation-reduction potential (ORP) cell **62** are mounted in contact with the solution flowing through the trough **20** from the first compartment to the second compartment.

In the treatment of hexavalent chromium contaminated waste solutions, the ORP meter is operative to control the injection of sulfur dioxide or equivalent reagents, such as sodium metabisulfite, through the injector **38** for admixture with the solution being recirculated through the horizontal branch **36**. The pH electrode **60** in turn is operative to control the addition of a mineral acid such as sulfuric acid from a reagent tank **64** mounted above the first tank **10** and provided with a remotely operable valved outlet **66** for intermittently or continuously discharging a controlled amount of acid into the first tank so as to maintain the pH at a level preferably at about 3.5 or below at which the acid sulfonation reaction progresses rapidly in accordance with the following equations:

$$SO_2 + H_2O \longrightarrow H_2SO_3$$

$$3H_2SO_3 + 2H_2CrO_4 \longrightarrow Cr_2(SO_4)_3 + 5H_2O$$

The foregoing two step reaction generally occurs in about ten minutes under appropriate pH conditions and the aqueous solution containing all of the hexavalent chromium in the form of trivalent chromic sulfate overflows into the rectangular duct of the second tank, where a second stage pH electrode **68** is operative to raise the pH to a level preferably within about 7 to about 8 by the addition of an alkali hydroxide, such as caustic, from a reagent tank **70** having a remotely operable valved outlet **72** which is disposed above the tank **12**. The rapid conversion of the trivalent chromic sulfate to chromic hydroxide proceeds in accordance with the following equation:

$$Cr_2(SO_4)_3 + 6NaOH \longrightarrow 2Cr(OH)_3 + 3Na_2SO_4$$

The solution discharged from the flanged outlet **58** in accordance with the foregoing treatment and flow cycle is at approximately a neutral pH and contains only nontoxic trivalent chromium hydroxide which can be removed by settling if desired or required. In accordance with the foregoing arrangement, it will be apparent that the elevation of the inlet compartment **24**, the trough **20**, the overflow chamber **44** and the discharge chamber **56** and attendant piping is at a level such that a natural flow of solution occurs through the two stages as a result and in direct proportion to the amount of contaminated waste liquid introduced through the inlet portion of the system. The control of this influx of liquid waste can be achieved such as by means of a flow control device **74**, as shown in Figure 4.6a which may comprise a metering pump or more usually, a remotely actuatable valve which on actuation prevents entrance of further waste solution in the event any chemical imbalance should occur in the first or second stage treatment, preventing discharge of solution from the apparatus until the imbalance is again corrected.

Typically, the volume of the aqueous solution contained within the first tank relative to the influx rate of waste solution is in the order of at least about 3:1 and preferably from around 10:1 to about 20:1 for dilute solutions of the type normally derived from water rinse treatments which contain up to 500 ppm, more usually about 100 ppm, toxic constituents. Accordingly, the first tank may have a volume of about 600 gallons with an influx rate of about 30 gal/min of waste solution, thereby providing an average residence time within the first treatment stage of about 20 minutes.

Similarly, the second tank is provided with a volume of about 600 gallons under these same circumstances such that a similar average residence time of 20 minutes is provided for during the second stage reaction. In addition to the foregoing, it is generally preferred that the rate of recirculation of solution from the second compartment to the first compartment be at least about two times the rate of influx of waste solution and preferably at least about 3:1 to about 4:1. Accordingly, at an influx rate of about 30 gal/min, the rate of solution recirculated through the horizontal branch **36** would preferably be in the order of about 90 to 120 gal/min. A detailed description of the control system is also provided.

Addition of Metal Salts as Exchange Agents for Chelates

According to a process described by *L.E. Lancy; U.S. Patent 3,682,701; August 8, 1972; assigned to Lancy Laboratories Inc.* used or waste wash water treatment solution from individual cyanide type plating lines that apply coatings of different metals on workpieces, having a chelating compound and a dissolved content of plating metal that have been picked up from drag-out on the workpieces, is treated with a salt of metals of the class consisting of calcium, barium, lithium, aluminum and magnesium. The salt acts as an exchange element in the chelating compound to replace a more noble dissolved metal content in the wash water treatment solution before it is reused to wash off drag-out from workpieces being processed in a different plating line. Freshly plated metal surfaces, especially zinc or cadmium surfaces, are very active, and the wash solution has a high pH of about 10 to 14 and contains a relatively high concentration of chlorine, about 50 mg to 4,000 mg/l. The problem particularly arises by reason of the fact that chelating or sequestering agents have come into use as an ingredient of the make-up in plating solutions as a compound added to the proprietary plating solution mix. Such an agent is thought to improve the plating process by adding brightness and smoothness. Typical agents are ethylenediaminetetraacetic acid (EDTA), nitrilotriacetic acid, ethylene glycol-bis(beta-aminoethyl ether)-N,N-tetraacetic acid.

It has been found that the presence of such an agent causes a complexing of the metal content of the solution and prevents such content from being precipitated out from the solution. Plating metal is thus retained in soluble form in the solution, where otherwise, under the pH conditions of about 10 to 14, the metal salt or salts used in the plating operation in view of such pH and presence of no other solubilizing agent would be insoluble and precipitate. However, the chelating agent has the characteristic of holding plating metals in solution as a chelate thereof.

The process as devised is carried out in such a manner as to satisfy the chelating compound's hunger for holding a before-used plating metal in solution, where otherwise such metal, under the solution conditions involved, would precipitate as an inert material. This hunger is satisfied by providing or adding one of a class of metals of low nobility that will serve as an exchange medium with the plating metal that is chelated or held by the chelating agent, whether it is copper, cadmium, tin, zinc, nickel, silver or other more noble plating metal. Such exchange agent is provided in the used wash solution in excess or over a stoichiometric amount to reduce the soluble more noble metal content to not greater than 4 to 5 mg/l which represents the optimum maximum that is permissible without resultant adverse reaction with the plated surfaces of a workpiece.

In a treating solution, adding calcium chloride in the amount of about 5 g/l (1.55 g/l of calcium content), with the pH and the chlorine concentration unchanged, the soluble copper concentration was reduced from about 3.56 mg/l to 0.34 mg/l. After making an addition to the treatment solution of the second solution example of 1.5 g/l of calcium chloride ($CaCl_2$) representing 0.465 g/l of calcium, analysis showed an unchanged pH of 13, an unchanged free chlorine concentration of about 2,800 mg/l but with a soluble cadmium concentration of only about 7 mg/l. To further prove the experiment, 500 mg/l of cadium in the form of cadmium cyanide was additionally added and, after replenishing the chlorine concentration consumed by the cadmium cyanide addition, and allowing an hour's reaction time, the reanalyzed same solution was found to contain only 3.2 mg/l of soluble cadmium in solution.

Electroless Plating

According to a process described by *D.R. Kamperman; U.S. Patent 3,770,630; November 6, 1973; assigned to Dart Industries Inc.* solutions that have been used for electroless plating a workpiece of nonconductor material, such as a resin, or for stripping metal from a workpiece are treated in such a manner that they may be discharged as nonpolluting waste into a natural body of water.

By way of example, a typical treatment system may employ a batch dumping of the process solution from either an electroless plating solution or metal workpiece stripping treatment. This involves first acidifying the solution to preferably a pH of about 5 and if highly alkaline in nature, to at least reduce the pH to about 8 or 9. Secondly, the batch waste solution should be diluted so that its metal content concentration is not more than 5 g/l. Thirdly, the solution should be heated to at least 120°F and preferably or as an optimum to about 140°F or higher. Finally, alkalinity is added along with an excess (above a stoichiometric amount) of sodium hydrosulfite so as to provide the solution with an excess of the latter of at least about 50 mg/l, but as an optimum higher, and up to about 500 mg/l.

The solution is then allowed to settle and the treated liquid is decanted, resulting in a content of less than one mg/l of dissolved metal. The solid waste is removed and discarded which, of course, is insoluble and thus suitable for solid waste disposal. The same process may be used for a chemical treatment rinse or for an integrated and recirculated treatment wash solution, with the described chemical conditions being maintained by either adding alkalinity to provide the desired higher pH after an acidic treating process or for acidic solution. An alkali metal salt such as calcium chloride will tend to precipitate and thereby reduce the pH; it can be used while maintaining the reducing conditions through continuous or intermediate hydrosulfite additions to maintain the necessary excess.

PULP AND PAPER

GENERAL PROCESSES

Closed Cycle Coking Process

H.V. Hess and E.L. Cole; U.S. Patent 3,717,545; February 20, 1973; assigned to Texaco Inc. describe a closed circuit process for treating waste liquors, such as spent sulfite liquors from wood digestion by coking the liquors in the liquid phase under pressure under turbulent flow conditions to form coke and a clear effluent of low chemical oxygen demand. In the process, a heat exchange zone, a coking zone, a steam and flue gas-producing zone, an absorbing zone, a wood digesting zone and a kilning zone are provided and interconnected by suitable conduits.

Pressurized spent liquor is sent to the heat exchange zone where it absorbs heat from the coking zone effluent before coke separation, thereby absorbing also heat content contained in the coke. The coke is discharged to the kilning zone where it is dried by hot flue gases produced by burning previously dried coke in the steam producing zone. Sulfur dioxide containing warm flue gas from the kilning zone is contacted in an absorbing zone with cool coker effluent to produce warm ammonium bisulfite solution which is passed to a wood digestion zone as well as cool flue gas substantially free of all air pollutants.

Sulfur-containing gases produced by coking and wood digestion are burned in the steam-producing zone to manufacture sulfur dioxide for recycling. Use of a closed system reduces to a minimum odors in the digestion section of a pulp plant since all the odorous sulfur compounds are burned and the sulfur dioxide is recovered.

With reference to Figure 5.1, spent sulfite liquor is pumped through a pump up to a pressure of around 1,000 psig and passed through a heat exchanger where it is in heat exchange with hot coker effluent and thence to a coker which is operated under turbulent flow conditions. Here the wood sugars and lignins present in the liquor are coked.

The total effluent from the coker consisting of water in liquid phase, suspended particulate coke **18** and a small amount of gas passes through a pressure reducing valve **4** to release a small amount of gas and then to the heat exchanger **2** where it is in heat exchange with the spent sulfite liquor feed. It should be noted that this arrangement allows for the transfer of the heat content of both the coke and the liquid effluent to the incoming feed stream. The coker effluent, after heat exchange, drops into the vessel **5** which is provided with a trap tray **20** as shown.

FIGURE 5.1: CLOSED CYCLE TREATING PROCESS FOR WASTE LIQUORS

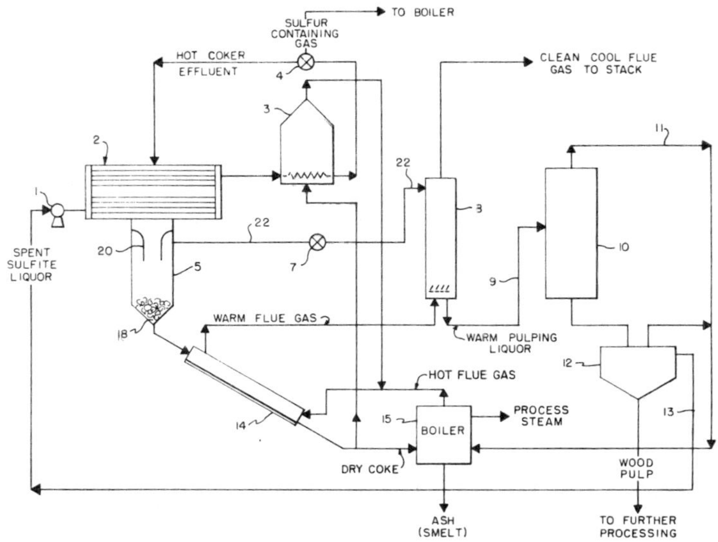

Source: H.V. Hess and E.L. Cole; U.S. Patent 3,717,545; February 20, 1973

The coke settles to the bottom of vessel **5** and coke-free, cooled coker effluent is withdrawn through line **22** and pressure reducing valve **7** where the pressure is reduced to atmospheric. Any gases emerging from pressure reducing valve **7** can be conveyed to the boiler. The cooled coker effluent at atmospheric pressure (optionally after NH_3 addition through line **22**) is passed to an absorber **8** where SO_2 is scrubbed from the relatively cool flue gas (about 300°F). A clean cool flue gas is produced and sent to the stack.

The liquid from the absorber (now warm pulping liquor) is sent (after fortification, if necessary, with the usual pulping chemicals) to the wood digester **10** through line **9.** In the digester wood chips are reduced to wood pulp and spent sulfite liquor. The gases produced will contain H_2S, methyl mercaptan, dimethyl sulfide, etc. and are released through line **11** and along with the gases produced from blow vessel **12** are sent to boiler **15** to burn the sulfur compounds to SO_2. In the blow vessel wood pulp is separated from the spent sulfite liquor which is recycled through line **13** as shown.

Wet coke from the separator vessel **5** is blown periodically by the pressure of the system as a dense slurry into the kiln **14**. In the kiln the wet coke is contacted countercurrently with hot flue gas from the boiler and emerges from the kiln as dry coke which is used to fuel the boiler and the coker heater. This contact at the same time cools down the flue gas for the SO_2 absorption step which is desirable.

Process steam and hot flue gas are produced in the boiler and all sulfur compounds are burned to SO_2. The flue gas then passes as a heat source to the kiln and eventually to the absorber **8** where the SO_2 is recovered by scrubbing with cool coker effluent to reconstitute the wood digestion liquor. The coke produced contains 6 to 7% sulfur and the sulfur is recycled for reasons of economics. In other metal based (Na, Ca, Mg) sulfite processes sulfur recovery is also necessary and in those cases any metal based chemicals adhering to the coke would be recovered as a result from the boiler.

This is true also of the kraft pulping which utilizes NaOH and Na_2S as pulping chemicals and the Soda process which uses NaOH. In the case of metal based pulping processes, the smelt would be properly treated for recycle to the wood digestion portion of the process. The process as described is most useful in the wood pulping industry where rather formidable problems exist with respect to both air and water pollution and also where it is necessary to recover chemicals both for reasons of economy as well as air and water pollution. The process allows for maximum recycle of water and pulping chemicals and essentially eliminates odor at least from the digester portion of a wood pulping plant. Since pulp and paper mills are heat deficient, the heat produced as process steam is welcome.

In the process, carbonization is substantially complete within a period of about 0.5 to 10 minutes. In a preferred example, the outlet temperature of treated liquid leaving the tubular coking zone is maintained within the range of about 550° to 650°F. The waste liquid undergoing treatment is maintained at elevated temperatures above 450°F for a sufficient period of time, e.g., one minute, to obtain substantially complete carbonization of organic compounds contained therein to water-insoluble solids or coke.

The coker used here is a tubular reactor. The flow of liquid in a tube reactor may be laminar or turbulent and for a given tube reactor the type of flow depends on the quantity of fluid flowing through the reactor. Turbulent flow is most desirable for coking as coke deposition on the reactor walls is minimized, thus assuring long operational life of the tubular reactor system. A useful relationship that substantially indicates the type of flow in a tube is the Reynolds number. The Reynolds number is defined by the relation $N_{Re} = DV\epsilon/\mu$ where N_{Re} is the Reynolds number; D is the inside diameter of tube, in feet; V is the average linear velocity, ft/sec; ϵ is the fluid density, lb/cu ft; and μ is the fluid viscosity, lb/ft-sec. Preferably Reynolds numbers above 4,000 are maintained in the tubular reactor. It is preferable that the tubes in the heater be sized to insure turbulent flow conditions and optimize heat transfer.

Hydropyrolysis and Use of Activated Carbon

A process described by *W.G. Timpe; U.S. Patent 3,762,989; October 2, 1973; assigned to St. Regis Paper Company* relates to the conversion of the nonfibrous materials in wood or other pulp source to activated carbon and the use of such carbon for the purification of the water utilized in pulping and papermaking operations. What has been found is that the major portion of the original organic chemical values in spent pulping liquor solids can be converted to a water-insoluble fraction by heating under pressure.

The process is applicable to the black liquor as it comes from the kraft digestor at a solids content of 16% by weight, or even lower, or indeed at any solids content up to, for example, 65% by weight, or even higher. Preferably, however, the hydropyrolysis step will be introduced into the white liquor regeneration and recovery system when the solids content of the black liquor is about 23 to 27% by weight, after removal of the tall oil salts.

In the process, the black liquor to be treated is heated under a pressure, usually autogenous, of from about 1,000 to 3,500 psi during a period of from about 10 minutes to 3 hours. The temperature of the hydropyrolysis step is from 490° to 700°F. The preferred conditions for purposes of economy and efficiency are 550° to 650°F, 1,300 to 2,700 psi pressure for a period of from 20 minutes to 1 hour.

During hydropyrolysis, the major portion of the organic chemical values separate as a water-insoluble precipitate. The inorganic constituents of the precipitate are principally sodium carbonate with minor amounts of sodium sulfate, sodium sulfide and sodium bicarbonate also present. The insoluble precipitate is separated by filtration or analogous process and the solution containing the dissolved inorganics is brought directly to the causticizing tank where the sodium carbonate is converted to NaOH by contact with calcium hydroxide. The aqueous solution from the causticizing tank contains mostly sodium hydroxide. Reaction of the alkaline solution with hydrogen sulfide (preferably the gas stream from thermal cracking) converts a sufficient quantity of sodium hydroxide to sodium sulfide to make a solution useful for pulping liquor. The following example illustrates the process.

Example: 10,000 gal of black liquor from a kraft pulping recovery operation is concentrated to a solids content of about 25% by weight and the tall oil soap removed. It is heated for about 30 min at a temperature of 600°F and a pressure of 2,000 psi in a hydropyrolysis reactor equipped with an agitator and a condensor. During the heating period a charry, black precipitate forms, and it is separated by filtration and washed with water. The washings and filtrate are combined. The precipitate which weighs approximately 6,000 pounds on a dry basis is separated into two equal fractions. One fraction is found to be an excellent fuel for the power boiler. The second fraction is heated in a Herreshoff furnace at 1675°F for 30 min in an atmosphere of carbon monoxide, carbon dioxide, steam and nitrogen to produce approximately 1,000 lb of water grade activated carbon.

A mixture of 100,000 gal of combined effluent from a kraft pulp and paper mill including liquors obtained from the pulping and bleaching operations together with water used in the papermaking is subjected to primary and secondary purification in the usual manner and then contacted with 200 lb of the activated carbon produced as described above by stirring the carbon in the effluent water at ambient temperature for 30 min. The water is tested and found to be of suitably high purity for reuse in a mill for pulp and papermaking purposes. The carbon is recovered by filtration and found to be an excellent fuel for the mill power boiler.

The vapor from the hydropyrolysis reaction is condensed with running tap water to recover a condensate from which water, phenol and cresols are separated. The uncondensed fraction of the vapor is passed over a bed of powdered cadmium sulfide catalyst in a tube reactor at a temperature of 1000°F to convert the major portion of the sulfur containing organic compounds in the gas stream to a mixture containing ethylene, methane and hydrogen together with approximately 600 lb of hydrogen sulfide.

The combined filtrate and washings from the hydropyrolysis reaction are causticized with calcium hydroxide in the usual manner to produce an aqueous solution containing a large amount of dissolved sodium hydroxide together with a lesser amount of sodium sulfide and sodium sulfate. The gas stream from the thermal cracking operation is passed through this solution to increase the concentration of sodium sulfide. The insoluble components in the gas stream pass through the solution and are useful as fuel in the power boiler. The hydrogen sulfide treated solution is suitable for use in the pulping process.

The process is repeated, except that before hydropyrolysis a stream of carbon dioxide and steam is passed through the concentrated black liquor until the pH is reduced to 9. The precipitate which forms is separated by filtration and washed with water. The washings are combined with the filtrate from the hydropyrolysis step. The precipitate is carbonized by heating at 900°F for 20 min and then converted to activated carbon.

R.E.S. Thompson; U.S. Patent 3,783,128; January 1, 1974 describes a process for the closed circuit processing of all waste products from forest product plants, without air or water pollution. Solid wastes, typically bark, wood chips, sawdust, or the like are carbonized to a char. The char is at least in part treated in a second furnace to produce activated carbon, char not so treated being briquetted and sold. The activated carbon is used as a filter medium for plant effluent liquors, which can be recycled after solids removal. The carbon and entrained solids are returned to one of the furnaces. Off-gas from all furnaces is passed to a boiler for steam generation, the spent gas being passed through an absorber, if necessary, before release.

Formaldehyde, Nitric Acid and Urea

J.N. Karnemaat; U.S. Patent 3,655,395; April 11, 1972 describes a process for treating waste materials containing relatively high amounts of solids, such as up to 60% by weight, or more, of solids. The process comprises the following steps: mixing the waste materials with formaldehyde to kill the bacteria and to allow the reducing properties of the formaldehyde to react with materials in the waste; then cautiously adding concentrated nitric acid and mixing same with the waste materials and formaldehyde in order to form a substantially homogeneous mixture, allowing time for the reduction of the nitric acid into fuming nitric, and for the oxidizing properties of the volatile NO_2 and other oxides to react with materials in the waste; and then adding urea to the homogeneous mixture and mixing same therein. After a period of polymerization, agglomeration then occurs.

The waste materials are thereby precipitated or agglomerated into a form in which they can be treated in any one of numerous manners, such as recovering of constituents for sale or reuse, use in recycling techniques, or the design of new products. The solids can, if desired, be disposed of by conventional landfill techniques. However, because they contain relatively large amounts of nitrogen and other valuable constituents, they can be used to make various useful products, such as animal feed supplements or fertilizers, or they can be further treated, as by ion exchange techniques, various chemical reactions and/or screening techniques, in order to recover various useful constituents. The following example illustrates the process.

Example: Process for Utilization of Paper Mill Wastes – The starting material was 100 g of paper mill wastes. 25 ml of formalin were added and heated to 60° to 80°C. 5 ml concentrated nitric acid were added (cautiously) and allowed to react for 10 min, or until the bubbling ceased. 50 g of prilled urea were added, the mixture was stirred and dried to 3 to 5% moisture. This material lends itself to fractionation and recovery of the basic ingredients by techniques such as differential solubilities, screening, density variations, etc.

This material also has certain properties which might be advantageous in the formation of new products without initial degradation. For example, the relatively high nitrogen content of the product coupled with moisture retaining qualities makes it suitable for use as an efficient soil conditioner. With humus and other fertilizer elements, including micronutrients, pesticides, etc., other specialty agricultural products can be designed. With the capability of adjusting the fiber content and the urea-formaldehyde resin polymer ratios, rigid and semirigid urea-formaldehyde resinous products can be obtained by increasing the amounts of formaldehyde and urea added.

Molecular Oxidation

A process described by *S.F. Galeano; U.S. Patent 3,655,343; April 11, 1972; assigned to Owens-Illinois, Inc.* pertains to a multistage tubular reactor by which the oxidation with molecular oxygen of the spent or weak or strong black liquor produced, for example, in a kraft pulping process, is efficiently and economically performed without any undesirable foam formation and without the concentrated accumulation of inactive sulfate type compounds. The oxidation of a spent or black pulping liquor produced in the conventional art known kraft pulping processes has been previously carried out by employing air as the oxidizing agent. In the prior art, oxidation is also being carried out on spent liquors produced from hardwoods on both the weak liquor, that is a spent liquor containing about 12 to 16% solids and on strong spent liquors, that is liquors containing 30 to 36% solids.

Generally, in the pulping installations employing the cooking of southern pine, there are only a few installations wherein the oxidation of a strong black liquor is practiced with air. Also, as far as is known, there does not appear to be any practice of the oxidation of the weak liquor due to excessive foaming. The oxidation of a spent liquor produced from a southern pine often is accompanied by a most objectionable characteristic, that is, the formation of excessive amounts of foam.

Different methods of air oxidation have been suggested by the prior art for controlling the foam, however, the methods have not received universal acceptance and the attempts have generally led to the fact of uncontrolled foam formation and attempts to use the foam in a better economical way. In view of this prior art discussion, the fact seemingly remains that the presently employed processes for the air oxidation of weak black liquor are not very feasible at this time. It has been suggested that a solution to this problem may reside in the employment of molecular oxygen for the oxidation of black liquor, especially the oxidation of a kraft weak black liquor from a kraft process using southern pine. However, so far this suggestion has not found acceptance in the pulping art because of the lack of a suitable means for effecting the oxidation. This process concerns a two-stage device for the molecular oxidation of a spent pulping liquor, mainly a black liquor. The two-stage device consists essentially of two reaction chambers connected one into the other as a series.

In using the device, the spent liquor is first atomized and turbulently mixed and intimately contacted with molecular oxygen to oxidize a high percentage of the spent liquor. The first oxidation stage occurs in the first reaction chamber. Next, the just oxygen treated liquor is again atomized, that is, for a second time, and again turbulently mixed and intimately contacted with molecular oxygen to further oxidize any unoxidized spent liquor remaining from the first oxidation in the first stage of the reactor. The second oxidation occurs in the second reaction chamber of the two-stage device.

Figure 5.2 depicts a two-stage reactor for use of oxygen absorption and chemical reaction in the oxidation of black liquor. A pipeline **20** is connected to source of spent liquor to be treated (not shown) and also connects to the first stage of a reactor **35** where the essentially bulk oxidation of the spent liquor occurs. The black liquor flow in the pipeline passes an oxidation reduction potential (ORP) cell or sulfide electrode **22** suitably mounted in the wall **23** of the feed pipeline. The pipeline is made of stainless steel or any other duct or conduit material that does not chemically react or is adversely affected by the spent black liquor.

The described cell **22** protrudes into the black liquor wherein, in operation, it relays a current signal through line **25** to an ORP Beckman 900 analyzer, for determining the sulfide content as ascertained by analyzer **24**. This analyzer provides essentially for a sulfide concentration signal, which is relayed to a primary control **26** which, in turn, is connected to a secondary controller or control **27**. The primary control is the first cascade of a two pen cascade controller. The secondary controller is an integral part of the control station **27** which contains the primary controller. The secondary controller is designed to interpret a signal which measures the flow of liquor in the pipeline. The flow of liquor is ascertained by an orifice meter **28** which is pneumatically connected through lines **29** to a pressure differential cell **30**. The pressure differential cell, in turn, is connected to a transducer, which converts the pneumatic signal to a current signal to be supplied through electrical line **31** to the secondary controller.

FIGURE 5.2: OXIDATION APPARATUS

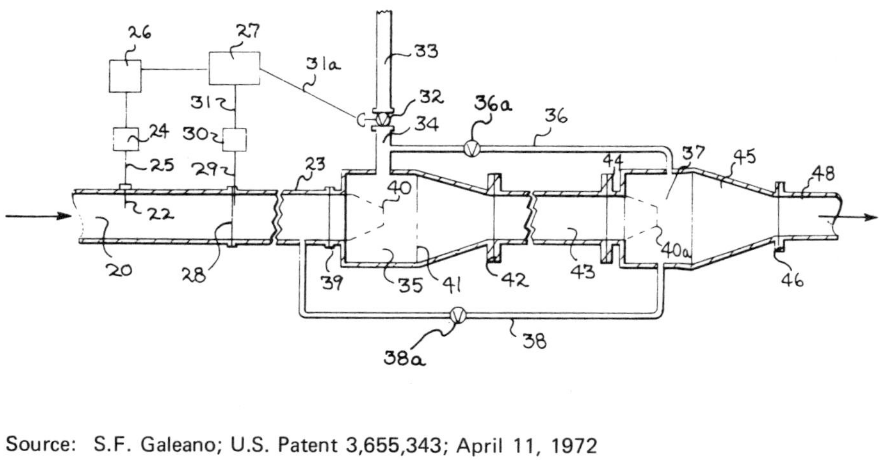

Source: S.F. Galeano; U.S. Patent 3,655,343; April 11, 1972

The secondary controller is connected through an electrical line **31a** to a conventional control valve **32** such as a double seated, diaphragm, cage in line type, which is part of an oxygen feedline **33.** Thus, in operation thus far, spent liquor flows in the pipeline past cell **22** which measures the amount of sulfide in the spent liquor while the flow rate of the spent liquor in the pipeline is also being measured by an orifice meter.

The two data are evaluated by the signals sent to the primary and secondary controllers which send a command signal to valve **32** through line **31a** for letting a now predetermined amount of oxygen enter into the system by a tee pipe **34** which permits some of the oxygen to enter the first stage reactor **35** and also permits some of the oxygen to travel in a conduit or pipe **36** suitably equipped with an in line gate valve **36a** to a second stage reactor **37.** The first stage reactor is connected to the second stage reactor by a pipeline **43,** and pipeline **20** extends into a flange **39** and enters the first stage reactor shortly thereafter.

The pipeline narrows and terminates with a nozzle **40** attached thereto for atomizing the spent liquor in the first stage reactor. The first stage reactor has a joint **41** for converging into a flange **42** which leads into a pipe section **43** through a flange **44** and into second stage reactor **37.** The length of the pipeline **43** is determined by use of the equation previously indicated. The second stage reactor is constructed in a manner similar to the first stage reactor by having a converging section leading into a flange **46** and then into a pipe **48** for transporting oxidized black liquor to storage tanks, not shown. The second stage reactor is also externally connected to inlet pipeline **20** through a circulating conduit **38** suitably equipped with an in line gate valve **38a.** The circulating conduit circulates the spent liquor to aid in the oxidation of the spent liquor. In operation, the pressure in pipeline **20** forces the liquor into conduit **38** for circulation into reactor **37.**

In actual operation, the tubular reactor is operated at a high Reynolds number, above 500,000 to guarantee turbulent conditions in the reactor. The relative dimensions of a typical reactor are, for example, as follows. The nozzle diameter is 0.5 times the pipeline diameter; the reaction chamber, for example, **35,** has a diameter 1.8 times the pipeline diameter; the length of the reaction chamber is 5.5 times the pipeline diameter; and the length of the converging zone is 3 times the pipeline diameter. The diameter of the oxygen line **36** and spent liquor lines **43** to the second stage reactor are 0.33 times the respective diameters of the incoming oxygen pipelines **33** and the incoming liquor pipeline which latter line has a diameter of any predetermined size, for example, according to the installation of 4, 6, 8 and 10 inches, etc.

The oxygen absorption process and the subsequent chemical reaction takes place in the reactor in three phases. In the first phase, oxygen molecules are transferred to the liquid black liquor surface. The liquid interphase has a definite thickness, usually a few molecules. Next, the second phase, oxygen must pass through this layer by diffusion. Finally, in the third phase, oxygen diffuses into the bulk of the spent liquor and the oxidation reaction takes place.

The speed of the phases can be increased by turbulence. In the first stage, a fraction of the black liquor, that is, 80 to 90% of the total flow in incoming pipeline **20,** is atomized at nozzle **40** at a high Reynolds number above 500,000. The necessary oxygen is injected at **34** for oxidation, about 1.05 times the required stoichiometric amount being used. Intimate mixing and diffusion of the spent liquor takes place in chamber **35,** with oxidation taking place therein and in tubular section **43.**

Section **43** leads to the second reactor where the remaining fraction of unoxidized liquor is introduced and atomized by nozzle **40a.** Oxygen, for the oxidation of the remaining liquor is introduced by line **36.** The mixing again takes place in the second stage reactor **37,** and the liquor passes out in pipe **48.** The process can be run on a limited quantity of liquor or on a continuous basis.

Wet Air Oxidation System

According to a process described by *L.A. Pradt; U.S. Patent 3,661,778; May 9, 1972;*

assigned to Sterling Drug Inc. raw liquor is introduced to a heat exchanger and then through a temperature control valve to a reactor and then to a separator. A flash tank is provided which receives oxidized liquor from the separator at an intermediate pressure resulting in flash steam used for preheating in the heat exchanger.

As shown in Figure 5.3, raw liquor and atmospheric air enter the system at **10** proceeding at **12** to the heat exchanger **14** from which condensate may be derived as shown. From the heat exchanger the preheated liquor or sludge proceeds to the reactor which has an inlet **16**. After passing through the reactor and through line **18** into separator **20**, the gases are emitted as at **22** as before for energy recovery but the oxidized liquor which exits from the separator through bottom valve **24** to a flash tank **26** (see Figure 5.3b) is reduced to an intermediate pressure in the approximate range of 10 to 20 atmospheres.

FIGURE 5.3: WET AIR OXIDATION SYSTEM

a.

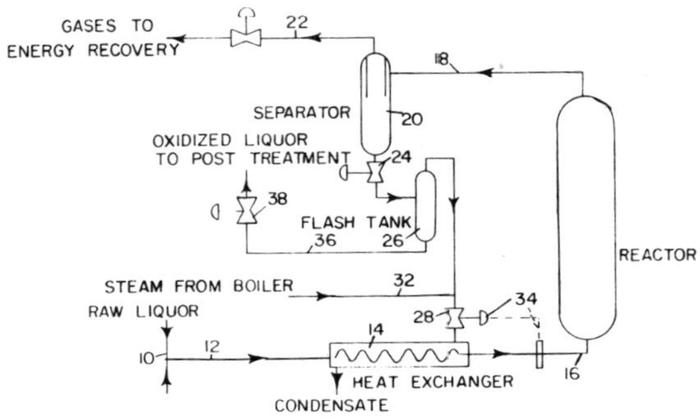

Flow Diagram

b.

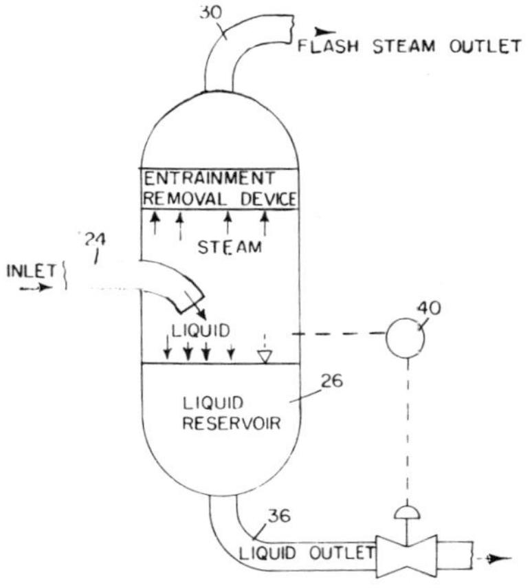

Flash Tank

Source: L.A. Pradt; U.S. Patent 3,661,778; May 9, 1972

This evolves flash steam as a result of the pressure reduction, and this steam is used for preheating in the heat exchanger, passing from the top of the flash tank, through the valve at **28**, to the heat exchanger **14**. Additionally, steam from the boiler may be introduced into line **30** from the flash tank **26** to valve **28**, as at **32**. The valve is controlled by a temperature control valve **34** controlling steam flow to the shell side of the heat exchanger.

The intermediate pressure in the flash tank depends upon the required reactor inlet temperature. For example, if the required reactor inlet temperature is 180°C, the flash steam temperature must be higher than this or about 200°C, fixing the intermediate pressure and flash tank at about 15 atmospheres. The reactor inlet temperature at **16** is easily controlled by the temperature control valve which controls steam flow to the shell side of the heat exchanger. If the flash steam evolved is insufficient this can be made up by supplying steam from the boiler. If excess flash steam is evolved, this excess can be directed to the steam mains for general plant use.

The quantity of flash steam can be calculated in the usual way considering the initial temperature of the oxidized liquid and the final pressure to which this liquid is reduced or flashed. For practical purposes the oxidized liquid can generally be assumed to be water, although large quantities of dissolved salts may have some effect on the temperature-pressure relationship of the vapor of this liquid. The flash tank is in essence a device to separate the flash steam from the remaining liquid after pressure reduction.

Design of such a tank is generally known in the art. This tank could be designed, for example, to give a liquid hold-up time of 3 min, an upward flash steam velocity of 3 to 4 fps and could be equipped with a mesh type or baffle type of entrainment removal device in the top. The tank would also be equipped with a device to control the level of the liquid in the tank. This is a conventional device indicated at **40**, and connecting the interior of the tank to the outlet, as shown in Figure 5.3b in dotted lines.

While liquor and air can be mixed before admission to the heat exchanger as shown and described, liquor only can be passed through the heat exchanger and air admitted after the heat exchanger if this should become desirable. Oxidized liquor from the flash tank can be sent directly to post treatment as at **36, 38** as shown, or can be used for additional preheating of raw liquor if desired and convenient. The oxidized liquor can be flashed to an even lower temperature and the resultant flash steam evolved used for heating instead of the oxidized liquor.

Continuous Air Oxidation of Sour Water

D.E. McCoy, R.M. McEachern and R.M. Dille; U.S. Patent 3,761,409; September 25, 1973; assigned to Texaco Inc. describe a continuous process for the liquid phase, noncatalytic, air oxidation of sour water in which the sulfur compounds in the water are converted to nonpolluting, nonoxygen-demanding compounds. The process involves heating the water and oxidizing the heated water with countercurrent or cocurrent flow of the water and oxidizing medium. The process rapidly converts objectionable sulfides and intermediate sulfur compounds such as thiosulfates, tetrathionates, polythionates, sulfites and polysulfides to sulfate.

Figure 5.4a shows schematically a case where oxidation is carried out on a countercurrent contact tower while Figure 5.4b illustrates the process as carried out in a cocurrent contact tower. Referring to Figure 5.4a, feed water, the pH of which has been adjusted as required, is flowed through line **10** to feed pump **12** which pumps it under pressure to feed effluent heat exchanger **14** and thence to heater **20** and oxidizer tower **22**.

The heater can be of any conventional construction and can be fired or steam heated. The oxidizer tower can be packed with Raschig rings, berl saddles, pall rings, or other suitable packings, or can contain bubble or sieve trays. The tower serves to contact the hot sour water with the oxidizing medium air, oxygen or a mixture in any proportion. This medium is introduced via line **24** and compressor **26** which brings it to systems pressure and then

FIGURE 5.4: CONTINUOUS PROCESS FOR THE AIR OXIDATION OF SOUR WATER

a.

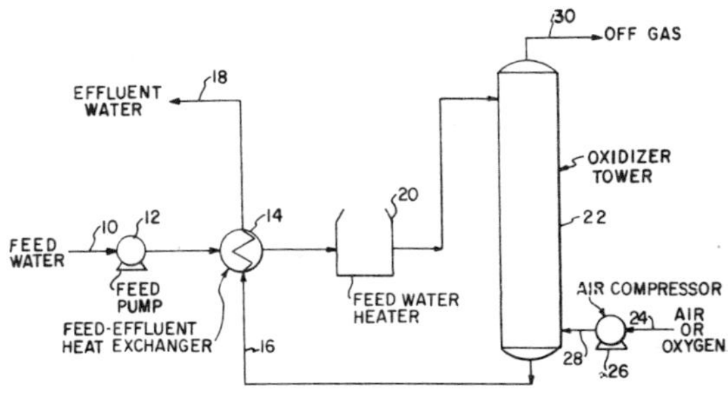

Countercurrent Contact Tower

b.

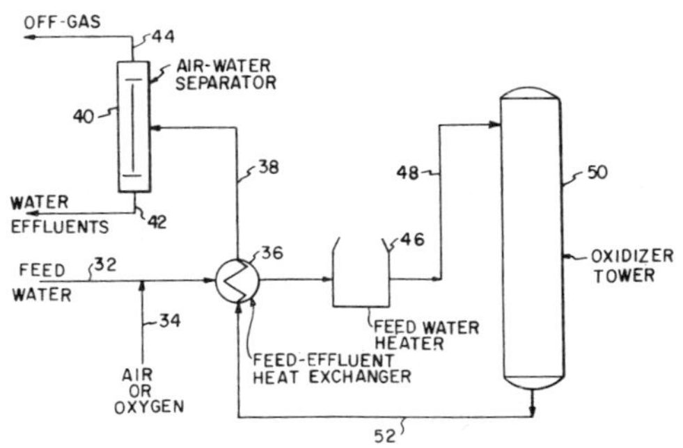

Cocurrent Contact Tower

Source: D.E. McCoy, R.M. McEachern and R.M. Dille; U.S. Patent 3,761,409; September 25, 1973

through line **28** in countercurrent relationship with the feed. The sulfur compounds react with the oxygen to form sulfate as shown in Equations 1 and 2 for sulfides and thiosulfates, respectively.

$$(1) \qquad H_2S + 2O_2 \rightarrow SO_4^{=} + 2H^+$$

$$(2) \qquad S_2O_3^{=} + 2O_2 + H_2O \rightarrow 2SO_4^{=} + 2H^+$$

The amount of oxygen required for complete oxidation to sulfate will depend upon which sulfur compounds are present and their relative amounts. To oxidize sulfides, an oxygen to sulfur weight ratio of 2 is required to produce sulfate. Oxidation of thiosulfate requires half as much oxygen as do sulfides. An excess of oxygen is desirable to assure complete and rapid oxidation of the sulfur compounds. Unused oxygen and inerts in the gas stream pass out of the top of the tower through line **30**. The water effluent from the bottom of the oxidizer passes through line **16** and the feed effluent heat exchanger and is discharged through line **18**, thereby preheating the feed. In Figure 5.4b the process is illustrated as operating with cocurrent downflow of oxidizing medium and water through the oxidizer tower.

In this case, sour water is fed through line **32** and mixed with the oxidizing medium supplied through line **34** before flowing into heat exchanger **36** where it is preheated to between 200° and 500°F by contact with hot effluent arriving from the oxidizer tower through line **52**. The preheated water is heated to operating temperature in heater **46** and passes into oxidizer tower **50** which preferably is packed with Raschig rings, berl saddles, pall rings or other suitable packing.

The oxidized effluent still containing off-gases passes into air separator **40**. The off-gas, mainly oxygen depleted air, is taken off through line **44** for discharge to the atmosphere or routed to a flue gas stack, and the water effluent is taken off through line **42**. The water has a reduced BOD and COD and may be discharged safely to a receiving water depending upon the concentration of other contaminants in the water. Specific examples of water treatments according to the process follow. The analysis of the untreated sour water is given in Table 1 below.

TABLE 1: UNTREATED SOUR WATER ANALYSIS

$S^=$	7,495	ppm as S
$S_2O_3^=$	320	ppm as S
$S_4O_6^=$	257	ppm as S
$SO_4^=$	44	ppm as S
NH_3	4,600	ppm as N_2

The analysis of partially oxidized sour water is shown in Table 2 below.

TABLE 2: PARTIALLY OXIDIZED SOUR WATER ANALYSIS

$S^=$	0	ppm as S
$S_2O_3^=$	5,440	ppm as S
$S_4O_6^=$	146	ppm as S
$SO_4^=$	2,563	ppm as S
NH_3	5,150	ppm as N_2

Example 1: The analysis of the oxidized water after treating the sour water of Table 1 with 100% excess of stoichiometric air requirement for the oxidation of sulfide to sulfates at 500°F, 750 psig and 6.4 Liquid Hourly Space Velocity (LHSV) was the following.

$S^=$	0	ppm as S
$S_2O_3^=$	64	ppm as S
$S_4O_6^=$	0	ppm as S
$SO_4^=$	5,139	ppm as S
Total S	5,346	ppm as S
Percent conversion to $SO_4^=$	96%	

Example 2: The analysis of the oxidized water after treating the sour water of Table 1 with 100% excess of stoichiometric air requirement for the oxidation of sulfides to sulfates at 450°F, 450 psig and 3.8 LHSV was the following.

$S^=$	0	ppm as S
$S_2O_3^=$	32	ppm as S
$S_4O_6^=$	0	ppm as S
$SO_4^=$	6,864	ppm as S
Total S	7,369	ppm as S
Percent conversion to $SO_4^=$	93%	

Example 3: The analysis of oxidized water after treating the sour water of Table 2 with 100% excess of stoichiometric air requirement for the oxidation of sulfides to sulfates at 400°F, 300 psig and 3.8 LHSV was the following.

$S^=$	0 ppm as S
$S_2O_3^=$	64 ppm as S
$S_4O_6^=$	736 ppm as S
$SO_4^=$	5,745 ppm as S
Total S	6,542 ppm as S
Percent conversion to $SO_4^=$	88%

Example 4: The analysis of partially oxidized water after treating the sour water of Table 2 with 100% excess of stoichiometric air requirement for the oxidation of thiosulfates to sulfates at 500°F, 750 psig and 5 LHSV was the following.

$S^=$	0 ppm as S
$S_2O_3^=$	96 ppm as S
$S_4O_6^=$	0 ppm as S
$SO_4^=$	7,411 ppm as S
Total S	7,340 ppm as S
Percent conversion to $SO_4^=$	100%

Extraction Technique for Wood Preservative Pollutants

M.W. Mar; U.S. Patent 3,715,306; February 6, 1973 describes a method using pyridine and a solvent selected from the group consisting of an aromatic halogen compound and an aliphatic halogen compound to extract wood preservative pollutants from an aqueous mixture. The following examples illustrate the process.

Example 1:

Component	Volume (ml)
Aqueous solution of pollutants	300
Carbon tetrachloride	60
Pyridine	1.0

In the first aqueous mixture of water plus pollutants, the pollutants had a boiling point in the range of approximately 150° to 185°C. The aqueous solution of pollutants was extracted, in a batch extraction process, at room temperature, by shaking the mixture of carbon tetrachloride and pyridine with the aqueous solution for about 5 sec. Then, the resulting mixture was allowed to separate into a lower layer of carbon tetrachloride, pyridine and pollutants and an upper layer of water plus reagents plus pollutants.

In the lower layer of carbon tetrachloride plus pollutants, the pollutants had a boiling point in the range of approximately 145° to 185°C. The ratio of pollutants in the first carbon tetrachloride layer to the aqueous solution was approximately 0.54 to 54%. In the water layer, containing pollutants and a small amount of pyridine and carbon tetrachloride, the pollutants had a boiling point of approximately 150°C. The ratio of pollutants in the first water layer to the ratio of pollutants in the original aqueous solution of pollutants was approximately 0.45 or 45%.

Then, the first water layer was once again extracted with a mixture of fresh carbon tetrachloride, about 60 ml. This batch extraction was at room temperature and for about 5 seconds and accomplished by shaking the mixture. There resulted a layer of carbon tetrachloride, pyridine and pollutants, and the pollutants had a boiling point in the range of about 150° to approximately 165°C. The ratio of the pollutants in the second carbon tetrachloride layer as contrasted to the pollutants in the first carbon tetrachloride layer was approximately 0.24 or 24%. The water layer had pollutants which boiled at approximately 150°C. The ratio of pollutants in the second water layer as compared with the pollutants in the original aqueous solution of pollutants was approximately 0.33 or 33%. From the foregoing it is seen that the carbon tetrachloride and pyridine layer extracted over one-half of the pollutants from the aqueous solution of pollutants. Also, the resulting water layer, from the second extraction contained a lower concentration of pollutants. Further, it is seen that when pyridine is absent, the transfer of pollutants from the aqueous solution to the carbon tetrachloride is lessened.

Example 2: The aqueous solution used was the same as in Example 1.

Component	Volume (ml)
Aqueous solution of pollutants	300
Carbon tetrachloride	60
Pyridine	1.0

In this batch process the aqueous solution of pollutants with the carbon tetrachloride and pyridine was boiled for about 15 min under a reflux condenser so that the carbon tetrachloride was introduced back into the boiling mixture. In the first extraction, after the mixture was allowed to cool, the pollutants had a boiling point in the range of about 145° to 185°C. The ratio of the pollutants in the carbon tetrachloride to the pollutants in the original aqueous solution and pyridine mixture was approximately 0.61 or 61%.

The water layer had pollutants which had a boiling point in the range of about 150°C. The ratio of the pollutants in the water layer as compared with the original aqueous solution of pollutants was 0.31 or 31%. Then, the water layer after the first extraction was boiled and refluxed for about 15 min with a fresh solution of 60 ml of carbon tetrachloride. Then, this mixture was allowed to cool and to separate into a lower layer of carbon tetrachloride, pyridine and pollutants and an upper layer of water, pyridine and pollutants.

In the layer of carbon tetrachloride the pollutants had a boiling point in the range of about 150° to approximately 165°C. The ratio of the pollutants in the second layer of carbon tetrachloride as compared to the first layer of carbon tetrachloride was 0.47 or 47%. The pollutants in the second water layer had a boiling point in the range of approximately 150°C. The ratio of the pollutants in the second water layer as compared with the pollutants in the original aqueous solution of pollutants was 0.14 or 14%; and, the ratio of pollutants in the second water layer as compared with the first water layer was 0.47 or 47%.

Alum Mud Treatment

R.R. Fuller; U.S. Patent 3,740,363; June 19, 1973; assigned to Gulf States Paper Corporation describes a waste effluent treatment which involves contacting a waste effluent, e.g., pulp and paper mill effluent, with a metal salt reagent, preferably alum mud. Treatment decolorizes the effluent and precipitates a substantial portion of the organic content. The precipitate and sludge is dewatered, then calcined, and the reagent regenerated from the ash for use again in a cyclic process. The decolorized effluent is biooxidized, preferably in a multistage biooxidation sequence and is now sufficiently pure for recycle purposes. The purified effluent may be bleached prior to recycle. The following examples illustrate the process.

Example 1: Various samples of the effluents from an unbleached kraft mill, and a bleached kraft, coated board mill were secured. Examples of the effluents tried are as follows: total effluent each type mill, pulp mill effluent of unbleached kraft mill, paper mill effluent of unbleached kraft mill, strong waste board mill (pulp mill less beach plant), effluent board mill less strong waste, caustic stage bleach plant, etc. Liter aliquots of these samples were treated with varying quantities of alum mud. Addition of 1 to 5% alum mud precipitates the color bodies from most effluents. The amount of precipitating material required for best results varied with the concentration of precipitatable organics in the effluent sample.

Simple reduction of pH to pH 5 was found to be a reliable guide for controlling the alum mud addition. The concentration of black liquor and/or bleach plant caustic effluent in the total effluent influenced the degree of color removal. Chlorination stage effluent, or a low pH mixed effluent, i.e., below pH 4, would not react to give adequate color removal; but addition of lime equivalent to about 1 lb/1,000 gal prior to treatment with alum mud gave the color reduction desired in most instances.

Example 2: Using a sample of strong waste of 6 to 7 pH, (a) varying amounts of alum mud were tried, and 1 to 2% was found to give the desired color removal; (b) alum mud addition was held constant at 1%, and various dosages of alum tried; (c) clay and alum in various ratios were tried; and (d) alum and sulfuric acid were tried in various ratios. The best results were obtained with alum mud (a) or alum mud and alum (b).

Example 3: Various metal salts and acids were tried without regard to quantity on a pulp mill effluent having a pH of 10.

Magnesium nitrate	Precipitate, color reduction poor
Lithium chloride	No change
Ferric chloride	Excellent precipitate, color reduction, but not good enough
Zinc acetate	Excellent precipitate, good color reduction
Magnesium sulfate	Precipitate, little color reduction
Aluminum chloride	Excellent precipitate, excellent color reduction
Barium chloride	Precipitate, little color reduction
Zinc chloride	Good precipitate, good color reduction
Cadmium sulfate	Good precipitate, color reduction not good enough
Lead acetate	Good precipitate, good color reduction
Copper sulfate	Good precipitate, good color reduction, but the supernatant is light blue
Calcium nitrate	Good precipitate, color reduction not good enough
Sulfuric acid	Precipitate, poor color reduction
Hydrochloric acid	Precipitate, poor color reduction
Ferrisul	Excellent precipitate, excellent color reduction, supernatant tends to have blackish tinge
Activated alum	Good precipitate, good color reduction

Some of these materials cause no pH reduction, some reduce to pH 6.5 and stop, other such as alum gradually reduce the pH and pH can be used to control the reaction. Acids reduced pH to 3.5 to attain precipitation, which is too low a pH. Some of the salts, e.g., lead, are not compatible with a secondary treatment step. The sludge volume remained at 25 to 35% of the solution after 24 hours settling.

Example 4: Effluent discharged to the river from a mechanically aerated lagoon of a plant producing bleached tissue, bleached board, and paper. This was an integrated mill producing its pulp by the sulfate process. The following are the characteristics of the effluent: 5-day BOD 190 ppm; oxygen demand 1,583 lb/million gallons; total solids 11,621 lb per million gallons; inorganic solids 7,483 lb/million gallons; organic solids 4,139 lb/million gallons; color, grayish black. Sodium aluminate was added to a liter of this effluent to pH 7.5, and alum mud added to pH 5 with slow agitation. Agitation was continued for 5 min and the precipitate allowed to coagulate and settle.

Characteristics of supernatant are as follows: 5-day BOD 71 ppm; oxygen demand 568 pounds per million gallons; total solids 10,064 lb/million gallons; inorganic solids 9,013 pounds per million gallons; organic solids 1,051 lb/million gallons; and color, comparable to tap water. The reduction in 5-day BOD was 64.12%, reduction in organics was 74.6%, removal of color was complete, and the effluent was comparable to tap water.

Example 5: The strong waste from the pulp mill of a bleach kraft coated board mill was treated for a 2 day period. Liquid 50% alum was metered into the effluent flow just ahead of the pump moving the effluent to a storage lagoon. Air-dry alum mud solids were fed at an approximate rate at the same point. The effluent at the start of the trial had an APHA color of 2,500; a COD of 960 ppm; and a 5-day BOD of 1,080 ppm. The effluent flow rate varied from 1.6 to 2.3 million gallons per day. pH was recorded continuously and also checked every 30 min. Samples were taken hourly of the treated effluent. Settling, precipitate (sludge) volume, pH, COD and color were determined.

In Phase 1 was added 12.5 lb air-dry mud solids per minute and control treated pH in the range 5.0 to 5.5. Input pH was 11 and alum feed rate was 2.6 to 2.8 gpm. In the treated effluent pH was 5.2, sludge volume after two hours was 25 cm, color was 120 and COD was 450 ppm.

In Phase 2 was added 8 lb air-dry mud solids per minute. The alum feed rate was 2.1 to 2.8 gpm. In the treated effluent pH was 4.9 to 5.2, sludge volume after 2 hr was 24 to 29 cm, color was 120 to 150, and COD was 430 to 520 ppm.

In Phase 3 was added 6 lb air-dry mud solids per minute. The alum feed rate was 1.7 to 2.8 gpm. In the treated effluent the pH was 5.1 to 5.6, sludge volume after 2 hr was 23.8 to 26.5 cm, color was 100 to 200, and COD was 440 to 520 ppm.

In Phase 4 was added 6 lb air-dry mud solids per minute and the pH range aim point was raised to 6.0 to 6.5. The alum feed rate was 2.0 gpm. In the treated effluent the pH was 6.3, sludge volume after 2 hr was 28.6 cm, the color was 1,000 and COD was 620 ppm.

In Phase 5 the caustic stage effluent was diverted from bleach plant into a strong waste flow. The air-dry mud solids feed was kept at 6 lb/min. In the treated effluent the pH was as follows.

pH	Alum added, g.p.m.	Color	COD, p.p.m.
5.2	3.7	375	360
4.7	2.8	625	400
4.9	3.4	250	330

The sludge volume after 2 hr was 28 cm and the 5-day BOD was 555 ppm. In Phase 6 Phase 3 was repeated and the treated effluent had a pH of 5.2, sludge volume after two hours was 25.5 cm, the color was 100 and the COD was 480 ppm.

In Phase 7 alum was added to a pH of only 5.0 to 5.5 and the alum usage was 2.3 to 2.7 gpm. In the treated effluent the pH was 4.7 to 5.1, color was 150 to 250 and COD was 420 to 570 ppm. During this work part of the alum was replaced by sulfuric acid. Acid addition was in the range of 0.1 to 0.4 gpm of 66° Bé acid. End results were not changed.

Example 6: The precipitates from laboratory and from Example 5 were filtered from the supernatant, dried at 105°C, ashed in a muffle furnace and the ash extracted with sulfuric acid. A typical analysis of these precipitates and extraction was as follows: heat value dry material, 6,777 Btu; ash content dry material, 37.77%; aluminum oxide content of ash, 35.5%; easily convertible oxide of ash, 34.7%; and recovery of Al_2O_3 present, 97.75%.

Ultrafiltration

A process described by *H.A. Fremont; U.S. Patent 3,758,405; September 11, 1973; assigned to U.S. Plywood-Champion Papers, Inc.* comprises the continuous method of removing and disposing of organic color bodies found in aqueous effluents from kraft pulp manufacturing operations, adjusting the pH of the effluents to from about neutrality to about 9, subjecting the effluents to ultrafiltration to form an aqueous permeate and a retentate containing substantially all of the color bodies in the effluents and having a solids concentration of at least about 15% and heating the retentate in the presence of oxygen at a temperature and for time sufficient to oxidize the color bodies to colorless inorganic salts and colorless inorganic gases which can be safely disposed of. Referring to Figure 5.5a, there is shown a flow diagram depicting the process in broad terms. The process comprises adjusting the first stage caustic extraction effluent by addition of acid. This pH adjustment can be accomplished using any acid, such as sulfuric, and by adding the acid and effluent to a conventional in line mixing device or a pipe of sufficient length to provide complete mixing. The rate of acid addition is controlled by conventional in line pH probes located after such mixing means. Depending upon the temperature of the effluent, it may be necessary to cool it to a temperature below 125°F prior to processing the ultrafiltration units in order to avoid damage to the membranes in the cells.

FIGURE 5.5: COLOR REMOVAL FROM KRAFT MILL AQUEOUS EFFLUENTS

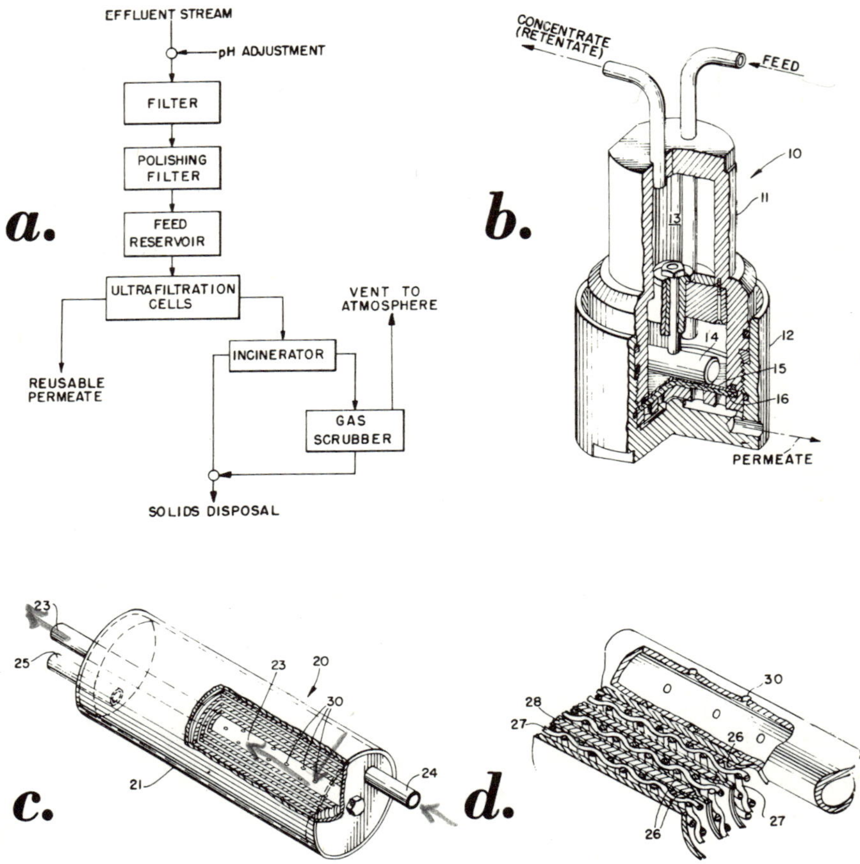

(a) Flow Diagram
(b) Cut Away View of Ultrafiltration Cell
(c) Cut Away View of Spiral Wound Filtration Cell
(d) Enlarged View of Membrane Module

Source: H.A. Fremont; U.S. Patent 3,758,405; September 11, 1973

After the pH adjustment (and temperature adjustment if necessary), the effluent is preferably filtered through a nonmolecular filter which will remove the suspended solids. This can be any conventional filter of the rotary vacuum drum type, gravity type, or sand bed type. In addition, polishing filters can be provided to insure the removal of the very small suspended solids in the order of microns. It is desirable to use one or both of these preliminary filtering operations since this helps to minimize any possible fouling of the ultrafiltration membranes by such solids.

The thus filtered effluent is then pumped at the required pressure through ultrafiltration cells, more particularly described below, having membranes of a pore size to retain the color bodies in the effluent while permitting lower molecular weight materials, such as inorganic salts, to pass therethrough. In order to have a steady source of material for the ultrafiltration cells it is preferred to place the effluent either before or after coarse filtration into a surge tank or feed reservoir.

The permeate from the ultrafiltration cells can be recycled for further use and the concentrate or retentate is burned in any incinerator adapted to burn high solids liquids. The residual ash, volatile acids, such as hydrochloric acid, and other volatiles are removed from the off-gases by scrubbing prior to venting. This will act to eliminate any possible or potential air pollution. The solid wastes removed from the gas scrubber are combined with the solids formed by incineration and can be disposed of in any conventional manner. Thermal oxidation is important in order to render the color bodies colorless by the oxidation that occurs during combustion. The colorless inorganic solids formed by this oxidation also require very little, if any, biological oxygen.

In accordance with this process, it has been found that regulation of the pore size to retain the high molecular weight organic color bodies and permit the lower molecular weight organics and inorganics to be passed through the membrane into the permeate has many advantages. For the color bodies ordinarily found in kraft mill effluent, a membrane with a pore size of from about 0.01 to 0.05 micron is used. This pore size can be varied dependent upon the size of the color bodies in the particular stream being treated. Such regulation of pore size permits high efficiency in retention of the organic color bodies and allows concentration to very high volumetric reduction ratios since the inorganic salts which comprise the bulk of the dissolved solids are not retained in the concentrate.

This avoids both a buildup of a high osmotic pressure in the concentrate which would inhibit concentration to high solids levels, and fouling resulting from increased concentrations of both organics and inorganics, e.g., residual calcium salts. Furthermore, a concentrate is obtained in the ultrafiltration process which contains primarily organic materials and disposal of this concentrate by incineration is substantially facilitated by the removal of inorganics, especially chlorides. It is possible to consider use of this purified organic concentrate as a suitable base for the production of by-product chemicals.

The ultrafiltration cell used can be any commercially available with a wide variety of membranes available for use. A conventional stirred cell with a diaphragm type membrane is depicted in Figure 5.5b. Such membranes are readily conventionally available and can be used as set forth in Figure 5.5b, or tubular or spiral wound membranes can be utilized. It is preferred to use spiral wound membranes such as shown in Figure 5.5c and 5.5d.

As is known, the membranes are usually modified cellulose acetate films approximately 100 microns in thickness with an active (dense) surface layer of approximately 0.2 micron. These membranes are manufactured by dissolving cellulose acetate in a mixture of solvents and then casting the mixture as a thin film. The solvents are evaporated and the film gelled. The surface of the film from which the evaporation first took place becomes the dense, active layer. The gelled film (or membrane) is then treated with hot water and is thereby caused to shrink to its final state. Referring to Figure 5.5b, there is shown a stirred ultrafiltration cell **10** having an outer jacket **11** and cap **12** defining an interior passage **13**, with a magnetically controlled stirrer **14**, attached to the interior of the outer jacket in the passage. A feed inlet and concentrate outlet are provided at the top of the cell and a permeate outlet at the bottom of the cell. Spaced below the stirrer and disposed between the feed inlet and permeate outlet in the passage is membrane **15** supported on a sintered disc **16.**

The operation of the cell will be largely evident from the description given. The effluent is pumped through the feed inlet into the passage. The bulk of the water and smaller molecular weight material will pass through the membrane and sintered disc and be discharged through the permeate outlet. The color bodies and minor amount of water are discharged through the retentate outlet. The stirrer is operated by magnetic means (not shown) at the bottom of the cap.

As to Figures 5.5c and 5.5d, there is shown a spiral wound cell **20** comprising an outer jacket **21** and spiral wound membrane module **22** in the jacket with a perforated hollow tube **23** having perforations **30** and extending through the cell. An effluent inlet **24** and retentate outlet **25** are also provided. The cell has been shown partially cut away to illustrate the spiral wound nature of the membrane module.

The enlarged sectional view of the membrane module as shown in Figure 5.5d, shows the membrane **26** and spacer **27** wound about the perforated hollow tube **23** in the form of a spiral so that there are alternate layers of membrane **26** and spacers **27**. Support means **28** are also provided. The spacer acts to provide passageways for the retentate. During operation, the effluent stream, under pressure, is introduced into the cell **20** through inlet **24** and passes axially through the membrane **26** and support means **28**. The color bodies cannot pass through the membrane **26**, but pass through the spacer **27** and exit out the bottom of the cell through retentate outlet **25.**

The remainder of the effluent permeates the membrane **26** and travels through the support **28** spacer **27** into the central tube through perforations **30.** The permeate exits the cell through the bottom of tube **23.** The spiral wound membrane module **22** is formed by first forming a sandwich of two sheets of membrane **26** separated by a porous support or backing material **28.** The backing material supports the membrane under operating pressure and provides a path for leading off the permeate. One side of the sandwich is lapped over and sealed to the perforated tube **23** to form the spiral shown in Figure 5.5c. The mesh spacer **27** is introduced between the facing membrane surfaces to provide passageways for the retentate.

In the examples that follow the figures given for color bodies (color units), total solids, total dissolved solids, total suspended solids were determined according to the analytical procedures set forth in *Standard Methods: Water and Wastewater,* 12 edition (1965), APHA, pp 127, 244 and 247. With respect to the color bodies (measured as Pt-Co units) absorbance at 420 m correlated with absorbance of a standard solution is the measure used.

The total solids were determined by gravimetric analysis of the raw effluent sample with drying at 105°C; the total dissolved solids were measured similarly but on the filtered effluent; total suspended solids were gravimetrically analyzed based on weight gain of the filter cell used to filter a measured volume of the raw effluent; and ash was also gravimetrically analyzed with drying at 600°C.

Example 1: A series of first stage caustic extraction filtrates from a kraft pulping operation having a pH of about 11.7 were all adjusted to pH 6.8 with sulfuric acid and after coarse filtration (including passing through a polishing filter having a 5 micron pore size) were passed through an ultrafiltration cell having 32 sq ft of membrane area (HF-35 and HT-00, Eastman Kodak Co.). The filtrates were passed through the membranes at a circulation rate of 2 gal/min. The results obtained at various operation conditions are shown in Table 1. The material balance calculations and resultant composition analysis are set forth in Table 2.

TABLE 1

Run	Membrane type	Percent total solids[1]	Operating conditions	Volumetric reduction	Average flux (gfd.)	Percent color* removal
1	HF-35	0.46	92 p.s.i. and 25° C	108X	18	83.9
2	HF-35	0.46	125 p.s.i. and 25° C	116X	17.5	82.4
3	HT-00	0.79	185 p.s.i. and 39° C	22X	22	90.5

[1] In caustic extraction filtrate.
*Based on color in neutralized effluent. Percentage would be higher if based on original effluent color and pH (11.7).

TABLE 2

	Effluents	Retentate	Permeate	Retentate and permeate
Run 1:				
Total Material, lbs	1,297	13.7	1,285	1,298.7 lbs.
Total dissolved solids, p.p.m.	4,600	82,000	3,710	5.88 lbs.
Color, Pt/Co units	13,300	780,000	2,150	1.35×10^7 lbs.-units.
Run 2:				
Total material, lbs	1,597	12.1	1,585	1,597.1 lbs.
Total dissolved solids, p.p.m.	4,600	102,700	3,370	6.60 lbs.
Color, Pt/Co units	10,900	1,330,000	1,860	1.91×10^7 lbs.-units.
Ash, p.p.m.	3,030	21,600	....	
Run 3:				
Total material, lbs	1,125	56.5	1,068	1,124.5 lbs.
Total dissolved solids, p.p.m.	7,940	60,500	6,000	9.72 lbs.
Color, Pt/Co units	21,500	460,000	2,050	2.62×10^7 lbs.-units.

The total solids in each run of retentate is over 15% and because of the high organic content the solids are readily incinerated. It is preferred to have a retentate with a high organic content and a solids concentration above 20% to minimize the cost of thermal oxidation.

Example 2: The described process can be combined with reverse osmosis and a run was made with an effluent as used in Run 1 of Example 1 except that the operating conditions were 94 psi and 45°C, the volumetric reduction was 20X and the average flux 22.4 gfd. After such ultrafiltration the 20X concentrate (permeate) was further concentrated to 90X by reverse osmosis (RO) using a conventional RO cell with 32 square feet of membrane area (high salt rejection, 97% NaCl at 600 psi). This cell had both high salt and very high color retention. From 20X to 90X, the buildup in salt concentration was not such as to create an inoperatively high osmotic pressure and permitted the reverse osmosis to procede easily and effectively. The material balance calculations resulting from both treatments are set forth in Table 3.

TABLE 3

Run 4	Effluents	Retentate	Permeate	Retentate and permeate
Total material, lbs	1,373	15.4	1,357	1,372.4 lbs.
Total dissolved solids, p.p.m.	3,920	86,400	3,120	5.56 lbs.
Color, Pt/Co units	10,600	76,500	1,820	1.43×10^7 lbs.-units.

A significant advantage of the process is that it permits a once-through continuous removal of the color bodies and materials having a high BOD. This eliminates the holding tanks and extensive equipment required in batch and recycle procedures.

Electrolysis

M. Ichiki and M. Ishii; U.S. Patent 3,793,174; February 19, 1974; assigned to Mitsui Mining & Smelting Co., Ltd., Japan have found that the ligninsulfonate contained in wastewater can be separated easily and efficiently by subjecting the wastewater to electrolysis using a ferrous anode plate.

The matter of primary importance according to the process is that the electrodes, especially the anode, should consist of iron. The electrolysis can perform purification of the wastewater to some extent, even when an anode plate made of aluminum is employed. Through experiments, however, it has been found that employment of a ferrous anode plate results in an exceedingly superior efficiency of purification in comparison with an anode plate made of other materials. The following example illustrates the process.

Example: Each of aqueous solutions containing 500 ppm, 1,000 ppm, 2,000 ppm, and 5,000 ppm of sodium ligninsulfonate respectively, was taken individually into a 5 liter cell provided with a ferrous anode plate. Then, each of aqueous solutions was subjected to electrolysis with an electric current of 5A and cell voltage of 4.0 v, while air was blown into the cell by an air pump, so as to separate insoluble substances from the mother liquid by floating them.

During the electrolysis, an examination was carried out at regular intervals of time for observing the removal of sodium ligninsulfonate from the mother liquid by checking the COD of the liquid by sampling the limpid liquid staying in the lower portion of the cell. The results are shown in the table below.

Waste water before the start of electrolysis		COD of the limpid liquid staying in the lower portion of the cell after the start of electrolysis (ppm) after the start of electrolysis in minutes								
Concentration of sodium ligninsulfonate (ppm)	COD (ppm)	10	20	40	60	90	120	240	300	360
500	398	200	113	63	42	27	16	6	8	6
1000	840	800	760	170	85	60	38	19	17	25
2000	1524	—	—	468	312	264	64	53	39	39
5000	3520	3520	3440	2880	960	824	792	190	103	92

The amount of removed sodium ligninsulfonate differed in accordance with the concentration thereof within the wastewater. When the wastewater contained less than 1,000 ppm of sodium ligninsulfonate, the COD value was decreased to $\frac{1}{10}$ or so by 1 hr electrolysis. In addition, the foregoing was also observed, when the aqueous solution contained calcium salt or potassium salt of ligninsulfonic acid.

Recovery of Elemental Sulfur

P. Urban; U.S. Patent 3,728,439; April 17, 1973; assigned to Universal Oil Products Co. describes a multistep process for the selective treatment of an aqueous stream containing a water-soluble sulfite compound for the purpose of converting the sulfite compound to elemental sulfur in a highly selective manner. Another purpose for the treatment procedure is to lower the total sulfur content (i.e., total amount of sulfur contained in the aqueous stream in any form, calculated on an elemental basis) of the aqueous input stream to the point where it can be safely discharged into a suitable sewer without causing a major pollution problem, or alternatively, reused in the process or industrial activity which originally produced it.

The method involves a four step process for treating an aqueous stream containing a water-soluble sulfite compound in which the sulfite compound is first selectively reduced to the corresponding thiosulfate compound, a portion of this thiosulfate compound is then selectively reduced to the corresponding sulfide compound and the resulting sulfide compound is then used to selectively reduce the remaining portion of the thiosulfate compound to form the corresponding polysulfide compound. Elemental sulfur is then easily recovered from the polysulfide compound via a conventional decomposition technique. Referring to Figure 5.6, an input water stream containing ammonium sulfide and bisulfide in an amount of about 10 weight percent thereof is continuously introduced into the system via line **1**

FIGURE 5.6: TREATMENT OF SULFITE WASTE TO PRODUCE ELEMENTAL SULFUR

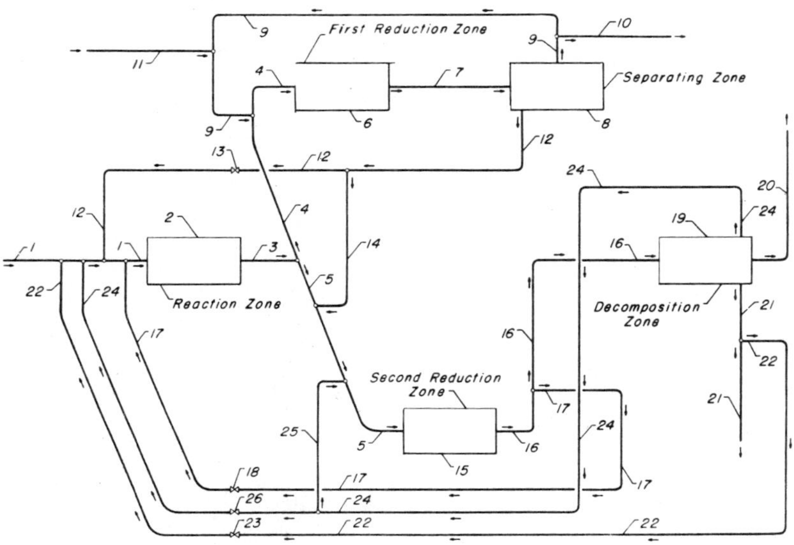

Source: P. Urban; U.S. Patent 3,728,439; April 17, 1973

and passed to reaction zone **2**. Also introduced into this zone is a reducing agent selected from the group consisting of ammonium hydrosulfide, H_2S, elemental sulfur, ammonium polysulfide and mixtures of these reducing agents. In the case where the reducing agent is elemental sulfur it is a portion of the sulfur product of subsequently described zone **19** which is charged to zone **2** via lines **21, 22** and **1** when valve **23** is in the open position and valves **26, 18,** and **13** are in the closed position.

Another source for the reducing agent for zone **2** is a portion of the ammonium polysulfide produced in subsequently described zone **15**. When polysulfide is the reducing agent, it is charged to zone **2** via lines **16, 17** and **1** when valve **18** is in the open position and valves **13, 26** and **23** are in the closed position. Yet another source for the necessary reducing agent for zone **2** is a portion of the sulfide-containing effluent stream produced in zone **8.** It is charged to zone **2** via lines **12** and **1** when valve **13** is in the open position and valves **18, 26** and **23** are in the closed position.

The final internal source for the reducing agent utilized in zone **2** is a portion of the H_2S-containing overhead stream from zone **19**. It is passed to zone **2** via line **14** when valve **26** is in the open position and valves **13, 18** and **23** are in the closed position. The preferred source for the necessary reducing agent is a portion of the H_2S-containing overhead stream produced in decomposition zone **19**. Of course during start up of the process, an exterior source for sulfide can be provided or it can be generated in zone **6** from the input water stream.

Regardless of the source of the reducing agent, it will be charged to zone **2** in an amount sufficient to react with all the sulfite compound present in the input water stream to produce thiosulfate. For example, in the preferred case where the reducing agent is H_2S, it is typically charged to zone **2** in an amount above the stoichiometric amount of 0.5 mol of sulfide per mol of sulfite, with good results typically obtained at a mol ratio of 1:1. Ordinarily during start up of the system an external source for the necessary reducing agent will be necessary; however, once the plant is started up and lined up, all of the necessary reducing agent will be obtained from internal sources. Zone **2** is a conventional thermal reaction zone designed in the preferred case to effect intimate contact between the input water stream and the H_2S-containing gas stream and in the other cases between the input water stream and the particular aqueous stream containing the reducing agent.

In the case shown in Figure 5.6, the zone is shown operating in a concurrent flow mode. In the preferred case, the temperature within zone **2** is maintained by conventional means at about 65°C and the pressure is about 50 psig. The flow rate of the reactants is adjusted to result in a residence time of about 3 min. These conditions with the preferred H_2S-reducing agent result in the conversion of 95% of the ammonium sulfite and bisulfite contained in the input water stream to ammonium thiosulfate, with less than 2% converted to undesired ammonium sulfate.

An effluent stream containing ammonium thiosulfate and minor amounts of unreacted ammonium sulfite, bisulfite and of unreacted ammonium sulfide or hydrosulfide is withdrawn from zone **2** via line **3** and split into two portions at the junction of line **3** with lines **4** and **5**. The first portion comprises about two-thirds of the effluent stream and is charged via line **4** to the first reduction zone, zone **6**. The second portion is charged to zone **15** via line **5**. Also introduced into zone **6** is a reducing gas stream which essentially comprises carbon monoxide. It enters zone **6** via lines **11, 9** and **4**.

After the process is started up, the reducing gas stream entering zone **6** is composed of two portions, one portion being makeup reducing gas which enters the system via line **11**, and the other portion being recycled unreacted reducing gas which is recovered in zone **8** and recycled via line **9**. Zone **6** contains a fixed bed of a catalyst comprising 10 to 12 mesh particles of activated carbon having a cobalt sulfide component combined in an amount sufficient to result in the catalyst containing 2 weight percent of cobalt. The amount of carbon monoxide introduced into zone **6** is sufficient to provide a mol ratio of carbon monoxide to ammonium thiosulfate charged thereto of about 5.5:1. By conventional means zone **6** is operated at a temperature of 200°C, a pressure of 500 psig, and a liquid hourly space velocity of 1 hr^{-1}.

An effluent stream is then withdrawn from zone **6** via line **7** and passed to separating zone **8** where a gas phase is allowed to separate from a liquid phase. Zone **8** is operated at substantially the same temperature and pressure as zone **6**. The gaseous phase is withdrawn via line **9**. It contains unreacted carbon monoxide, carbon dioxide and minor amounts of H_2O and H_2S. A portion of this gas stream is vented from the system via line **10** in order to prevent the build-up of CO_2 in the flow circuit between zones **6** and **8**, and the remaining portion is recycled to zone **6** via lines **9** and **4**. The liquid phase formed in zone **8** contains ammonium hydrosulfide, ammonium carbonate, unreacted ammonium thiosulfate, ammonium hydroxide and minor amounts of ammonium polysulfide and undesired ammonium sulfate.

An analysis of this liquid phase shows that 99% of the thiosulfate charged to zone **6** is converted therein at a selectivity for sulfide of about 95% and with less than 2% being converted to ammonium sulfate. It is withdrawn from zone **8** via line **12**. In all cases, a substantial portion of this liquid phase is passed via lines **12, 14** and **5** to a second reduction zone **15**. In accordance with one mode of operation of zone **2**, a portion of this liquid effluent stream from zone **8** is passed via line **12** through valve **13** and line **1** into zone **2** where the sulfide component constitutes the reducing agent for use in zone **2**.

Turning now to zone **15**, it is a conventional thermal reactor designed to effect intimate contact between the two liquid streams. It is preferred to operate zone **15** with a substantial excess of sulfide reactant and ordinarily this additional sulfide reactant is obtained from decomposition zone **19** via lines **24, 25** and **5**. In effect the additional sulfide reactant circulates around a flow circuit between zone **15** and **19** which is formed by lines **16, 24, 25** and **5**. In the particular case illustrated in Figure 5.6, sufficient H_2S is obtained from this second source to insure that the mol ratio of sulfide reactant to thiosulfate reactant within zone **15** is 7.5:1. Ordinarily during start up of the process, the necessary excess sulfide reactant can be manufactured in zone **6** until sufficient sulfide accumulates in this sulfide circuit, thereafter a portion of thiosulfate reactant can be diverted to zone **15** via lines **3** and **5** and the normal operation of zone **15** commenced. Zone **15** is maintained by conventional means at a temperature of about 200°C and a pressure of about 650 psig. The flow rates of the input streams are, moreover, adjusted so that the residence time of the reactants charged thereto is approximately 6 minutes.

An effluent stream is then withdrawn from zone **15** via line **16** and is found to contain ammonium polysulfide, unreacted ammonium thiosulfate, unreacted ammonium sulfide, ammonium carbonate, ammonium bicarbonate, ammonium hydroxide, and ammonium sulfate. An analysis of this stream indicates that 99% of the thiosulfate charged to zone **15** is converted to ammonium polysulfide. Likewise, it indicates that less than 0.5% of the total sulfur-containing reactants charged to zone **15** are converted to the undesired ammonium sulfate.

At the junction of line **16** with line **17,** the effluent stream for zone **15** can be divided into two portions depending upon whether or not valve **18** is in the open position. When valve **18** is open, a portion of this polysulfide-containing stream is passed to zone **2** via lines **17** and **1** in order to furnish at least a portion of the necessary reducing agent therefor. Ordinarily, in this last mode of operation, the portion passed to zone **2** is approximately 10 to 25% of the effluent stream from zone **15.** In the case where valve **18** is in the closed position the entire amount of the effluent stream from zone **15** is charged via line **16** to decomposition zone **19.**

In zone **19** the ammonium polysulfide-containing water stream is decomposed to yield H_2S, elemental sulfur and a treated water stream containing ammonium carbonate, ammonium bicarbonate, ammonium hydroxide and minor amounts of undesired ammonium sulfate. Preferably, zone **19** is a distillation zone which is maintained at conditions including a temperature in the range of 100° to about 200°C and a pressure of about 0.1 to 10 atmospheres effective to form an overhead gas stream containing H_2S, H_2O, and minor amounts of NH_3 which is withdrawn therefrom via lines **24,** and an aqueous side stream, which is the product treated water stream and is withdrawn from the system via line **20.**

In order to prevent carry-over of NH_3, it is a preferred practice to inject a portion of this side stream into the top of the distillation column in order to maintain it at low temperatures and to scrub NH_3 out of the overhead stream. In most cases, it is advantageous to accelerate the polysulfide reaction in the bottom of zone **19** by conventional means such as a steam coil or conventional reboiler.

The overhead stream from zone **19** is withdrawn therefrom via line **24,** and at least a major portion of it is recycled to zone **15** via lines **24, 25** and **5.** As previously explained, in one mode of operation of zone **2**, a portion of this overhead stream from zone **19** is passed via lines **24** and **1** back to zone **2** when valve **26** is in the open position. It is to be noted that this overhead stream can be, if desired, cooled by conventional means to form an aqueous solution containing ammonium hydrosulfide and ammonium hydroxide which then can be charged to zone **15** and **2** in the manner previously indicated.

When the temperature utilized in the bottom of zone **19** is less than the melting point of sulfur, the sulfur product of the polysulfide decomposition reaction will be present in the form of a slurry of solid particles in an aqueous stream withdrawn from the bottom of zone **19.**

This sulfur-containing bottom stream can then be subjected to any of the techniques taught in the art for recovering a solid from a liquid such as filtration, settling, centrifuging and the like in order to recover the elemental sulfur. In the case shown in Figure 5.6, decomposition zone **19** is operated at a bottoms temperature greater than the melting point of sulfur so that liquid sulfur forms in the bottom of the zone. In particular, excellent results are obtained when zone **19** is operated at a bottoms temperature of 140°C, a pressure of 38 psig and with 20 to 30% of the liquid feed being taken overhead. In this case, the liquid sulfur formed therein will collect at the bottom of zone **19** and be withdrawn via line **21**. In one mode of operation of zone **2**, a portion of this sulfur product from zone **19** will be passed to zone **2** via lines **21, 22** and **2**, when valve **23** is in the open position. The treated water stream withdrawn from zone **19** via line **20** will contain only a minor amount of sulfide; for example, at the particular conditions specified above for zone **19**, this side stream will contain only about 800 ppm of sulfide. An overall weight balance on the total system indicated that greater than 90% of the ammonium sulfite and bisulfite charged to this system via line **1** is converted to elemental sulfur, with substantially less than 5% converted to undesired ammonium sulfate.

FLOCCULATION

Oxyaminated Polyacrylamides

E. Reinwald and J. Galinke; U.S. Patent 3,707,465; December 26, 1972; assigned to Henkel & Cie GmbH, Germany describe a process of clarification of aqueous suspensions consisting of contacting an aqueous suspension with an oxyaminated polyacrylamide in order to form a floc and separating clarified water from the floc. The oxyaminated polyacrylamides can be used as flocculating agents for the aqueous and preferably acidic suspensions which are obtained in the paper industry. Polyacrylamides are utilized as a starting material for the preparation of the oxyaminated polyacrylamides. Preferably a polyacrylamide polymerized in a redox system is used. Suitable polyacrylamides have, for example, average molecular weights of between 2 and 10 million.

The oxyamination of polyacrylamide may be carried out by a process similar to that for oxyalkylation. It has been proven advantageous to add to the mixture of polyacrylamide and oxyamination agent slight amounts of an alkaline catalyst such as an alkaline oxyalkylation catalyst. It has been found sufficient if catalytic amounts of about 0.1 to 2%, based on the polyacrylamide of sodium hydroxide or potassium hydroxide are added.

Optionally it may be appropriate to add to this reaction, proceeding in an aqueous phase, also smaller amounts of water-miscible solvents as dissolving intermediaries for the oxyamination agent. Here, for instance, dioxane, methanol, ethanol, acetone and the like are of interest. The oxyamination is suitably carried out at slightly raised temperatures, between about 40° to 100°C and requires about 15 min to 4 hr. The following examples illustrate the process.

The starting materials were prepared as follows. 30 g of acrylamide were dissolved in 170 milliliters of deionized water and freed of oxygen by agitation for 20 min in a strong stream of nitrogen. Then 3 mg of sodium sulfite and 5 mg of potassium peroxydisulfate, dissolved in a little water, were added and after adjusting the pH of the reaction solution to 6.5 with sodium hydroxide, it was heated to 35°C. After 4 hr of polymerization at 35°C, the polymer solution, which has a monomer content of 3 to 5%, could be reacted directly with the aminoalkylene oxides. The specific viscosity of 0.1% of the isolated material dissolved in a 1 N sodium nitrate solution was 2.3.

200 g of a 15% polyacrylamide solution were reacted with 0.6 g of 1-diethylamino-2,3-epoxypropane and 2.5 ml of a 2 N sodium hydroxide solution. The reaction solution was stirred for 1 hr at 60°C. Subsequently, the reaction product was precipitated by the addition of ethanol, suction filtered and washed with ethanol. The specific viscosity of the isolated material, dissolved in a 1 N sodium nitrate solution, was 2.7.

In the same way the reaction with 1.2 g of 1-diethylamino-2,3-epoxypropane and with 0.65 g of 1-dibutylamino-2,3-epoxypropane was carried out. In a further charge the same amount of the polyacrylamide was reacted with 0.7 g of trimethylammonium-2,3-epoxypropane chloride. These oxyaminated polyacrylamides had specific viscosities of 2.1, 2.3 and 2.4, respectively.

Example 1: Flocculation of a Pulp Suspension – The effectiveness of the oxyaminated polyacrylamides was tested on wood pulp suspensions. For the determination of the water removal behavior from wood pulp suspensions, the following process was used. Dry fibrous material was distributed homogeneously in water and flowed with a constant flow rate on a filter cloth of known dimensions and porosity.

The speed with which the water penetrates through the filter cloth is a measure for the water removal behavior. The filter cloth is appropriately placed in a cylindrical vessel which in its lower part has in different heights, two discharge openings. In a slow rate of water removal (rate of filtration), the large part of the liquid leaves through the lower opening. By the addition of flocculating agents the rate of water removal is increased and thus the part of the liquid which flows through the upper opening.

100 g of ground mechanical wood pulp were suspended in 5 l of tap water and pulverized in an impact apparatus. 1 l of the suspension thus obtained was diluted with 9 l of water and adjusted to a pH of 6 or 4.3 with aluminum sulfate. 1 l of the thus obtained pulp suspension was flocculated with the addition of 4 ml of a 0.1% solution of the flocculation agent in water in the Schopper-Riegler apparatus. The following Schopper-Riegler values (reduction of degree of fineness, leaflet V/7/61 of the Verein der Zellstoff und Papier-Chemiker und Ingenieure) were measured.

Mol Percent	Polyacrylamide Oxyaminated with	Schopper-Riegler Values pH 6	Schopper-Riegler Values pH 4.3
1	1-Diethylamino-2,3-epoxypropane	–40	–36
2	1-Diethylamino-2,3-epoxypropane	–39	–34
1	1-Dibutylamino-2,3-epoxypropane	–37	–29
1	1-Trimethylammonium-2,3-epoxy-propane	–39	–30
	Nonoxyaminated polyacrylamide (comparison test)	–20	0

Example 2: Flocculation of a Bentonite Suspension – 1.7 g of bentonite and 3.7 g of sodium chloride were added to 550 ml of distilled water of 50°C. To this suspension each time 0.001 g of flocculating agent was added. The suspension was then poured into a pipe of 4 cm diameter and 35 cm height which was kept constantly at 50°C by a heating jacket. The sedimentation speed in cm/sec is given in the following as a measure for the flocculation effect.

Mol Percent	Polyacrylamide Oxyaminated with	Cm/Sec
1	1-Diethylamino-2,3-epoxypropane	1.4
2	1-Diethylamino-2,3-epoxypropane	1.3
1	Trimethylammonium-2,3-epoxy-propane chloride	1.3
	Nonoxyaminated polyacrylamide (comparison test)	0.7

Example 3: Flocculation of a Pulp Suspension – One liter samples of the pulp suspension described in Example 1 were adjusted to a pH of 6 or 4.3 by means of ferric sulfate. Using polyacrylamides oxyaminated with diethylamino-2,3-epoxypropane, dibutylamino-2,3-epoxypropane and trimethylammonium-2,3-epoxypropane chloride as flocculation agents the same Schopper-Riegler values were obtained as in Example 1.

Oxyalkylated Polyacrylamides

E. Reinwald and J. Galinke; U.S. Patent 3,707,466; December 26, 1972; assigned to Hendel & Cie GmbH, Germany describe a process of clarification of aqueous suspensions consisting of contacting an aqueous suspension with an oxyalkylated polyacrylamide in order to form a floc and separating clarified water from the floc.

In the clarification of aqueous suspensions, it has been known to improve the speed of sedimentation or filtration of aqueous suspensions by the addition of flocculation or filtration aids which are based on natural products, such as starch. In addition, synthetic high polymers such as polyacrylamides have also been used as effective flocculating or filtration aids, particularly for the clarification of aqueous suspensions as they are obtained in the paper industry. The advantage of polyacrylamide lies in a higher effectiveness and particularly in the uniform quality. A certain disadvantage of polyacrylamides is, however, the extensive dependence upon the pH of the suspension in order to effect good flocculation.

It has been found that the pH dependence can be widely reduced, if oxyalkylated polyacrylamides are used as flocculating agent for aqueous suspensions. With a particular advantage the oxyalkylated polyacrylamides can be used as flocculating agents for the aqueous and preferably acidic suspensions which are obtained in the paper industry. The amounts of addition of the flocculating agents may vary widely upon the suspension to be flocculated and are generally between 0.02 and 2 g, based on 100 g of the solids. The flocculating agent may be added as solids or in an aqueous solution. It has further been found that such oxyalkylated polyacrylamides are particularly effective if they contain oxyalkyl groups with 3 or 4 carbon atoms.

As starting material for the preparation of the oxyalkylated polyacrylamides, for instance, an acrylamide polymerized with a redox system in a known manner is used. Suitable polyacrylamides have, for instance, average molecular weights of between about 2 to 10 million. The oxyalkylation of polyacrylamide is also effected in a known manner with alkylene oxides. Preparation methods for oxyalkylated polyacrylamides are, for instance, described in the German Patent 925,130.

The effectiveness of the oxyalkylated polyacrylamides was tested on wood pulp suspensions. For the determination of the water removal behavior from wood pulp suspensions the following process was used. Dry fibrous material was distributed homogeneously in water and flowed with a constant flow rate on a filter cloth of known dimensions and porosity. The speed with which the water penetrates through the filter cloth is a measure for the water removal behavior.

The filter cloth is appropriately placed in a cylindrical vessel which in its lower part has, in different heights, two discharge openings. In a slow rate of water removal (rate of filtration) the large part of the liquid leaves through the lower openings. By the addition of flocculating agents the rate of water removal is increased and thus the part of the liquid which flows through the upper opening. The following examples illustrate the process.

In the examples the preparation of the starting material is as follows. 30 g of acrylamide were dissolved in 170 ml of deionized water and freed of oxygen by agitation for 20 min in a strong stream of nitrogen. Then 3 mg of sodium sulfite and 5 mg of potassium peroxydisulfate, dissolved in water, were added, and after adjusting the pH of the reaction solution to 6.5 with sodium hydroxide, it was heated to 35°C. After 4 hr of polymerization at 35°C, the polymer solution which has a monomer content of 3 to 5%, could be reacted directly with the alkylene oxides. The specific viscosity of 0.1% of the isolated material dissolved in a 1 N sodium nitrate solution, was 2.3.

200 g of a 15% polyacrylamide solution were charged together with 3 g of sodium hydroxide in an autoclave. After adding 15 g of propylene oxide, a pressure of 3 atmospheres was set with the aid of nitrogen. The autoclave was heated to 80°C and after 30 min the reaction was stopped. The contents of the autoclave were stirred into excess ethanol. The addition product precipitated. It had a specific viscosity of 2.0.

As described above, 30 g of polyacrylamide in a 15% aqueous solution was reacted with 15 g of butylene oxide with addition of 2 g of sodium hydroxide. Also in this case the reaction was carried out during about 30 min at 80°C. The isolation of the addition product was effected as described above. The adduct had a specific viscosity of 1.3.

Example 1: Flocculation of a Pulp Suspension – 100 g of mechanical wood pulp were suspended in 5 l of tap water and pulverized in an impact apparatus. One liter of the suspension thus obtained was diluted with 9 l of water and adjusted to a pH of 6 or 4.3 with aluminum sulfate. One liter of the thus obtained pulp suspension was flocculated with the addition of 4 ml of a 0.1% solution of the flocculation agents described below in water in the Schopper-Riegler apparatus. The following Schopper-Riegler values (leaflet V/7/61 of the Verein der Zellstoff-und Papier-Chemiker und Ingenieure) were measured.

Flocculation Agent	Schopper-Riegler Values pH 6	pH 4.3
Polyacrylamide	–20	0
Propoxylated polyacrylamide	–38	–32
Butoxylated polyacrylamide	–38	–26

Example 2: Flocculation of a Pulp Suspension – One liter samples of the pulp suspension described in Example 1 were adjusted to a pH of 6 or 4.3 by means of ferric sulfate. Using propoxylated polyacrylamide or butoxylated polyacrylamide as flocculation agents the same Schopper-Riegler values were obtained as in Example 1.

Example 3: Flocculation of a Pulp Suspension – One liter samples of the pulp suspension of Example 1 were adjusted to a pH of 6 or 4.3 with hydrochloric acid. Using polyacrylamide or propoxylated polyacrylamide as flocculation agents the following Schopper-Riegler values were measured.

Flocculation Agent	Schopper-Riegler Values pH 6	pH 4.3
Polyacrylamide	0	0
Propoxylated polyacrylamide	–3	–2

Dispersed Ammonia Gaseous Phase

According to a process described by *J.W. Gooch; U.S. Patent 3,733,269; May 15, 1973* suspended solids and inorganic and organic waste materials are agglomerated, coagulated and extracted from waste effluent streams of packing and rendering plants, paper and pulp mills, food processing plants, wood preserving plants, rice processing plants and other similar mills or industrial facilities. Typical waste materials include those which are generally organic in nature such as wood and wood fibers, animal and vegetable tissues including oils, fibers, blood, saccharides, urea, carbohydrates, fats, sugars, fatty acids, cellulose, dyes, and other similar material.

Also typically found within a waste effluent stream are human and animal excrement, hard and soft garbage, animal and vegetable organisms, and other biological and mineral products. Those portions of the waste material which exceed the density of the liquid component of the waste stream by a large degree will quickly settle out and may be removed by conventional mechanical separation techniques. In addition, those portions of the waste material which are lighter to a large degree than the liquid component of the waste stream will float to the surface of the stream and may also be easily removed by mechanical separation techniques.

After removal of that portion of the waste material which may be easily removed by mechanical techniques, there remains a large amount of finely divided dispersed, suspended or emulsified waste material of a density closely approximating the density of the liquid component of the stream, usually a water solution. The finely divided material may be economically and efficiently removed from the liquid component by this process. In one example of the process, a reagent solution is prepared in a metastable form and introduced into the waste stream in such a manner that the reagent is released from the solution and dispersed within the waste stream in the form of a very finely divided and microscopic gaseous phase. The reagent, in gaseous phase, is selectively absorbed on the surfaces of the suspended, dispersed or emulsified waste material to selectively react with the surfaces of the waste material and produce surface characteristics which induce agglomeration and coagulation of the waste material. Upon agglomerating and coagulating, the apparent density of the material is increased by the expulsion of liquid from within the waste material and by the reduction in apparent surface area for the waste material. The agglomerated and coagulated particles of waste material then settle out of the waste material stream and agglomerate and coagulate into large easily removed portions of waste material.

Upon the removal of these portions, by decanting, filtration, centrifuging or the like, a reasonably clear liquid remains which may be returned to the mill, plant or industrial facility for further use or the liquid may be flowed to and deposited in a river, lake or other by-water without any harmful or objectionable effect thereon provided the original mineral content of the liquid component is acceptable. The waste material may be accumulated, dried and optionally sterilized for further use as fertilizer, animal feed, or energy source for burning in the generation of power. In addition, it should be noted that the extracted waste material is capable of and useful for subsequent and further use as a surface active cationic flocculating agent in the treatment of waste material streams.

Suitable reagents include those compounds or compositions which may be spontaneously released from a liquid media as a gas to become attracted to a particulate surface in the liquid media upon which an activating reaction takes place. Preferred reagents include ammonia, hydrogen sulfide, halogen compounds, carbon dioxide, sulfur dioxide and ozone. A particularly preferred group of reagents include ammonia, chlorine, bromine and fluorine with especially superior results being achieved through the use of gaseous ammonia. In one example, a supply of saturated ammonia solution, for example, saturated ammonia water, is prepared with the solution heated sufficiently to cause the chlorine to become and remain metastable.

Temperatures from in excess of 80° to about 700°C have been found to produce suitable results and are recommended for the operation of the process. Superior results have been achieved through the use of a temperature of about 85° to about 250°C. In addition, the supply of metastable saturated ammonia solution should be maintained in a pressure-tight system with an increase in pressure maintained in the system over that pressure which is required to suppress premature release of the ammonia gas. Suitable pressures have ranged up to 6 atmospheres with superior results being achieved at a pressure of 2 atmospheres.

The metastable saturated ammonia solution is introduced into a waste material stream which is at a lower temperature and pressure than the ammonia solution. The shock created by this introduction, together with the differential between the pressures of the waste stream and ammonia solution and the reduction in partial pressure of ammonia, causes the ammonia gas to be released and dispersed throughout the waste stream in a gaseous phase of microscopic bubbles, i.e., microbubbles.

The microbubbles generally are within the range of 3.72 A to 0.1 mm in diameter and because of their small size quickly migrate and disperse throughout the waste stream. These microbubbles have a low tendency to coalesce or dissolve in the waste stream and therefore maintain substantially the same size as they travel through the liquid. Further, flow around each bubble is approximately laminar, lessening localized disturbances and mixing.

It is preferable that a short contact time be provided during which the microbubbles may be thoroughly dispersed throughout the waste stream. As each microbubble encounters particulate solid waste material, the bubble of ammonia gas is selectively adsorbed upon only the surface of the particulate material to selectively react with the material only at that point on the surface. The reaction produces an ammonium salt having a dipole charge at the point of adsorption. The dipole charge is sufficient to create a charged site on the surface thus causing that area of the surface to become a cationic flocculating agent. Thus, surface active cationic flocculating agents are produced in situ from waste material by this process. The following examples illustrate the process.

Example 1: A viscous sludge of cresol, phenolic acids, wood fiber and other contaminants of wood origin is taken from the bottom of a settling basin of a wood preserving plant and diluted with water to one part sludge per four parts water. The pH of this solution is measured at 4.0. A metastable ammonia water solution is prepared and heated to a temperature of about 90°C in an airtight solution preparation tank by mixing ammonia gas with distilled water. The ammonia solution is then gradually pumped under 2 atmospheres of pressure into the bottom of a dispersion tank, containing approximately 500 ml of the waste solution at 20°C, until 0.15 g of ammonia has been added.

Microbubbles of ammonia gas form within the dispersion tank and the waste material begins to agglomerate and coagulate as a precipitate soon thereafter. The waste solution is clear and is removed by decanting. The supernatant liquid is found to have a pH of 6.8. The chemical oxygen demand is determined for the diluted waste solution and the supernatant liquid for comparison purposes. It is found that the diluted waste solution has a COD of 170,000 mg/l and the treated supernatant liquid has a COD of 600 mg/l.

Example 2: A quantity of paper mill waste is mixed with water to form a suspension and the larger particles separated therefrom, leaving a solution containing smaller particles. A 231 ml sample of the solution is found to have 0.27 g of solids therein. A chlorine water solution is prepared at 200°C and the sample of waste solution is treated therewith as in Example 1. A brown precipitate quickly forms at the bottom of the tank. The water remaining is found to be free of solids.

Example 3: A solution of rice steep waste from a rice processing plant is pumped at 21.1°C into the bottom of a dispersion tank at a rate of 500 cc/min. A metastable ammonia water solution is prepared in a preparation tank at a temperature of 99.8°C and a pressure of 2 atmospheres. The metastable ammonia water solution is then pumped into the dispersion tank with the rice steep waste solution at a rate sufficient to add 0.12 g of ammonia per minute.

The flow of both reagent and rice waste continued and gave consistently coagulative results with sedimentation occurring in 5 min. 95% of total solids are removed by gravitational sedimentation in 15 min, total settleable solids in 45 min. The supernatant was then centrifuged to reduce the amount of fine suspended particles. The centrifuged supernatant is then flowed through an activated carbon bed to remove all color from the liquid. The following results are obtained upon testing.

Analysis of Rice Waste

	Sample		
	No. 1	No. 2	No. 3
pH	4.8	6.9	6.6
Chemical oxygen demand, mg/l	2,360	450	
Biochemical oxygen demand, mg/l	–	–	30

In the above table, Sample No. 1, rice waste was untreated; Sample No. 2, rice waste was treated, resulting in gravitational sedimentation; and in Sample No. 3, rice waste was treated and centrifuged.

Example 4: A 100 ml sample of dye waste solution including azo dyes, phenolphthalein, and methyl orange dye is found to have 0.0124 g of solids therein. The sample is treated with a chlorine water solution as in Example 2. A white precipitate forms at the bottom of the dispersion tank. All visible solids are precipitated from the water.

DECOLORIZING

Rare Earth Salts

M. Ghassemi, H.L. Recht and E.V. Kleber; U.S. Patent 3,736,255; May 29, 1973; assigned to North American Rockwell Corporation describe a method for chemically removing coloring matter from water by treating the coloring matter containing water with a trivalent rare earth salt, generally as a mixture of such salts, to form an insoluble rare earth precipitate and then separating the precipitate from the water. The rare earth precipitant salt, such as lanthanum chloride or lanthanum sulfate, may be regenerated by first heating the rare earth precipitate to form rare earth oxides and to drive off the organic components by converting them to carbon dioxide and water. The rare earth oxide may be used directly in acidic waters or converted to a soluble rare earth salt by using acid such as dilute hydrochloric or sulfuric acid.

Lime has been found to aid trivalent rare earth ions in the removal of coloring matter from wastewater. The lime results in improved settling characteristics of the flocs which form during coloring matter precipitation and results in improved compaction and dewaterability of the rare earth precipitates. The lime can be added to the colored wastewaters either before, after or at the same time that the rare earth precipitant is added. The quantity of lime added to the colored waters is usually considerably less than is employed in the massive lime treatment process (U.S. Patent 3,120,464).

In the case of acidic waters, lime is particularly useful in raising the pH of the acid wastes which generally makes the trivalent rare earth precipitant more effective. In addition to lime or other calcium(II) coagulants, conventional iron(III), aluminum(III) and polyelectrolyte coagulants may be used in conjunction with rare earth precipitants to effect color removal and improve the characteristics of the precipitates. The following examples illustrate the process.

Example 1: A 2 gal sample of alkaline bleach waste from a kraft pulping mill was diluted 6 to 1 (6 parts diluent to 1 part alkaline bleach waste) with distilled water. 500 ml of the diluted waste were placed in a 1 l beaker which was then placed in a stirring apparatus. While stirring at 90 rpm, 10 ml of 0.05 M lanthanum nitrate [$La(NO_3)_3$] solution were added to the beaker. An automatic titrator was used to maintain the pH at 6 by the addition of dilute sodium hydroxide or hydrochloric acid.

After 2 min of stirring at 90 rpm, the stirring rate was slowed to 20 rpm and the stirring was continued at this rate for 10 minutes. After the stirring was completed, 20 min of quiescent settling was allowed after which a portion of the supernatant was removed for turbidity analysis. The turbidity analysis was conducted on a standard turbidiometer calibrated to read in Jackson Turbidity Units (JTU) [see *Standard Methods for the Examination of Water and Wastewater,* 12 edition, American Public Health Association, New York (1965), p 312].

The supernatant was filtered through Whatman No. 42 filter paper and the color of the filtrate was determined at a wavelength of 420 mμ using a Lumetron Colorimeter (Model 402E, Photovolt Corp.). The residual turbidity of the supernatant following 20 min of quiescent settling was 1.6 JTU. The optical density of the supernatant after filtration was 0.24 which represented a 92% reduction in color based on the optical density of the original diluted sample of alkaline bleach waste.

Example 2: Diluted alkaline bleach waste from the 2 gal sample in Example 1 was treated in accordance with the procedure of Example 1 except that the automatic titrator was used to maintain the pH at 4. The residual turbidity of the supernatant following 20 min of quiescent settling was 4.2 JTU. The optical density of the supernatant after filtration was 0.46 which represented an 84% reduction in color based on the optical density of the original diluted sample of alkaline bleach waste.

Example 3: Eight milliliters of a 0.1 M lanthanum nitrate [$La(NO_3)_3$] solution were added to a 250 ml sample of alkaline bleach waste from a kraft pulping mill which had a color at pH 6.2 of 7,600 platinum-cobalt (Pt-Co) units. The sample was placed in a stirring apparatus and stirred rapidly at 90 rpm for 2 min. The stirring rate was then reduced to 20 rpm and the stirring was continued at this rate for an additional 10 min.

After the stirring was completed, the pH of the sample, which was 6.3, was measured and the sample was allowed to settle for 20 min. Following the 20 min period of quiescent settling, the supernatant was filtered through Whatman No. 40 filter paper and the color of the filtrate was determined at a wavelength of 420 mμ using a Lumetron Colorimeter (Model 402E) which had been calibrated against platinum-cobalt color standards. The color of the supernatant at a pH of 6.7 after filtration was 410 platinum-cobalt units which represented a 94% reduction in color.

Lime and Carbon Dioxide Treatment

According to a process described by *H.C. Croom; U.S. Patent 3,736,254; May 29, 1973; assigned to Owens-Illinois, Inc.* waste effluent liquid from a paper mill is intimately mixed with a water slurry of lime and a water slurry of lime mud, composed principally of calcium carbonate, under alkaline conditions. The resulting mixture is then subjected to clarification separation of formed precipitants. The liquid effluent overflow is carbonated by the bubbling of carbon dioxide gas through to precipitate calcium carbonate and residual color-imparting entities and to effect adjustment of the pH to approximately neutral conditions whereby the resultant liquid is adapted for discharge.

Referring to Figure 5.7, a holding tank **11** contains mixed waste liquid from mill and storage ponds normally requiring undesirable out-of-mill settling treatment. Reference **12** designates a holding tank for a supply of hydrated lime $Ca(OH)_2$, e.g., calcium hydroxide. Reference **13** designates a holding or supply tank for washed white liquor clarifier mud from the caustic room which is an adjunct to most pulping operations and which consists essentially of calcium carbonate. Liquid mixing tanks are shown at **15** and **16**, while **17** identifies a rather large clarifier and **18** and **19** identify vertical carbonation towers in between which is positioned a settling tank **20.** The above cited equipment comprises the essential apparatus vessels for carrying out the method of reducing or removing color bodies from waste pulp liquid effluent.

In accordance with the process, a proportion of the raw waste liquid is pumped via pump **22** through line **23** leading from holding tank **11** to the interior regions of mixing tank **15**; the latter being provided with interior baffle **15a** and an appropriately driven mixer **15b**. Simultaneously, an amount of hydrated lime is pumped via pump **25** through line **26** leading from lime holding tank **12** into the interior of mixing tank **15**, as shown. Also at the same time, a separate amount of raw waste is pumped via pump **27** through line **28** leading from the raw waste liquid holding tank **11** to mixing tank **16** which, like tank **15**, contains interior baffle **16a** and an appropriately driven bladed stirrer **16b.**

Also simultaneously, an amount of lime mud consisting essentially of calcium carbonate is pumped via pump **29** through line **30** into the interior of mixing tank **16** where the two liquids, namely, the raw waste liquid and the slurry of calcium carbonate, are intimately mixed. In tank **15,** the amount of raw waste liquid and an amount of hydrated lime are intimately mixed together. The slurry mixtures from tank **15** and tank **16** are respectively pumped via pump **31** and pump **32** through lines **33** and **34** to an in line mixer **40** from which the merged mixture of the two streams, contained respectively in lines **33** and **34**, pass via line **42** to the clarifier tank **17**, as shown.

The tank **17** includes an appropriately driven stirrer **17a** and an array of baffles **17b**. Bottom discharge line **44** leads to a precipitate or sludge storage while line **45** at the top of the clarifier tank carries effluent to the first carbonation tower **18.** Carbon dioxide is fed through pipe **46** laterally into the bottom of first carbonation tower **18.**

The carbonation tower may be packed or provided with suitable interior structure as to provide efficient gas/liquid contact between the CO_2 gas and the liquid effluent or overflow entering through pipe **45.** Carbonated liquid is passed via line **48** as urged by pump **49** to settling tank **20.** Settled solids (inclusive of residual calcium carbonate and color bodies) emerge from the bottom through line **20a** as controlled by valve **20b.** Overflow effluent from settling tank **20** passes via line **50** to carbonation tower **19** which is

FIGURE 5.7: REMOVAL OF COLOR FROM PAPER MILL LIQUID WASTE

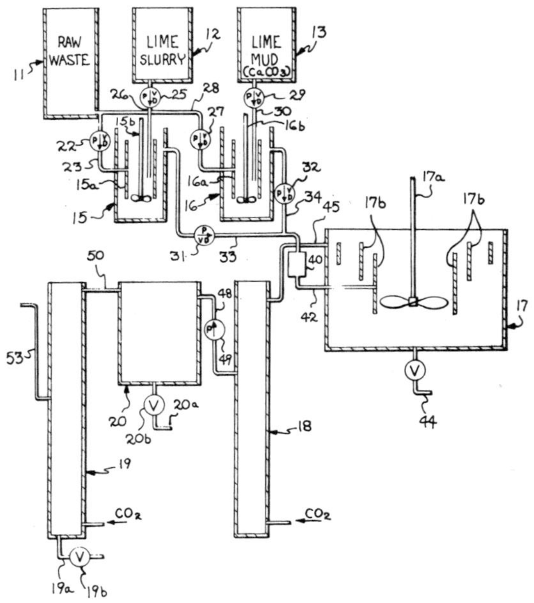

Source: H.C. Croom; U.S. Patent 3,736,254; May 29, 1973

constructed similarly to carbonation tower **18.** Carbon dioxide gas is introduced to the carbonation tower near the bottom through line **52.** Settled solids are exited through line **19a** as controlled by valve **19b.** Finally treated discharge waste leaves carbonation tower **19** through exit line **53.**

In carrying out the process, an amount of raw waste was treated in accordance with the sequence of steps set forth above. The raw waste liquid, for example, that in holding tank **11**, measured 1,500 to 2,000 true color units as determined by the known platinum-cobalt method of determining color value which is described in the text *Standard Methods for Determination of Water and Wastewater,* 12 edition. An amount of raw waste was mixed with an amount of a lime slurry, while at the same time an amount of raw waste was intimately mixed with an appropriate amount of lime mud comprising essentially calcium carbonate.

The two resulting mixtures were mixed in an in line mixer identified by **40.** The in line mixer may be a physical impeller mixer, a jet action mixer or an educator type of mixer providing an intimate contact and mixing of the two streams. The mixed liquid in mixer **40** and line **42** was determined to contain about 200 to 250 parts per million of lime as $Ca(OH)_2$ and 250 to 400 parts per million of lime mud calculated on the basis of $CaCO_3$.

This proportion is eminently satisfactory and is achieved by determination of the concentration of hydrated lime and calcium carbonate in feed tanks **12** and **13** and appropriate control of pumps **22, 25, 27** and **29.** The pH of the liquid mixture carried in line **42** and as introduced to clarifier **17** measured at the high end of the pH scale; that is, in the neighborhood of about 11.0 to 13.0. While in the clarifier, precipitation of calcium carbonate and conjoint agglomeration separation of color bodies from the liquid slurry occur particularly at the high pH.

Overflow effluent from the clarifier was passed to the carbonation tower from which it proceeds to settling tank **20** and thence to secondary carbonation tower **19.** Precipitation of residual calcium carbonate and conjoint agglomeration separation of residual color bodies also occur in the carbonation towers and particularly in the settling tank. Discharge liquid exiting from the secondary carbonation tower through pipeline **53** was found suitable for discharge to a source body of water such as the lake, river or stream.

Thus, the discharge water was marked by efficient and marked color removal in the range of 85 to 95% and a measured pH of approximately neutral due to the residence time and contact with carbon dioxide in the carbonation towers and settling tank. Liquid in the first carbonation tower **18** had a measured pH of about 11.0 to 12.0. It will be appreciated that higher concentrations of hydrated lime, calculated on the basis of $Ca(OCH)_2$, and lime mud, calculated on the basis of $CaCO_3$, will be indicated with raw waste liquid possessing color values higher than that recited hereinabove.

Generally, the raw waste liquid, the hydrated lime slurry and the lime mud slurry are desirably combined in such relative proportion and amounts that the calcium carbonate ($CaCO_3$) provided by the lime mud slurry is present in the total slurry in an amount which is at least equal to the hydrated lime, calculated as $Ca(OH)_2$, provided by the hydrated lime slurry. The lime mud usage desirably and significantly reduces the amount of hydrated lime that might otherwise be required to counteract the level of color bodies in the raw waste liquid to be treated.

It is a collateral and subsidiary feature of the process that the color removal method herein outlined is compatible with conventional chemical and semichemical pulp operations, requiring no expensive modifications or changes. The lime slurry, of course, can be prepared from fresh lime or burned lime could be used. Lime mud is readily available and its use contributes to the advantages in economy.

The precipitate and/or sludge from settling tank **20** is conveniently passed via line **20a** to the clarifier, although this is not specifically shown. Sludge precipitate collecting at the bottom of the clarifier passes through line **44** to a sludge storage and is usually conveniently processed through a kiln such as a lime recovery kiln for recovery of the chemical values.

Macroreticular Aromatic Resins

A process described by *J. Paleos; U.S. Patent 3,652,407; March 28, 1972; assigned to Rohm and Haas Company* relates to the decolorization of wood pulp bleach effluents by utilizing a macroreticular aromatic or aliphatic adsorbent resin. In the process, the caustic extraction effluent stream which normally has a pH of about 9 to 11 is first given a 2% lime treatment to reduce the color of the stream from 2,000 to 500 color units on the Co-Pt color scale. The lime reduces the color bodies by flocculation and/or precipitation of the bodies. This is then followed by a filtration step to dewater the effluent slurry.

The resulting still brown-colored filtrate is then mixed with the chlorination effluent which has a pH of about 2 to 3 and the mixture adjusted to a pH of about 2 to 4. The color of this mixture is an intense yellow and has an optical density of about 0.43 to 0.6 measured at a wavelength of 500 mμ or an optical density of about 0.2 to 0.6 measured at 560 mμ. Optical density is measured on the Spectrophotometer Beckman DU instrument. An optical density of 0.1 measured at 500 mμ corresponds to about 60 color units on the cobalt-platinum scale and an optical density of 0.1 measured at 560 mμ corresponds to 86 color units on the Co-Pt color scale.

This intensely yellow colored effluent mixture is then passed through a bed of an aromatic or aliphatic macroreticular adsorbent, of the type more fully described below, where virtually all of the remaining color can be removed. The overall schematic process, except for the regenerating and/or burning steps described hereinafter, is shown in the diagram on the following page. In an example, the caustic extraction filtrate and chlorination filtrate are mixed at approximately a 1 to 5 ratio after filtering the excess lime.

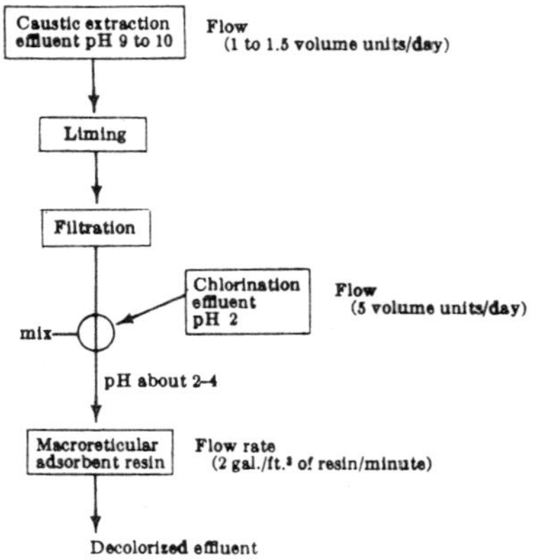

Loading of this stream onto a mass or bed of macroreticular resin adsorbent in a column is performed at a flow rate between 1.0 and 2.0 gal/cfm. The macroreticular adsorbent in this example is a synthetic insoluble cross-linked polystyrene polymer or resin used in the form of 20 to 50 mesh beads and used in a column operation known in the art. The adsorbent polymer has a surface area of about 330 m^2/g, a porosity of about 0.44 ml/ml and an average pore diameter of about 85 to 95 A. The optical density (0.550) of the influent is measured at an adjusted pH of 7.0 using a Beckman DU at 560 $m\mu$. Bed volume fractions are collected periodically and the optical density of these fractions is measured to obtain percentage color removal and percentage leakage data.

A typical exhaustion cycle showing effluent optical density and percentage color removal and leakage values is given in Table 1 below. Average color removal as calculated on the basis of the influent color is about 80% after the bed has been in extended use. During its initial use the bed removes as much as 96% or more of the color. The balance, or color not removed, is reported as percent leakage.

TABLE 1

BV*	Effluent Optical Density	Percent Color Removal	Percent Leakage
5	0.02	96	4
20	0.12	78	22
30	0.12	78	22
45	0.11	80	20
55	0.11	80	20
60	0.12	78	22
65	0.12	78	22

*BV - bed volume

Regeneration of the macroreticular adsorbent used in the example above can be accomplished in a number of ways. Ordinarily, most effective regeneration is accomplished with dilute solutions of sodium hydroxide (about 0.2% NaOH to 1% NaOH or higher). An additional advantage of the process is the ability to utilize as the regenerant various in-plant streams. For this purpose one can use as regenerants streams known in the art as white liquor or weak wash, analyses of which are given below in Table 2.

A further advantage of this process is the convenient elimination of the color disposal problem, i.e., the regenerant stream, after regeneration of the exhausted macroreticular adsorbent resin, which will then contain the color bodies, can be burnt, either alone or together with excess white liquor and/or with black liquor in a suitable furnace. Also, a very substantial portion of the caustic can be recovered from the furnace, if desired, and reused as a regenerant. Thus, the regenerant may be essentially a no cost item.

TABLE 2: ANALYSIS OF EFFLUENT STREAMS USED FOR REGENERATION OF MACRORETICULAR RESIN ADSORBENT

	White Liquor (g/l)	Weak Wash (g/l)
Total Na_2O	92.6	20.8
NaOH as Na_2O	62.5	13.0
Na_2S as Na_2O	18.7	5.1
Na_2CO_3 as Na_2O	11.3	2.8
NaOH as NaOH	80.6	16.7
Na_2S as Na_2S	23.6	6.4
Na_2CO_3 as Na_2CO_3	19.4	4.8

COAL, ORE AND SAND PROCESSING

COAL

Reverse Osmosis-Neutralization Process

A process described by *R.D. Hill, R.C. Wilmoth and R.B. Scott; U.S. Patent 3,795,609; March 5, 1974; assigned to the U.S. Environmental Protection Agency* relates to treatment of mineral matter contaminated water such as for example, acid mine drainage and in particular involves a procedure for separating contaminated waters into a purified water product and a relatively small volume of inert sludge. The water product has a low enough mineral content for direct discharge into natural waterways. The sludge is sufficiently inert and water-insoluble for permanent disposal, e.g., landfill.

The method involves reverse osmosis treatment of mineral contaminated water to concentrate the dissolved minerals into a small fraction of the feed stream and concomitantly to separate out at least about 50% of the water content of the feed stream as a purified water product stream capable of being safely discharged into the adjoining surface waters. In the instance of acid mine drainage, pH adjustment of the purified water product will often be required before discharge to surface waters.

The concentrated fraction or brine from the reverse osmosis treatment is chemically treated to precipitate mineral matter therefrom. In the instance of acid mine drainage, for example, chemical treatment may be with inexpensive neutralizing agents such as lime, limestone, sodium carbonate. The pH adjustment made in such brine is intended to achieve a pH level for optimum solids precipitation, rather than neutrality. Thereafter, this brine stream is clarified to remove the precipitated solids as a sludge product of the process, and the supernatant brine recycled back for reverse osmosis treatment.

Desirably the ultimate products of the process are a purified water stream and a wet sludge sufficiently inert for permanent disposal as, for example, by landfill. As a practical matter some brine blowdown may have to be discharged as a coproduct to avoid continued buildup of some nonprecipitating mineral constituent to harmful levels. Not more than about 20% of the feed water will be so discharged, and usually less than about 5%. The sludge product may contain up to about 2% of the feed water. Desirably, the purified product stream will contain at least about 90% of the entering waters.

In the instance of acid mine drainage, the principal dissolved mineral material present in the water are: iron, calcium, sulfate, magnesium, manganese, aluminum, and some traces of other metals. The usual ranges of acid mine drainage waters is set forth in Table 1.

TABLE 1: MINE DRAINAGE COMPOSITION*

Constituent	Range
pH	1.5 - 6.5
Iron, total, mg/l	0 - 10,000
Acidity, mg/l	0 - 50,000
Aluminum, mg/l	0 - 2,000
Sulfate, mg/l	100 - 50,000

*Mine drainage also contains various amounts of Ca, Mg, Mn, Cu, Zn, Cd, Ni, Pb and As.

Referring to Figure 6.1, the contaminated acid mine drainage waters entering from a source **10** may have to be pretreated to insure suitability for separation by reverse osmosis. A typical pretreatment for acid mine drainage is oxidation of ferrous ions present in the contaminated water in oxidizer **14**, chemically as by ozonation, chlorination, or biologically. In addition, disinfection to control bacteria and other organisms that cause membrane fouling may be required in disinfecter **15**, chemically, e.g., by chlorination, or by ultraviolet light. Suspended solids present in the contaminated mine waters are removed by passage through a filter system **16**.

If the drainage waters are already in an oxidized state and do not contain membrane fouling organisms, the feed stream may be passed by way of line **12** directly to filter system **16**. In any event, the filtered mineral contaminated water is pressurized by pump **18**, then passed through reverse osmosis unit **20** where the water is separated into a product water stream which passes out of the osmosis unit through line **22** and a brine stream which leaves through line **24**.

The purified product water in line **22** is low enough in mineral content for discharge directly to a stream **26**. If the product water is intended for direct domestic or industrial use, it may require further treatment like neutralization in pH adjustment unit **28** and disinfection in unit **29** before discharge via line **30** to its intended domestic or industrial use. The purified product water stream leaving the reverse osmosis unit **20** by way of line **22** will constitute at least 50% of the water feed entering reverse osmosis unit **20**. However, since an internal brine recycle forms an important part of the system as a whole, the purified product water will constitute at least about 80%, and normally more than 90% of the water input to the system from inlet **10**.

The brine stream which leaves osmosis unit **20** through line **24** will then comprise less than 50%, and usually less than 25%, of the combined feed water and recycled brine entering reverse osmosis unit **20**. The brine enters into (partial) neutralizer **34** where it is reacted with suitable alkaline material, e.g., limestone or lime, fed from hopper **32** under agitation from mixer **36**. Typically acid mine wastes are relatively high in iron, in aluminum and sulfate and partial or full neutralization with inexpensive neutralizing agents like lime or limestone will cause a considerable proportion of these contaminants to precipitate from the brine. In some instances where the ferrous ion content of the feed stream is high it may be desirable to place a ferrous-to-ferric oxidation step at this stage of the process. In such instances, the slurry and precipitate pass to an oxidation unit **35** where ferrous ion is converted to ferric iron by aeration or some other oxidation process.

The slurry of brine and precipitate passes from the neutralizer **34** or oxidizer **35** via line **38** into clarifier **40**, where the sludge is separated out from the neutralized brine stream and removed by way of line **42** to disposal. The supernatant brine is recycled from clarifier **40** through line **44** back to repeat passage through reverse osmosis separation. In the flow sequence illustrated by the drawing, recycled brine from line **44** is mixed with the feed water just ahead of filter **16**.

In order to separate out as much of the contaminants as is reasonably possible in the form of an inert solid coproduct without upsetting the system, the reaction in neutralizer **34** is adjusted to a pH which is optimum for the system as a whole. In the instance of acid mine drainage, neutralization to a pH of about 4.5 is believed to be optimum. If the pH were raised much nearer to neutrality some danger exists that the neutralized brine recycled from clarifier **40** by way of line **44** will cause instantaneous iron precipitation when admixed with incoming acid mine drainage waters. Any such precipitate might foul filter **16** or the membranes inside reverse osmosis unit **20**.

FIGURE 6.1: REVERSE OSMOSIS-NEUTRALIZATION PROCESS

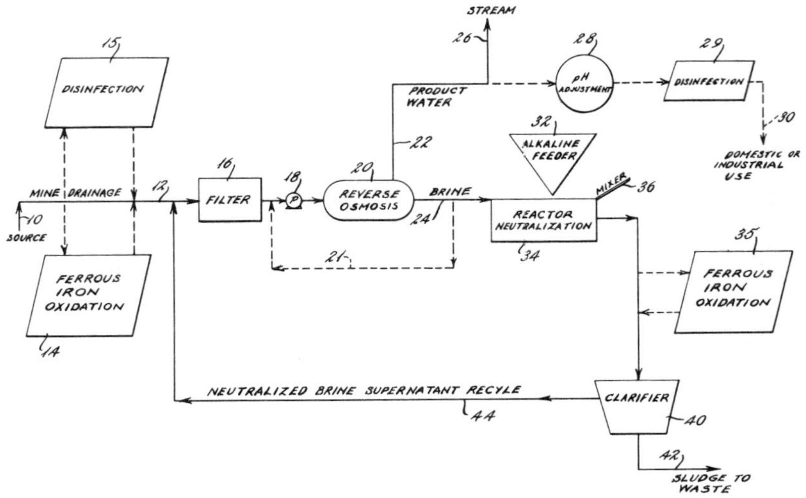

Source: R.D. Hill, R.C. Wilmoth and R.B. Scott; U.S. Patent 3,795,609; March 5, 1974

In the flow sequence illustrated, the only products of the separation process are the purified product water leaving in line **22** and the waste sludge leaving line **42**. Allusion has already been made that the purified product water preferably constitutes in excess of 90% of the water feed entering from source **10**. The balance of the water, which may be less than 2%, leaves as part of the sludge product from the clarifier **40** by way of line **42**. The high water recovery rate and the separation out of mineral contaminants as an essentially inert sludge demonstrates how efficiently the procedure of the process can treat contaminated drainage waters.

Such complete separation of the drainage waters depends upon the nature of the mineral contaminants, and the equilibriums attained in the system. Thus the precipitation which occurs in neutralizer **34** must reduce the mineral content of the brine and desirably reduce the content of all mineral components of the brine. Also, precipitation should not occur elsewhere in the system such as, for example, in the mixture of recycled brine with feed water, nor should the concentration of any mineral component in the brine build up to the point where its presence interferes with reverse osmosis separation. Obviously the mineral concentration levels at various stages of the system in the recycled brine can be adjusted by providing for some brine blowdown. This prevents buildup of relatively minor mineral constituents to interfering levels.

The refractory nature of some drainage waters may require that up to 20% of the water feed will be removed as brine blowdown. Preferably less than 5% will be so removed. Some process variations may be employed to minimize brine blowdown such as, for example, periodically reducing the purified water recovery rate in the reverse osmosis to a level low enough to clear any scale deposited therein at high recovery rates. Also the recycled brine may be maintained as a separate stream during reverse osmosis separation thereof, or admixed with feed water after partial separation of the feed water has been effected.

The reverse osmosis unit employed in the tests described by the following examples was a 10,000 gpd rated unit (Gulf Environmental Systems, Inc.) which contained 750 ft^2 total of membrane area. In order to insure that suspended solids did not plug the small brine channels in the reverse osmosis unit, sand filters and 10-micron cartridge filters were used to remove suspended solids from the raw feedwater and from the recycled neutralized brine.

In order to have sufficient brine flow in the reverse osmosis unit available for these tests it was necessary to recycle a substantial proportion of the brine directly from the outlet of the reverse osmosis unit back to the inlet (through direct recycle line **21**). Reverse osmosis units of the type employed in the test work of the following examples are capable of efficiently separating acid mine drainage waters into a concentrated brine stream and a relatively pure product water. Below tabulated is a representative reverse osmosis run.

TABLE 2: TYPICAL PERFORMANCE DATA AT 74.4% WATER RECOVERY

	Raw feed	Product water	Percent rejection
pH			
Conductivity, micromhos/cm	2.7	4.4	
Acidity ($CaCO_3$), mg./l.	1,350	55	97.9
Ca, mg./l	604	4	99.8
Mg, mg./l	115	2	99.3
Al, mg./l	38	0.9	99.2
Fe, mg./l	39	3.1	97.3
SO_4	153	0	100
	936	4.2	99.8

The same reverse osmosis unit was also operated at 91 ± 1% for extended periods of time. Some calcium sulfate scaling occurred but could be removed by operating the unit at 50% recovery for short periods of time to flush them from the membrane. Preferred practice of the process involves operating the reverse osmosis unit at high recovery rates, e.g., rates in excess of 80% water recovery.

Example 1: The pilot plant configuration used in the tests is the system illustrated in Figure 6.1, with brine recycle line **21** recycling brine directly from the outlet line of the reverse osmosis unit back to the inlet line where it mixed with the feed; the neutralized brine was added to the feed water downstream of the filter. The raw acid mine drainage, 5.02 gallons per minute, was pumped through a rapid sand filter and a 20-micron filter to remove suspended solids. The filtered water was then blended with 0.40 gpm of recycled neutralized brine and with the direct brine recycle.

The mixture was pumped to the reverse osmosis unit, which unit was operated at 91% water recovery. The 4.93 gpm of product water was sampled and discharged from the system. The brine flow was split with 3.60 gpm (88%) being recycled directly for blending with the filtered raw water (to maintain sufficient brine flow through the reverse osmosis unit).

The remaining 0.49 gpm (12%) of the brine flow was passed to a 50-gallon reactor where sufficient lime was added to raise the pH to 4.4. The neutralized water was then passed from the reactor to a 1,200-gallon upflow clarifier where the suspended solids were removed as sludge. The supernatant neutralized brine (0.40 gpm) was then passed through a sand filter and recycled back for blending with the raw mine drainage.

The run lasted a total of 77 hours; the details thereof and the results are tabulated in Table 3 shown on the following page.

TABLE 3: SUMMARY REVERSE OSMOSIS, NEUTRALIZATION TEST

	Raw feed	Blended feed water	Brine	Recycled neutralized brine	Product	Tube[1] rejection, percent	Overall[1] rejection, percent
Conductance, mmhos./cm	945	3,450	7,890	2,950	212	94	78
pH	2.9	2.7	2.4	4.4	3.3	----------	----------
Acidity ($CaCO_3$) mg./l	372	1,360	2,880	72	106	93	72
Ca (as Ca) mg./l	76	438	900	1,025	2.1	99	97
Mg (as Mg) mg./l	32	152	335	457	1.1	99	97
SO_4, mg./l	606	3,240	7,050	3,740	7.3	99	99
Fe, mg./l	55	269	560	3.4	1	99	98
Al, mg./l	21	106	231	12	0.85	99	96

[1] Tube rejection is the percent decrease in concentrate based on the water the membranes are actually receiving, e.g., blended feed—product ×100/blended feed. Overall rejection is the efficiency of the whole process, e.g., feed water—product water ×100/feed water.

NOTE:
Raw feed water flow, g.p.m. 5.02.
Product water flow, g.p.m. 4.93.
Brine flow vessel 5, g.p.m., 4.09.
Brine flow recycle, g.p.m. 3.60.
Brine flow to neutralization plant, g.p.m. 0.49.
Neutralized brine flow recycle, g.p.m. 0.40.
Neutralized brine flow to waste (sludge), g.p.m. 0.09.
Percent of neutralized brine resulting as sludge, 22.5.
Water recovery reverse osmosis unit, percent 90.9.
Water recovery total process, percent 98.3.
Average pressure, p.s.i. 603.
Average flux corrected to 10° C., gal./sq. ft./day 10.08.
Average water temperature, ° F. 47.

The overall 98.3% water recovery illustrates the extent to which the process can result in but two discharge streams, i.e., product water and a small quantity of sludge which is relatively inert. Attention is directed to the nearly 50% removal of sulfate from the brine and the rather small increase in the brine calcium content even though over 2,800 mg/l of lime as calcium carbonate was added to neutralize the acidity of the brine. Large amounts of calcium sulfate were being removed by the neutralization process.

Example 2: In this set of tests the same arrangement of equipment was employed as in the run of Example 1. The raw acid mine drainage treated in the runs of this example have the following characteristics:

Acidity ($CaCO_3$), mg/l	600	Aluminum, mg/l	40
Total iron (95% ferric), mg/l	100	Sulfate, mg/l	900
Calcium, Ca, mg/l	100	pH	2.7
Magnesium, Mg, mg/l	40	Conductance, micromhos/cm	950

A preliminary run was made with the reverse osmosis unit alone for 100 hours at a 91% recovery rate while the attendants were on duty and at 85% at night during the unattended hours. During the course of this run some evidence was noted of membrane fouling, probably by calcium sulfate. A temporary reduction of the recovery rate to 50% at the expiration of the run seemed to restore the capacity of the unit, apparently by a flushing effect. The details of this preliminary run are tabulated below (Table 4).

TABLE 4: SUMMARY OF 91% WATER RECOVERY TEST

	Raw water	Blended[1] feed water	Brine	Product	Tube[2] rejection, percent	Overall[2] rejection, percent.
Conductance mmhos./cm	1,190	4,210	9,600	248	95	79
pH	2.7	2.2	2.0	3.4	----------	----------
Acidity ($CaCO_3$) mg./l	633	2,584	5,914	116	96	82
Ca (as Ca) mg./l	266	1,206	2,756	7.5	99	97
Mg (as Mg) mg./l	134	750	1,671	2.8	99	98
SO_4, mg./l	810	4,024	9,542	16.8	99	98
Fe, mg./l	110	528	1,190	2.8	99	97
Al, mg./l	35	172	398	1.1	99	97

[1] Blended feed was the mixture of raw water and recycled brine actually pumped to the reverse osmosis unit.
[2] Tube rejection is the percent decrease in the concentrate based on the water the membranes were actually receiving, e.g., blended feed—product ×100/blended feed. Overall rejection is the efficiency of the whole process, e.g. feed water—product water ×100/feed water.

NOTE:
Average water recovery, percent 91.2.
Range of water recovery, percent 90.6-93.1.
Average pressure, p.s.i. 606.
Average feed water temperature, ° F. 44.
Water flux rate, corrected to 10° C., gal./sq. ft./day: Start of run, 11.6, end of run 9.8, range 9.5-13.4.
Length of run, hours 100.

Thereafter a first run was made with neutralization of the brine product to pH 4.7 and recycle of the clarified neutralized brine.

Neutralization reduced the acidity, aluminum, calcium, iron and sulfate sufficiently to demonstrate the possibility of long continued operation with, however, some occasional blowdown to avoid a continued buildup of some materials which do not precipitate at the acid pH such as, for example manganese. Some membrane fouling seemed to be occurring and the reverse osmosis unit was flushed every 20 to 24 hours to restore the flux rate therein. The total system of water recovery was 99%, i.e., 99 gallons of product water and 1 gallon sludge. The details of this test are presented below in Table 5.

TABLE 5: SUMMARY OF TEST 1

	Raw water	Blended [1] feed	Brine	Neutralized brine recycled	Product water	Salt [2] rejection percent
Conductance, mmhos./cm	1,720	5,160	10,000	3,680	335	95
pH	2.7	2.2	2.0	4.7	3.4	
Acidity, mg./l	647	2,680	6,030	462	130	95
Magnesium, mg./l	38	271	547	312	2	99
Calcium, mg./l	100	473	1,130	755	3	99
Aluminum, mg./l	38	187	405	65	1	99
Iron, mg./l	117	516	1,210	2	2	99
Sulfate, mg./l	982	4,680	11,100	2,160	21	99
Alkalinity, mg./l	0	0	0	1	0	

[1] Blended feed was a mixture of raw water, recycled brine and recycled neutralized brine. This water was pumped to the reverse osmosis unit.
[2] Salt rejection percent=blended feed concentration minus product water concentration ×100 divided by blended feed concentration.

NOTE:
Reverse osmosis unit water recovery, percent 91.
Total water recovery "Neutrolosis" system, percent 99.
Pressure, p.s.i. 602.
Water temperature, ° F. 58.
Raw water flow, g.p.m. 5.02.
Product water flow, g.p.m. 4.97.
Brine water flow, g.p.m. 0.48.
Neutralized brine recycled, g.p.m. 0.43.
Averaged water flux, gal./sq. ft./day at 77° F. 15.4.
Length run, hours 99.6.

A second test was conducted, this time with the reverse osmosis unit operated at 82% water recovery. The product water recovery was 98.7 or essentially the same as in the previous test. However, fouling of the membranes in the reverse osmosis unit seemed to be less than in the previous test and less flushing was necessary. The salt rejection was approximately 99% for all multivalent ions. The product water quality was better because a less concentrated feedwater was introduced. The details of Test 2 are tabulated below in Table 6.

TABLE 6: SUMMARY OF TEST 2

	Raw water	Blended [1] feed	Brine	Neutralized brine recycled	Product	Salt [2] rejection, percent
Conductance, mmhos./cm	1,660	3,000	5,640	2,500	213	93
pH	2.7	2.5	2.3	4.9	3.7	
Acidity, mg./l	585	1,150	2,430	175	36	97
Alkalinity, mg./l	0	0	0	7	0	
Calcium, mg./l	133	460	939	845	2	99
Aluminum, mg./l	43	91	189	30	1	99
Iron, mg./l	101	214	459	7	1	99
Sulfate, mg./l	1,340	3,010	6,380	3,490	18	99
Magnesium, mg./l	32	118	254	131	1	99

[1] Blended feed was a mixture of raw water, recycled brine, and recycled neutralized brine. This water was pumped to the reverse osmosis unit.
[2] Salt rejection percent=blended feed concentration minus product water concentration ×100 divided by blended feed concentrate.

NOTE:
Reverse osmosis unit water recovery, percent 82.
Total water recovery "neutrolosis" system, percent 97.7.
Pressure, p.s.i., 600.
Water temperature, ° F., 61.
Raw water flow, g.p.m., 4.83.
Product water flow, g.p.m., 4.77.
Brine water flow, g.p.m., 1.31.
Neutralized brine recycled, g.p.m., 1.25.
Average water flux, gal./sq. ft./day at 600 p.s.i. at 77° F., 12.3.
Length run, hours, 130.

Combination of Anionic Flocculants

M.F. Werneke; U.S. Patent 3,717,574; February 20, 1973; assigned to American Cyanamid Company describes a process for flocculating colloidal coal particles in aqueous alkaline medium. It has been found that substantially all of the colloidal coal is flocculated when the effluent is rendered acid and is treated with two specific water-soluble anionic flocculants. One flocculant is a low molecular weight polymer which is strongly anionic. It is at least about 50 mol percent composed of acrylic acid linkages and has a molecular weight between 50,000 and 500,000.

The second flocculant is a weakly anionic vinyl polymer which is of very high molecular weight. It is at least 95 mol percent composed of acrylamide, contains a sufficient number of vinyl acid linkages so that the polymer possesses a net anionic charge, and has a molecular weight in excess of 5 million. The higher its molecular weight, the greater is its efficiency. By the use of these polymers in combination it has been possible in the laboratory to flocculate substantially all of the colloidal coal in a typical aqueous medium, so that the supernatant phase (the medium remaining after the flocculated coal has settled) is substantially as clear as pure water. The following examples illustrate the process.

Example 1: The following illustrates the process as applied to the flocculation of colloidal coal in the effluent from the fluidized transport of bituminous coal after the over-sized particles have been removed by centrifugation. The effluent is a jet black mobile fluid containing 3.35% by weight of colloidal bituminous coal particles and has a pH of 7.35.

In each instance, the test is performed by taking a 50 cc sample of the suspension, adjusting the pH to 6.5 by addition of dilute sulfuric acid and then adding a solution of a 30:70 molar ratio acrylamide:acrylic acid copolymer having a molecular weight of 200,000 followed by an aqueous solution of a 99:1 acrylamide:acrylic acid copolymer having a molecular weight of 12 to 15 million. The test tube is inverted after each addition to ensure uniform distribution of the polymer.

The test tube is then allowed to stand for 2 hours at the end of which time the height of the supernatant layer is measured and also the turbidity of this supernatant layer is determined by means of a laboratory turbidimeter. Control tests are run on the suspension as supplied, on the suspension with addition of sulfuric acid (but without addition of polymer) and with addition of sulfuric acid and each of the polymers. Results are as follows.

	pH of Suspension		Amount of Polymer Added, ppm		Supernatant Layer After 2 Hours	
Run #	Start	After H_2SO_4	Low	High	Height, inches	Turbidity, ppm
A	7.4	-	(**)	(**)	0	Black
B	7.4	6.5	(**)	(**)	0	Black
C	7.4	6.5	20	(**)	1.5	400
D	7.4	6.5	(**)	20	0.25	Black***
E	7.4	(*)	12	8	2.5	Black***
1	7.4	6.5	12	8	2.5	150
2	7.4	6.5	14	6	2.25	90

*No acid added.
**No polymer added.
***Solids content in excess of 1,000 ppm.

The results show that while a practically useful amount of clarification was achieved in the best of the control runs (Run C) far superior clarification was achieved by the interaction of the two polymers.

Example 2: The procedure of Run 2 of Example 1 is repeated except that sufficient of a 4% aqueous solution of a water-soluble anionic starch is added with the low molecular weight polymer in amount sufficient to supply 30 ppm of the starch based on the total weight of the coal suspension. After standing 2 hours the height of the supernatant layer is 2.5" and the turbidity is 50 ppm. The starch thus acts as an auxiliary flocculant.

Acrylamide-Acrylic Acid Copolymer Latexes

F.S. Spoerle and E.W. Palmer; U.S. Patent 3,790,476; February 5, 1974; assigned to Nalco Chemical Company describes an improved method of thickening or dewatering solids from industrial wastes comprising the addition of an aqueous solution of a water-soluble vinyl addition polymer formed by the inversion of a polymeric latex added to the industrial wastes under floc forming conditions and thereafter allowing the solids to settle from the industrial waste to provide a clear aqueous supernate. The polymeric latex is produced by the steps of:

(A) Forming a water-in-oil emulsion from:
   - (1) water which contains a water-soluble ethylenic unsaturated monomer, thereby producing a monomer phase which has a concentration of from 30 to 70% by weight of the emulsion;
   - (2) an inert hydrophobic liquid;
   - (3) a water-in-oil emulsifying agent in a concentration of from 0.1 to 10% by weight;
   - (4) a free radical initiator;

(B) Heating the emulsion under free radical forming conditions to polymerize the water-soluble ethylenic unsaturated monomer forming a polymer which is contained in the emulsion; and

(C) Polymerizing the monomer in the water-in-oil emulsion to produce a polymeric latex.

As indicated, this process produces a polymeric latex which comprises a water-in-oil emulsion which contains dispersed therein a finely-divided water-soluble vinyl addition polymer.

The preferred copolymers of this process are the copolymers of acrylamide with either sodium acrylate or dimethylaminoethyl methacrylate. The preferred ratio of acrylamide to sodium acrylate is 70:30 while the preferred ratio of acrylamide to dimethylaminoethyl methacrylate is 95:5. When aqueous solutions of the monomers are used, they can vary widely in monomer content. Proportions between 70 and 30% by weight monomer and correspondingly 30 to 70% water are used, depending upon the monomer and the temperature of polymerization. The ratio of monomer phase to oil phase is also widely variable, advantageously between 30 and 70 parts of the former to between 70 and 30 parts of the latter by weight. A monomer phase to oil phase ratio of about 70 to 30 is preferred.

In illustration of the marked improvement of this process, the thickening and dewatering ability of the inverted polymer latex was measured and compared to that of dry solid polymers dissolved in water. This study was performed on two different types of industrial wastes, a coal mining slurry having a 5% suspended solids and a pulp mill sludge having 4% suspended solids. The coal mining slurry was obtained from a coal processor from West Virginia while the pulp mill sludge was obtained from a paper manufacturer in a midwestern state. The results are shown in the following tables.

Coal Mining Slurry

Run No.	Polymer	Polymer type	Dosage, percent	Solids settled, percent	Improvement, percent
1	Acrylamide-acrylic acid	Solid	0.01	15	-----
2	do	do	0.05	21	-----
3	do	do	0.10	38	-----
4	do	do	0.30	52	-----
5	do	Latex	0.01	29	93
6	do	do	0.05	42	100
7	do	do	0.10	57	50
8	do	do	0.30	73	40
9	Acrylamide-DMAEM	Solid	0.05	19	-----
10	do	do	0.10	28	-----
11	do	do	0.30	39	-----
12	do	do	0.70	52	-----
13	do	Latex	0.05	31	65
14	do	do	0.10	49	75
15	do	do	0.30	62	58
16	do	do	0.70	85	63

Where: DMAEM is dimethylaminoethyl methacrylate. Acrylamide-acrylic acid ratios are 70:30.

Pulp Mill Sludge

Run No.	Polymer	Polymer type	Dosage, percent	Solids settled, percent	Improvement, percent
1	Acrylamide-acrylic acid	Solid	0.1	15	
2	do	do	0.5	32	
3	do	do	1.0	55	
4	do	do	2.0	73	
5	do	Latex	0.1	29	93
6	do	do	0.5	48	50
7	do	do	1.0	69	26
8	do	do	2.0	91	25
9	Acrylamide	Solid	0.5	13	
10	do	do	1.0	35	
11	do	do	1.5	48	
12	do	do	2.0	72	
13	do	Latex	0.5	26	100
14	do	do	1.0	48	40
15	do	do	1.5	61	29
16	do	do	2.0	90	26

NOTE.—Acrylamide-acrylic acid ratios are 70:30.

The effect of the polymeric latex as compared to the aqueous solution of the dry solid can be observed from these results. It is clearly shown that the polymer latex shows superior results to the dry solids.

Use of Waste Steam to Produce Salt-Free Condensate

W.J. Didycz and D. Glassman; U.S. Patent 3,790,448; February 5, 1974; assigned to U.S. Steel Corporation describe a frontward flow and backward flow method of and apparatus for purifying waste fluid containing wastewater and corrosive salts with high temperature steam from a continuous coke quenching apparatus to produce substantially pure salt-free condensate.

One method of purifying waste fluid, containing wastewater and corrosive salts produced in a coke producing apparatus and normally utilized to quench coke in a coke quenching apparatus, with waste steam from the coke quenching apparatus to produce substantially salt-free condensate, comprises the steps of:

(a) heating the waste fluid in heat exchange relationship with the steam to a first predetermined temperature at a first predetermined pressure;

(b) condensing the high temperature steam to form substantially salt-free condensate;

(c) vaporizing the heated waste fluid at the first predetermined temperature at the first predetermined pressure to vaporize a portion of the waste fluid and to concentrate the corrosive salts to form a concentrate, the first predetermined pressure being lower than the pressure of the waste steam from the coke quenching apparatus;

(d) condensing the heated vaporized portion of the waste fluid to form substantially salt-free condensate; and

(e) utilizing the substantially salt-free condensate in an industrial application, thereby eliminating the disposal of the waste fluid and the contamination of the atmosphere with the waste fluid if the waste fluid were employed to quench the coke in the coke quenching apparatus, and utilizing the waste steam for a useful purpose.

Coal Liquefaction

G.P. Masologites; U.S. Patent 3,661,774; May 9, 1972; assigned to Atlantic Richfield Co. describes a method for separating solid particles from a hydrocarbonaceous liquid by subjecting the solids containing liquid to force conditions which cause the solid particles to collect in a localized portion, introducing into that localized portion a displacement liquid which is substantially inert to the solids containing liquid and which has a specific gravity substantially greater than the solids containing liquid so that the displacement liquid takes the place of (displaces) at least part of the solids containing liquid in the localized portion and takes up solid particles that are migrating into the localized portion. In effect therefore, these particles, when moving under the imposed force conditions, move into the localized portion of the solids containing liquid and in that portion are transferred from the solids containing liquid to the displacement liquid. The displacement liquid, with solid particles dispersed therein is removed from the localized portion of the solids containing liquid.

There is separately removed a substantially solids free hydrocarbonaceous liquid, this liquid being of a higher quality since it is substantially solids free and also being amenable to further processing to obtain hydrocarbonaceous products such as naphtha. The solids containing displacement liquid can be processed for removal of the displacement liquid for reuse in the process or other disposal as desired, thereby leaving substantially only the solid particles for disposal as waste, for coking, and the like.

By following the process, the amount of liquid initially present in the solids containing liquid hydrocarbonaceous product which is made available for processing into naphtha, and the like is maximized because substantially only the solids are removed because of the use of the displacement liquid medium. Additionally, the particular displacement liquid employed can be chosen to be more easily removed from the solid particles by distillation, solvent extraction, filtering, centrifugation, and the like so that substantially all of the displacement liquid can be recovered for reuse.

Figure 6.2 shows a frustoconical chamber **1** having a right cylindrical top portion **2**. A solids containing liquid hydrocarbonaceous product is introduced into an intermediate portion of chamber **1** by way of pipe **3**. The liquid introduced by the pipe is introduced at a sufficiently slow rate to establish and maintain a substantially quiescent pool **4** of liquid. In pool **4** the particles in the liquid from the pipe are subjected to the force of gravity under conditions which allow the particles to freely migrate downwardly towards a bottom portion **5** of the chamber. Thus, the upper portion of the chamber, particularly cylindrical portion **2**, contains liquid from the pipe which is substantially free of solids and which can be drawn off by way of pipe **6**, preferably at a rate substantially the same as the rate of introduction of liquid from pipe **3**.

FIGURE 6.2: DISPLACEMENT LIQUID TECHNIQUE FOR SEPARATING SOLIDS FROM A LIQUID

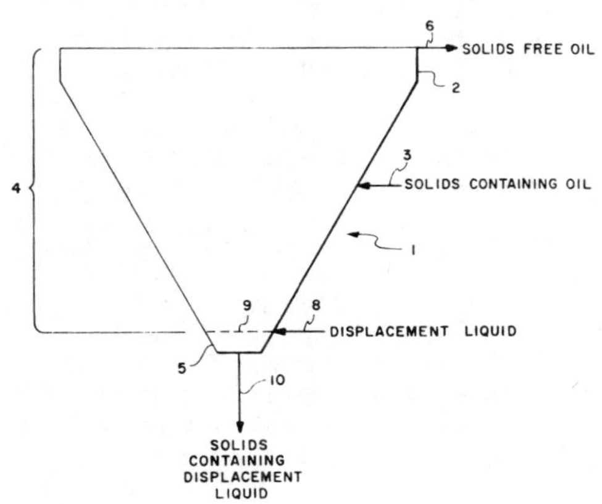

Source: G.P. Masologites; U.S. Patent 3,661,774; May 9, 1972

The substantially solids free hydrocarbonaceous liquid withdrawn by way of pipe **6** is a higher quality liquid because of the substantial absence of solid particles and can be further processed to produce naphtha. Bottom portion **5** of chamber **1** is a localized portion where the solid particles collect.

A displacement liquid is introduced into this localized portion by way of pipe **8**, the displacement liquid being substantially inert, i.e., nonreactive, to the hydrocarbonaceous liquid introduced by way of pipe **3**. This liquid has a specific gravity which is greater, e.g., at least 0.05 specific gravity units greater and preferably 0.10 specific gravity units greater than the specific gravity of the hydrocarbonaceous liquid introduced by way of pipe **3**. The displacement liquid is introduced by way of pipe **8** at a flow rate such that substantial mixing of the displacement liquid and the liquid from pipe **3** is avoided and an interface **9** is achieved between the displacement liquid and the liquid from pipe **3**. Thus, the displacement liquid is introduced so as not to disturb the quiescence of pool **4** and to establish a localized portion where solid particles migrating downwardly under the force of gravity move from the liquid of pipe **3** across interface **9** into the displacement liquid.

Solids containing displacement liquid is removed from the bottom of the chamber by way of pipe **10** for further processing such as flashing of displacement liquid from the solids thereby leaving the solids for coking, or other disposal as desired. The solids containing displacement liquid removed by way of pipe **10** is preferably removed at a rate substantially the same as the rate of introduction of displacement liquid by way of pipe **8** thereby setting up a dynamic equilibrium between the displacement liquid and localized portion **5** in the quiescent pool **4**.

Under these force conditions solids containing particles continually migrate from pool **4** into the displacement liquid in localized portion **5** while substantially solids free liquid from pipe **3** is continually removed by way of pipe **6**. All removal and addition of materials from and to chamber **1** are accomplished so that substantial hindrance of the gravity flow of solid particles downwardly into localized portion **5** is avoided.

Other known apparatus can be employed to accomplish the transfer of solid particles from the liquid of pipe **3** to the displacement liquid of pipe **8**. For example, instead of employing only gravity force conditions, other force conditions can be imposed in addition to or in lieu of gravity, e.g., centrifugal force, electrical force, magnetic force, and the like can all be employed alone or in any desired combination to give the solid particles from pipe **3** vectorial movement with a measurable velocity towards a localized or collecting portion of the chamber so that the particles, shortly after injection into the chamber will move towards this localized portion and under the momentum of this movement will transfer themselves from the liquid of pipe **3** across interface **9** and into the displacement liquid.

The displacement liquid is maintained in contact with the liquid from pipe **3** for a time sufficient for substantial amounts of solid particles to migrate from the liquid of pipe **3** into the displacement liquid under the prevailing force conditions.

From the above it can be seen that the particular force conditions applied to a given situation can vary widely depending upon the particular chemical and physical nature of the liquid in pipe **3** and the displacement liquid, the conditions of temperature and pressure imposed on the contents of chamber **1**, the interfacial forces between the liquid of pipe **3** and the displacement liquid, the desired rate of separation of solids from the liquid in pipe **3**, the desired rate of recovery of substantially solids free liquid by way of pipe **6**, the size, physical configuration and chemical nature of the solid particles in the liquid in pipe **3**, and the like.

However, the requirements of imparting to the solid particles a velocity in a desired direction, the velocity being sufficient to transfer the particles to the displacement liquid, and the displacement liquid being substantially inert as to the solids containing liquid and a substantially greater specific gravity than the solids containing liquid apply under all conditions of operation.

Example 1: The process is carried out in the apparatus as shown in Figure 6.2 where the solids containing liquid is a complex hydrocarbonaceous liquid obtained by the hydrogenation of Illinois No. 6 coal subdivided to all pass a 20 mesh sieve (Tyler), the hydrogenation being accomplished with molecular hydrogen at a temperature of about 850°F and a hydrogenation partial pressure of about 2,250 psia.

After flashing the light distillate oils from the hydrogenation product the residual liquid hydrocarbonaceous product obtained from the hydrogenation process has a boiling range starting from 600°F and has about 50% boiling above 1000°F and contains about 12 weight percent of particles composed primarily of cobalt molybdate hydrogenation catalyst particles, unconverted coal particles, ash, coke, and char, the particles having a largest cross-sectional dimension of from 1 to 50 microns.

The solids containing liquid is passed into chamber **1** and maintained therein with substantially no mixing at a temperature of about 650°F, which temperature gives the liquid a specific gravity of 0.943. Displacement liquid which is a 500° to 850°F oil product of the coal hydrogenation at 100°F and a specific gravity at that temperature of 1.04 is introduced into localized portion **5** at the rate of 5 gal/min.

Solids containing displacement liquid is removed from localized portion **5** by way of pipe **10** at the rate of 5.6 gal/min and contains 17 weight percent of solid particles as defined hereinabove in this example and 83 weight percent hydrocarbonaceous liquid, both weight percents being based on the total weight of the displacement liquid. Substantially solids free hydrocarbonaceous liquid, i.e., having less than 0.02 weight percent solid particles based on the total weight of the liquid, is removed by way of pipe **6** at the rate of 9.4 gallons per minute. Solids containing hydrocarbonaceous liquid is charged by way of pipe **3** at the rate of 10 gal/min. The charging rate of liquid from pipes **3** and **8** is sufficient to prevent turbulence or substantial stirring of any of the liquid in chamber **1**.

Example 2: The total liquid product from the coal hydrogenation of Example 1, having a boiling range starting about 1250°F and having about 42 weight percent boiling above 1000°F and containing about 10 weight percent of particles of the type described in Example 1, is passed into chamber **1** and maintained therein with substantially no mixing thereof at a temperature of about 150°F. This temperature gives the liquid a specific gravity of 1.15.

Displacement liquid which is an aqueous solution of aluminum chloride containing 35% aluminum chloride by weight, is introduced into localized portion **5** at a temperature of 80°F. The displacement liquid has a specific gravity of 1.28 at this temperature. The solids containing liquid is charged to chamber **1** at a rate of 11.1 gal/min and the displacement liquid is introduced into localized portion **5** at a rate of 1 gal/min.

Solids containing displacement liquid is removed from localized portion **5** by way of pipe **10** at the rate of 1.4 gal/min and contains 50 weight percent of solid particles as defined in Example 1 and 50 weight percent aqueous aluminum chloride solution, both weights being based on the total weight of the displacement liquid. Substantially solids free hydrocarbonaceous liquid, i.e., having less than 0.02 weight percent solid particles based on the total weight of the liquid, is removed by way of pipe **6** at the rate of 10.7 gal/min.

Fine Coal Recovery and Water Clarification Process

F.G. Miller; U.S. Patent 3,696,923; October 10, 1972; assigned to Bethlehem Steel Corp. describes a method for removing and recovering fine coal and coal-containing particles from a coal slurry and clarifying the tailings product sufficiently for reuse in a coal washing circuit or for disposal in adjacent environments without contaminating same.

The improved method for removing and recovering fine coal and coal-containing particles in a coal slurry includes treating the coal slurry in a coal cleaning system, which includes froth-flotation cells, to obtain a first froth product containing fine coal and coal-containing particles and a first tailings product containing gangue materials. The first tailings product is passed to waste. The first froth product is passed to at least one dewatering apparatus to recover the fine coal and coal-containing particles as a substantially dry cake product. A first effluent from the dewatering apparatus is substantially water containing a small percentage of coal and coal-containing particles. The first effluent is treated in a solids recovery system which includes froth-flotation cells to remove a substantial portion of the fine coal and coal-containing particles as a second froth product. The second froth product is recycled to the dewatering apparatus to recover the fine coal and coal-containing particles. A second tailings product from the solids recovery system is recycled in the coarse coal or fine coal washing circuit or is passed to waste in adjacent environments without contaminating same.

FIGURE 6.3: COAL RECOVERY AND WATER CLARIFICATION PROCESS

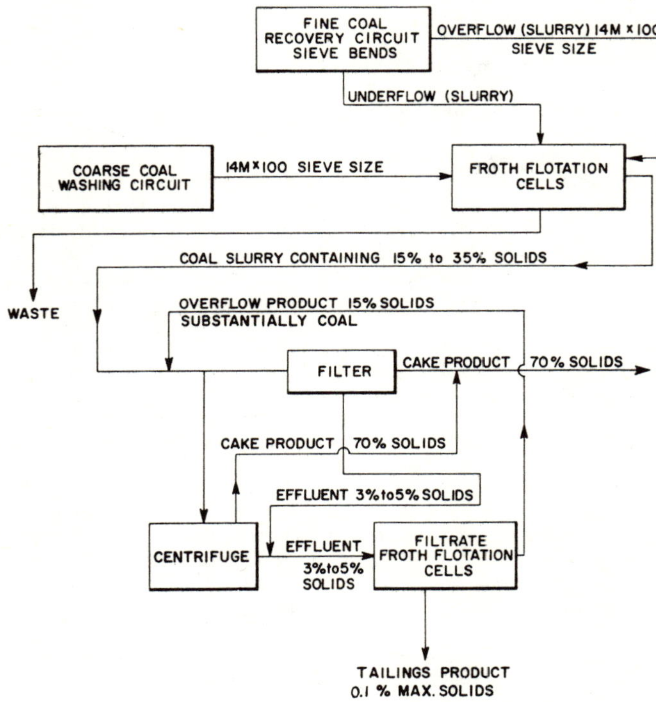

Source: F.G. Miller; U.S. Patent 3,696,923; October 10, 1972

As shown in Figure 6.3, raw coal is washed and screened in a coarse coal washing circuit and/or in a fine coal recovery circuit. The coal from these circuits contains gangue materials and must be beneficiated so that the fine coal and coal-containing particles in the resultant slurry can be recovered as a usable product.

The slurry or slurries from the abovementioned circuits are treated in a coal cleaning flotation system which includes froth-flotation cells. A suitable flotation and/or frothing agent, such as MIBC (methyl isobutyl carbinol), sodium xanthate, pine oil or fuel oil is added to the flotation cells to float the fine coal and coal-containing particles. It is preferred to use MIBC as the frothing agent. It has been found that about 0.1 lb to 0.4 lb of a flotation and/or frothing agent per ton of coal is used to float substantially all the fine coal and

.1lb -.4lb of MIBC/ton

coal-containing particles. The gangue materials are not affected by the flotation or frothing agent and sink in the flotation cells. The froth product produced in the coal-cleaning flotation system contains about 15% to about 35% solids. The froth product is dewatered in at least one dewatering device, such as a vacuum filter or a centrifuge. A cake product containing about 70% solids is formed in the dewatering apparatus. The cake product is recovered as usable coal. The gangue materials from the coal cleaning flotation system are passed to waste.

The effluent from the dewatering apparatus is a slurry containing 3 to 5% solids which are substantially all fine coal and coal-containing particles. A portion of the flotation and/or frothing agent added in the coal cleaning flotation system adheres to the surface of the fine coal and coal-containing particles and is present in the effluent slurry. The effluent is passed to a solids recovery flotation system which includes froth-flotation cells. An additional amount of a frothing agent can be added to the slurry in the flotation cells. Up to about 0.1 lb of frothing agent per ton of coal is added to float the fine coal and coal-containing particles.

Substantially all the fine coal and coal-containing particles in the effluent are floated and removed as a froth product. As in the coal cleaning flotation system, the frothing agent may be MIBC, sodium xanthate or pine oil but it is preferred to use MIBC. The froth product from the solids recovery flotation system contains about 15% solids. The froth product is recycled to the dewatering apparatus where the fine coal and coal-containing particles are recovered as a usable product.

The tailings product from the solids recovery flotation system is substantially clarified water containing not more than about 0.1% solids. The tailings product is used as makeup water in the coarse coal washing circuit replacing a portion of fresh water normally used in the circuit. Of course, the tailings product can be disposed of in adjacent environments without contaminating same.

In a specific example of the improved method, the overflow slurry from sieve bends in a fine coal recovery circuit was passed to a coal cleaning flotation system. About 0.2 lb of MIBC per ton of coal was added to the flotation cells in the system. A froth product containing 35% solids in the form of fine coal and coal-containing particles was produced in the flotation cells. The froth product was dewatered in a vacuum filter. Usable coal in the form of a cake containing 72% solids was recovered in the vacuum filter. The effluent from the vacuum filter containing 3% solids was passed to a solids recovery flotation system including froth-flotation cells.

About 0.1 lb of MIBC per ton of coal was added to the flotation cells. The froth product produced in the flotation cells contained 13% solids. The froth product was recycled to the dewatering apparatus. The tailings product produced in the solids recovery system was substantially clarified water containing 0.09% solids. The water was recycled in the coarse coal washing circuit.

ORE PROCESSING

Addition of Tailings to Slimes

C.C. Cook and E.M. Haynsworth; U.S. Patent 3,763,041; October 2, 1973; assigned to American Cyanamid Co. describe a process for enhancing the rate at which water is removed from the waste slimes of ore processing operations, particularly phosphate rock processing.

Since valuable ore deposits often occur in nature intimately mixed with a variety of less valuable or desirable constituents, it is a primary function of the ore processing industry to remove or separate as much of these extraneous constituents from the desired ore as possible. The flotation process developed in the early 1930s has proven to be a valuable tool for assisting in the removal of unwanted waste products from ores and is in wide usage today in a variety of ore processing operations. However, since the development and utilization of the flotation process by the ore processing industry, disposal of the waste products has presented a monumental problem. Basically these waste materials fall into two categories identified by those skilled in the ore processing art by the terms slimes and tailings, respectively.

Slimes and tailings are produced as waste products in a host of widely varying ore processing operations such as, for example, in copper mining, in the mining of heavy minerals such as titanium and rutile, and in virtually all nonmetallic mining such as the mining of phosphate, potash, feldspar, clays, and fluorspar.

A principal generator of copious amounts of slimes and tailings is the phosphate rock processing industry, a substantial portion of which is located in the state of Florida. Slimes and tailings from such an operation can be conveniently referred to as phosphate slimes and phosphate tailings. An extensive analysis and characterization of Florida phosphate slimes can be found in the Bureau of Mines Report of Investigations 6,163 by J.H. Gary et al which appeared in 1963 under the title *Chemical and Physical Beneficiation of Florida Phosphate Slimes.* A typical size distribution screen analysis (dry basis) and a typical chemical analysis (dry basis) of a Florida phosphate slime are presented below.

Screen Analysis

Mesh	+20	-20	-35	-48	-65	-100	-150	-200	-325
Percent by weight					100	99.85	99.65	99.20	98.30

Chemical Analysis

Constituent	Percent by Weight
P_2O_5	10 - 20
Bone phosphate of lime	21 - 42
Fe_2O_3	2 - 8
Al_2O_3	5 - 20
CaO	10 - 20
Insolubles (quartz)	17 - 23

Florida phosphate slimes typically contain approximately one-third divided quartz, one-third finely divided fluorapatite, and one-third finely divided clays (primarily attapulgite). Such slimes generally contain very large amounts of extremely fine solid particles. In the screen analysis given above, for example, it can be seen that 98.3% by weight of the solids have a particle size of -325 mesh. It should also be noted that the slimes may contain small amounts of tailings. In the above screen analysis, for example, the slimes contained 0.35% tailings solids. A typical size distribution screen analysis (dry basis) and a typical chemical analysis (dry basis) of Florida phosphate tailings are presented below.

Screen Analysis

Mesh	+20	-20	-35	-48	-65	-100	-150	-200	-325
Percent by weight	1.2	98.8	85.1	53.1	19.0	8.0	3.4	-	-

Chemical Analysis

Constituent	Percent by Weight
P_2O_5	0.5 - 2.5
Bone phosphate of lime	1.1 - 5.5
Fe_2O_3	0.2 - 0.5
Al_2O_3	0.1 - 0.3
CaO	0 - 3
Insolubles (quartz)	90 - 95

From the above screen analysis, it can be seen that 95.4% of the tailings solids have an average particle size within the range -20 + 150 mesh, a particle size range which is fairly typical of a Florida phosphate tailings. It should be noted that the tailings may contain small amounts of slimes as can be seen from the above screen analysis which shows a slimes solid content of 3.4%.

It is disposal of the tremendous amounts of slimes generated by ore processing plants which presents the greatest waste disposal problem and it is the solution of this problem to which this process is primarily directed. Slimes present a problem because they retain substantial amounts of water and consequently their fluidity even after years of settling and, as a result, possess substantially no bearing strength whatever. Even after settling for many years, the slimes settle to only 25 to 30% by weight solids and still possess a jelly-like consistency. As such, the disposal site in which they are deposited becomes virtually useless and is an obvious hazard to passers-by.

The slimes are typically disposed of by discharging them into excavations or, as is perhaps more common, into reservoirs or ponds which are formed by a constructed earthen dam work. The slimes are allowed to settle by gravity; the water which separates from the slimes during settling is usually recovered from the slimes settling pond for reuse in the ore processing plant.

Where dammed sites are utilized, the dams must be continuously maintained for when a dam fails, as they may occasionally do, the countryside and rivers surrounding the site are inundated with vast quantities of the jelly-like slimes resulting in pollution of the land and surrounding waterways. Moreover, when the slimes disposal area is filled, the site is nothing more than a liability. A crust may form on the surface of the site but a few inches below the surface, the slimes are a jelly-like mass, and obviously a series hazard to anyone crossing the site. Backfilling over the surface may permit use of the filled up disposal site for limited agricultural use but use of heavy equipment and building on the site is not possible.

In view of these disposal problems, this process involves producing a fertile reconstituted soil of acceptable bearing strength which comprises admixing slimes having a critical solids content with prescribed amounts of tailings. According to the process, it has been found that when slimes are mixed with tailings, the slimes dewater at a substantially greater rate than is achievable using gravity settling and, moreover, that the reconstituted slimes-tailings mixtures produced thereby are sufficiently fertile to support plant life and possesses acceptable bearing strength. By acceptable bearing strength is meant that the bearing strength of the reconstituted mixture will approach that of normal soil in the region from which the ore being processed originates. In general, an acceptable bearing strength means that the land will support animals and human beings, heavy equipment such as tractors and the like, and can be used for the construction of buildings and dwellings. More particularly, it has been found that if the slimes are simply mixed with tailings that:

(1) There is a substantial enhancement in the rate at which the water is released from the slimes. For example, whereas slimes containing 2% solids will settle by gravity to only 7% solids after 60 days, the addition of tailings to the slimes results in slimes containing 15% solids in 30 days and 20% solids in 60 days.

(2) Water is released from the slimes over a relatively short settling time in amounts substantially greater than was heretofore thought possible. For example, whereas slimes containing 2% solids after settling 60 days produced only 70% clear water, the same slimes upon addition of tailings produced about 80% clear water after 30 days and 90% clear water after 60 days.

The increased amount of water which the process allows to be rapidly recovered from the slimes has the two-fold advantage of making substantially more water available for recycle to the ore processing plant while simultaneously significantly reducing the acreage required for the slime disposal site.

Referring to Figure 6.4a the waste primary slimes **10** from the flotation pretreatment process steps in a phosphate ore processing plant and the waste secondary slimes **11** from the flotation step itself are admixed to form a dilute aqueous suspension of slime solids **12** which typically contains 1 to 2% solids. Usually about 90% of the total slimes produced are primary slimes with the remaining 10% being secondary slimes. All or a portion of the primary or secondary slimes may be used to produce stream **12**. It is, of course, possible to use only the primary or only the secondary slimes to make up stream **12** if desired.

Slimes stream **12** is fed to thickener **13** where it is typically thickened for 20 to 50 hours, and preferably 22 to 26 hours, to produce a solids content in the slimes of 3 to 6%. The thickener underflow **14** containing 3 to 6% solids and typically 3 to 4% solids, is then fed to tank **15** where it is mixed with waste tailings **16** from the flotation step. Tailings **16** are dewatered and typically contain about 20% water. The thickened slimes stream **14** is used to repulp tailings **16** in tank **15** to produce an aqueous slurry of a mixture of slimes and tailings **17** which is then pumped out of tank **15** to the waste disposal excavation or other suitable land reclamation site.

The respective quantities of streams **14** and **16** fed to tank **15** will, of course, depend on several factors such as, for example, the ratio of tailings to slimes solids desired in the landfill composition and provision of sufficient water to render stream **17** pumpable to the excavation site. Compositions for landfills preferably contain 30 to 40 parts by weight slimes, 60 to 70 parts tailings and 15 to 25% moisture. Generally, stream **17** will contain a solids content of from 15 to 45% and preferably 40 to 45%.

FIGURE 6.4: REMOVAL OF WATER FROM ORE PROCESSING SLIMES

a.

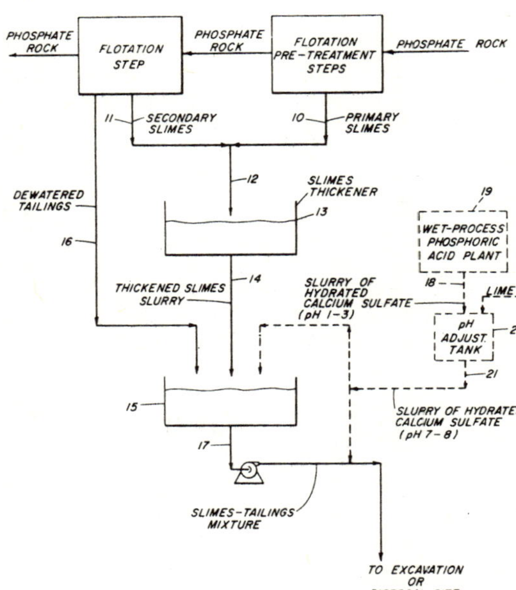

Process Flow Diagram

(continued)

FIGURE 6.4: (continued)

b.

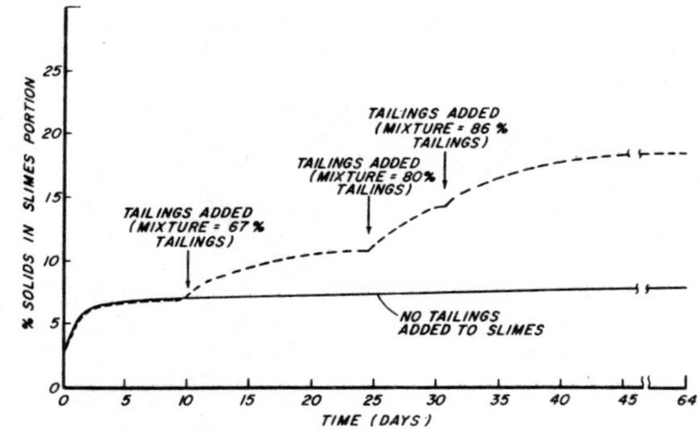

Enhanced Rate of Dewatering of Slimes Upon Addition of Tailings

c.

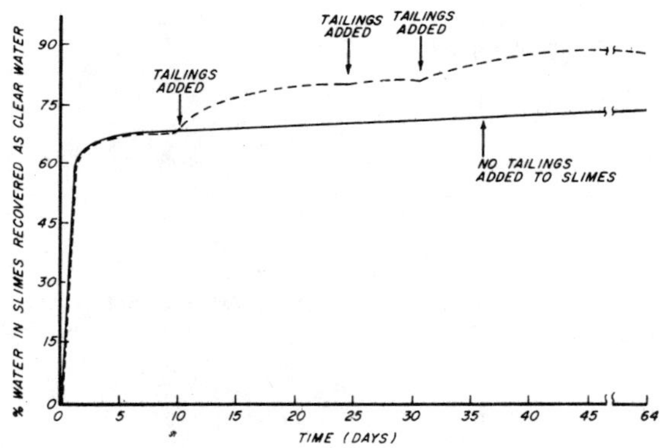

Enhanced Recovery of Clear Water from Slimes Upon Addition of Tailings

Source: C.C. Cook and E.M. Haynsworth; U.S. Patent 3,763,041; October 2, 1973

If it is desired to add hydrated calcium sulfate to the landfill composition, an aqueous slurry of hydrated calcium sulfate **18** from the wet process phosphoric acid plant **19** or from the calcium sulfate ponds of such a plant is fed to a suitable vessel **20** where the acid pH of stream **18** is rendered alkaline by treatment with lime. The alkaline calcium sulfate slurry **21** is thereupon added to tank **15** where, along with the slimes stream **14**, it serves to repulp tailings **16**. Optionally stream **21** can be injected directly into the pipeline which conducts the slimes-tailings mixture **17** to the excavation site where, due to the continuous churning and mixing which occurs in the stream flowing in the pipeline, the calcium sulfate mixes with the slimes-tailings mixture. The quantity of stream **21** required will, of course, depend on how much hydrated calcium sulfate is desired in the landfill composition. Stream **21** can contain a wide range of solids content, usually up to about 20%.

The slimes-tailings mixture **17** is then pumped to an excavation which is typically a ditch a half to 1 mile or longer in length, 40 to 60 feet deep, 30 to 50 feet wide at the base and from 100 to 150 to 200 feet wide at the top. Mixture **17** is introduced at one end of the ditch. Once mixture **17** is deposited in the ditch, a substantial portion of the slimes tend to physically separate from the tailings and flow away from the tailings. These slimes roll on ahead of the tailings accumulating ahead of the tailings where they are then given the opportunity to concentrate to the required solids content of 10 to 25%. Almost as soon as mixture **17** is deposited in the ditch, water is rapidly released from the mixture and proceeds by gravity down the ditch until it eventually reaches a dam at one end of the ditch. This dam will typically have an overflow weir which directs the overflow from the accumulated water into the fresh water return canal of the phosphate rock processing plant.

After those discharged slimes which have accumulated ahead of the tailings have become concentrated to the required 10 to 25% solids level, the discharge pipe for mixture **17** is advanced to discharge mixture **17** onto these concentrated slimes. As the mixture is discharged, most of the slimes discharged again separate from the tailings in the mixture and flow on ahead of the tailings. The discharged tailings, however, settle by gravity into the concentrated slimes and mix with these slimes to form a fertile landfill composition of acceptable bearing strength. In effect, these concentrated slimes remix with the slimes-tailings mixture as the deposited mixture advances along the ditch. The discharged slimes-tailings mixture is gradually advanced along the length of the ditch until the ditch is substantially filled with the landfill composition. Where desired, low check dams can be placed at intervals across the ditch to impede slimes flow and assist in concentrating the slimes solids to the 10 to 25% level where the tailings will mix with the slimes instead of merely displacing them or capping them.

It should also be noted that despite the fact that substantial portions of the slimes flow away from the tailings when the slimes-tailings mixture is deposited in the ditch, the tailings will retain as much as 1 to 3% by weight on a dry basis of the slimes even at the surface of the deposited tailings. This clay provides nutrients for plant growth and also acts as a sponge to retain the moisture necessary for plant growth and erosion inhibition. The following examples illustrate the process.

Example 1: Enhanced water release, increased solids content, marked improvement in the angle of repose and a significant increase in the bearing strength of slimes treated in accordance with the process are demonstrated by the following tests.

Five gallon samples of plant slimes from a Florida phosphate plant, containing 2.15% solids, are poured into 3' x 4'6" wooden troughs and permitted to settle. Percent solids in the slime portion, percent clear water, angle of repose and bearing strength of the solids is determined on a daily basis. Percent solids in the slimes is determined by evaporation of moisture from a weighed sample. This determination is made by heating the sample to 140°C until a constant weight is obtained and calculating the solids content therefrom. Percent clear water is determined by recovering the water from the trough in which the slimes sample is stored.

Angle of repose is the angle of incline of the stacked slimes determined by measuring (1) the distance from the leading edge of the slimes to the point of introduction, (2) measuring the depth of the slimes at the point of slimes introduction and (3) calculating the angle from the leading edge to the top of the stacked slimes.

Bearing strength of the slimes is determined by placing a cylindrical container of known area on the surface of the slimes and loading the container with fine shot. When the container breaks through the surface of the slime the container is removed, weighed and the bearing strength of the sample determined. Specified amounts of dry tailings are added to various slimes samples at spaced time intervals to determine what effect addition of the tailings to the slimes has upon the various properties being studied.

Data obtained are presented in the following tables. A portion of the data is represented graphically in Figures 6.4b and 6.4c. The data clearly show that each mixture of tailings with slimes not only significantly enhances the rate of dewatering of the slimes but also results in the recovery of substantial additional amounts of clear water from the slimes. For example, the data of Figure 6.4b indicate that whereas a dilute slimes (2% solids) had concentrated only to about 8% solids in 60 days, when tailings were added to the dilute slimes, a solids level of about 20% was achieved in somewhat less than 60 days. Similarly, while only about 70% of the water present in the dilute slimes was recovered as clear water after 60 days, 90% was recovered as clear water in the same time period in the case where tailings were added to the dilute slimes sample. This represents almost a 29% increase in the amount of water recovered from the slimes.

A comparison of the bearing strength data of the tables reveals that slimes have substantially no bearing strength whatever whereas slimes-tailings mixtures have appreciable bearing strengths provided that the slime used to prepare the mixture had a solids content of 10% by weight or higher. Results substantially similar to those obtained above were observed when the Florida plant slimes were replaced with Sydney Primary Slimes.

Slimes Settling Tests

Sample	Time (days)	Percent solids of slimes portion	Percent clear water	Angle of repose	Bearing strength of solids (lb./sq. in.)*
	0	2.15	0	1°	.02
	1	5.50	61	1°	.02
	2	6.15	65	1°	.02
	3	6.52	67	1°	.02
	4	6.72	68	1°	.02
	5	6.72	68	1°	.02
Plant slimes......	6	6.95	69	1°	.02
	7	6.95	69	1°	.02
	8	6.95	69	1°	.02
	9	6.95	69	1°	.02
	10	7.17	70	1°	.02
	20	7.40	71	1°	.02
	30	7.70	72	1°	0.02
	64	8.00	73	1°	0.02

*Weight required to break through surface.

Slimes-Tailings Mixture, Settling Tests

Sample	Percent of tailings added*	Time (days)	Percent solids of slimes portion	Percent clear water	Angle of repose	Bearing strength of solids (lbs./sq. in.)**
	0	0	2.15	0	1°	.02
	0	9	6.95	69	1°	.02
	2	10	7.17	70	1°	.02
	....	11	8.30	74	1.25°	.02
Plant slimes..............	....	16	10.2	79	2.25°	.03
	....	17	10.7	80	2.25°	.03
	2	24	11.3	81	4.00°	.15
	2	30	15.4	86	6.00°	.20
	....	64	19.5	89	20.0°	4.5

*Ratio of tailings to slime-solids dry basis.
**Weight required to break through surface.

Example 2: In the following land reclamation test, a cut or ditch, approximately 8,500 feet long and averaging 38 feet in depth and 136 feet in width at the top, was filled with approximately 1,219,038 tons of plant tailings and 87,382 tons of plants slimes in accordance with the process. Across the exit end of the cut there was erected a dam and an overflow weir leading to the fresh water return canal of the phosphate rock processing plant. Tailings from the phosphate processing plant were pulped with return water and pumped to the cut where they were deposited as landfill. Approximately 3,400 feet of the cut was filled in this manner before the process was then put into use. The waste disposal area at this time had good bearing strength but is not fertile.

At this point, the plant slimes thickener was run to produce maximum underflow density. The thickened slimes underflow containing 3.5 to 4.0% solids and typically 3 to 6% solids was used to repulp the dewatered plant tailings and the resulting mixture was then pumped to the disposal cut. The mixture contained a solids content of 15 to 18% by weight at this time. Due to the nature of the mixture the slimes tended to separate and run ahead of the tailings after the mixture was discharged into the cut. However, analysis of the soil samples showed that at least about 1 to 3% of slimes (dry weight) are trapped within the tailings despite separation of most of the slimes. This improves fertility and compaction as the clay contained in the slimes retains a substantial amount of moisture.

Since some initial separation of the slimes was evident small check dams were placed across the cut at intervals along its length. The dams impeded the advance of the densifying slimes permitting the clear water to proceed down the cut. When the separated slimes thickened to about 10 to 15% solids, mixing of these thickened slimes with the now advancing slimes-tailings mixture occurred. Such mixing improved the release of water from the slimes. This is evidenced by the fact that by utilizing the process, some 71 million gallons of water were recovered that would not have been recovered by using conventional tailings and slimes disposal systems.

The 1,219,038 tons of tailings processed would have required 560 acre feet (at 100 lb/ft^3) and the 87,382 tons of slimes treated would have required 252 acre feet, or a total disposal area of 812 acre feet (at 22.5% solids, 1.14 SG). The finished fill in this test actually required 595 acre feet or 217 acre feet less than would conventionally be employed. Thus, an additional 217 acre feet of water over and above that which would be recovered by conventional procedures were recovered by this process. This is equivalent to 70,709,667 gallons of additional water released and recovered.

After the cut was completely filled to ground level, six borings were made. The results are reported below in the following tables as percent slimes (i.e., –150 mesh material), and bearing strength. The results indicate that the landfill possesses good bearing strength. The entire site has been seeded with Bermuda grass and good growth is evident. The landfill has good bearing strength and supported a variety of heavy pieces of equipment such as, for example, D-8 tractors, within very short periods after the filling of the cut was completed.

Boring 1

Depth	% Passing 150 Mesh	Average Bearing Strength (tons/sq ft)
0.0 - 3.0	4.7	0.75
3.0 - 6.0	2.5	3.35
6.0 - 9.0	2.4	2.21
9.0 - 12.0	1.5	1.72
12.0 - 15.0	2.4	2.38
15.0 - 18.0	2.1	1.08
18.0 - 21.0	4.6	0.91
21.0 - 25.5	7.9	1.40
25.5 - 27.0	7.0	1.24
27.0 - 30.0	8.1	1.75
30.0 - 33.5	14.9	0.34
33.5 - 35.0	16.9	2.13

Boring 2

Depth	% Passing 150 Mesh	Average Bearing Strength (tons/sq ft)
0.0 - 3.0	2.2	0.75
3.0 - 6.0	3.0	1.40
6.0 - 9.0	4.1	0.26
9.0 - 12.0	3.9	0.59
12.0 - 15.0	22.3	0.42
15.0 - 17.0	3.2	1.08
17.0 - 18.5	12.2	0.26
18.5 - 22.0	7.4	0.26
22.0 - 25.0	11.8	0.26
25.0 - 28.0	21.0	0.51
28.0 - 31.0	17.9	1.08
31.0 - 34.0	24.9	1.32

Boring 3

Depth	% Passing 150 Mesh	Average Bearing Strength (tons/sq ft)
0.0 - 3.0	5.0	2.54
3.0 - 6.0	2.5	3.43
6.0 - 9.0	1.3	1.73
9.0 - 12.0	1.7	0.43
12.0 - 15.0	2.0	0.51
15.0 - 16.5	6.6	0.42
16.5 - 18.0	3.9	0.75
18.0 - 21.0	8.2	0.43
21.0 - 24.5	8.6	0.59
24.5 - 28.0	9.1	0.70
28.0 - 32.0	33.2	0.10
32.0 - 35.5	31.6	1.24
35.5 - 39.1	18.9	1.56

Boring 4

Depth	% Passing 150 Mesh	Average Bearing Strength (tons/sq ft)
0.0 - 3.0	4.1	2.78
3.0 - 6.0	4.0	5.30
6.0 - 9.0	1.5	2.30
9.0 - 12.0	1.9	1.24
12.0 - 15.5	3.2	0.50
15.5 - 19.0	5.3	0.75
19.0 - 22.5	3.9	0.51
22.5 - 25.5	19.0	1.42
25.5 - 28.5	10.0	3.43
28.5 - 31.5	9.3	3.05
32.0 - 33.5	13.1	3.02
33.5 - 36.4	6.1	4.81

Boring 5

Depth	% Passing 150 Mesh	Average Bearing Strength (tons/sq ft)
0.0 - 3.0	2.3	2.13
3.0 - 6.0	2.0	2.38
6.0 - 9.0	3.2	2.21
9.0 - 12.0	4.7	1.00
12.0 - 15.0	3.0	0.42
15.0 - 18.5	9.9	0.26
18.5 - 21.5	9.2	0.26
21.5 - 25.5	12.3	0.42
25.5 - 29.0	34.0	1.77
29.0 - 32.0	7.1	1.00
32.0 - 36.5	18.9	2.30

Boring 6

Depth	% Passing 150 Mesh	Average Bearing Strength (tons/sq ft)
0.0 - 3.0	6.0	1.89
3.0 - 6.0	4.1	3.52
6.0 - 9.0	7.2	3.53
9.0 - 12.0	6.1	3.27
12.0 - 15.0	9.5	3.51
15.0 - 18.0	4.7	1.56
18.0 - 22.0	2.9	0.67
22.0 - 25.5	21.5	1.97
25.5 - 28.5	6.0	1.24
28.5 - 32.0	23.1	1.07

Polyacrylamide Sulfate as Flocculant

A process described by *T. Liu and D.R. Stern; U.S. Patent 3,680,698; August 1, 1972;*

assigned to Occidental Petroleum Corporation relates to the settling and thickening of solid matter contained in a liquid, such as slimes, wastes, pulps and the like for recovery or disposal of the solid matter.

Slimes, for example, have been to the mining industry an aqueous suspension of fine particles that are difficult to dewater. Slimes typical to the phosphate industry appear as a waste from a beneficiation plant at a solids concentration of from 0.5 to 5%. These slimes generally contain particles finer than about 150 mesh and are often as rich as the original mined matrix. They are uneconomical to process with present technology and as a result, about a quarter to a third of the ore values are lost.

Normally about 1,000 to 2,500 lb are produced per ton of finished ore processed. Since water has been added, slime volume exceeds the volume of the matrix mined. Consequently, the volume required to store the slimes is greater than the mined out area. To contain the slimes, earthen dikes are built around a mined out area for long term storage.

Slimes have been accumulated almost since the inception of the phosphate industry, some 50 years ago. They have remained in suspension at a solids concentration usually less than 30%. The land has not been recoverable, remains blighted and blighted areas are constantly growing in size and number. The earthen dikes constructed to contain the slimes are costly and difficult to maintain, often breaking and causing serious stream pollution problems. Similar problems exist in other mining industries, such as the potash industry where slimes produced are also difficult to settle and practically impossible to filter.

It has been found that solids suspended in a liquid, such as slimes and wastes, can be more effectively compacted to a handleable state for recovery or disposal of the solid matter by the steps of adding to the liquid containing suspended fine solids, a directly prepared liquid coagulant reaction mass with agitation, preferably high shear agitation, in at least one stage to flocculate the solids; settling the flocculated solids, with slow agitation, to form at least a partially compacted solids mass and a clarified liquid; then separating the clarified liquid from the partially compacted solids.

Where desired, additional liquid may be extracted from the partially compacted solids by slowly agitating the partially compacted solids in contact with a porous filter, preferably sand. Agitation of the partially compacted solids while in contact with the porous filter materially improves the rate of flow of liquid through the porous filter and can promote the formation of a clarified overflow for removal by decantation.

The addition of sand to the solids which have been compacted to a solids concentration of at least 10% by weight, results in the formation of a fairly homogeneous mixture of solids and sand bound to some degree by the coagulant. This mixture is highly porous and rapidly releases retained liquid. The resultant dewatered mixture is also porous which allows land reclamation without at least the initial formation of impervious clay strata.

It has also been found that gypsum may be added with the coagulant reaction mass to promote flocculation and that aging a fresh liquid containing suspended solids will reduce the amount of coagulant required to settle the solids. Economy is also favored by the addition of the liquid coagulant reaction mass in several stages.

The liquid coagulant reaction mass may be obtained in a number of ways. Conveniently, it may be obtained by the polymerization of monomers of a polymeric coagulant in the presence of a suitable catalyst in a liquid compatible with the suspension to be treated. An example of such polymeric coagulants are the polyelectrolytes which may be prepared by the direct vinyl polymerization in an aqueous media of a polymerizable amount of at least one unsaturated monomer in the presence of a suitable polymerization catalyst therefore.

Illustrative of the monomers, which serve as basic building blocks for the synthesis of polymeric coagulants in an aqueous media there may be mentioned acrylamide and its hydrolysis derivatives; acrylic acid; methacrylamide and its hydrolysis derivatives; methacrylic acid; ethyleneimine; 2-ethylethyleneimine and propyleneimine. Vinyl carboxamide sulfate polymers are particularly preferred because of their high effectiveness and low cost of preparation. The following examples illustrate the process.

Example 1: To determine the effect of using a directly synthesized polyelectrolyte reaction mass as compared to commercial solid coagulants a phosphate slime containing about 8% solids was observed for its settling properties without the use of a flocculant and with the addition of 0.20% by weight of a commercial polyacrylamide coagulant, based on the total weight of the solids and an equivalent amount of an aqueous polymerized acrylamide sulfate reaction mass. Three samples were prepared and the volumes occupied by the settled solids observed at periods of time. The results are shown in the following table.

	Vol of Suspension (ml) at Time (hr)						Final Settled Solids (wt %)	
Flocculant	0	1/6	1/2	1	3	24	Natural	Centrifuged
No flocculant	100	96	87	81	69	51	8.9	18.8
Solid acrylamide polymer	100	44	42	42	41	41	10.8	25.0
Aqueous polyacrylamide sulfate	100	36	35	35	35	-	12.5	40.0

It was observed that compaction using the aqueous polyacrylamide sulfate reaction mass was greater and more rapid and that in the centrifugation step substantially greater than the unconditioned slime and the slime conditioned with the commercial polyacrylamide.

Example 2: Using a polyelectrolyte obtained by polymerizing a monomeric system containing predominantly acrylamide sulfate, using a potassium bisulfate polymerization catalyst, stagewise flocculation was studied. Three samples were prepared each containing about 8% by weight phosphate slime solids. One of the samples served as the control and flocculant was added in differing amounts to each of the remaining samples A and B with slow agitation. The degree of concentration of the solids was observed at the end of one-half after the first addition of flocculant, one-half hour after the second addition of flocculant and 20 hours after the third addition of flocculant. The control was compared after 24 hours standing. The results are shown in the following table.

Stagewise Addition of Flocculant (Slow Agitation)

		Solids Concentration (wt %)		
Amount Flocculant, ppm/part solids	Time, hr	Sample A	Sample B	Control
0	24.0	-	-	8.9
160	0.5	14.8	15.7	
320	0.5	18.8	20.3	
500	20.0	25.3	25.3	

Example 3: To illustrate the effect of sand dewatering, a polymeric acrylamide sulfate, prepared in accordance with Example 2, was added to a phosphate slime containing approximately 5% solids with slow agitation in an amount equivalent to 50 ppm parts solids. The mixture was then added to a bed containing fine sand and allowed to dewater. The results are shown in the following table where it was observed that a ninefold concentration in 2 days is a most unexpected and rapid sedimentation.

Aqueous Polyelectrolyte System	Feed Conc (wt %)	Final Conc (wt %)	Time (days)
Polyacrylamide sulfate	5	45	2

Example 4: Two agitated thickeners provided with slowly rotating agitators (0.5 rpm) were used for a large scale study of the settling behavior of slimes. The bottom third of one thickener was filled with graded sand for final dewatering, while the other was used for the initial thickening. A run was made using a situs prepared acrylamide sulfate flocculant in a total amount equivalent to 0.5 lb/ton slime solids. This run was made using 39,375 volumes of a phosphate slime of 8% solids content. In the first step the feed was mixed with 200 volumes of a 0.1% solution of the flocculant and left overnight to settle.

A total of 23,000 volumes of supernatant liquid was collected, leaving an underflow containing 19% solids. Two additional 200 volume portions of 0.1% flocculant were added to the slime in the thickener. Within 24 hours, an overflow of 3,500 volumes was collected. The thickened slime, containing over 25% solids, was then mixed with 150 volumes, of 0.1% flocculant and transferred to the second thickener which was lined with a sand bottom. After 16 hours, the solids content reached 38.6% and after a 24-hour duration in excess of 40%.

Modified Nucleoproteins as Flocculants

M.H. Yueh; U.S. Patent 3,681,283; August 1, 1972; assigned to General Mills, Inc. describes a flocculating agent including nucleoprotein material which is treated with an organic or inorganic salt. Typical examples of waste materials that may be treated according to the process include various suspended inorganic and organic wastes such as those produced by industrial plants, i.e., phosphatic slimes, silica, taconite tailings, soybean wastes, sewage sludge, clay and soil.

The nucleoprotein containing material is treated by dispersing in an aqueous medium along with water-soluble organic or inorganic salts, especially soaps and anionic detergents. The salts appear to at least partially solubilize the otherwise insoluble nucleoprotein materials. In so doing, it is theorized that the deoxyribonucleoprotein unwinds and swells, thus generally forming a viscous sol.

The nucleoprotein material used in the process may be obtained from any substance having a substantial amount of nucleoprotein present, preferably avian blood, microbial material or fish milt. Other sources would include wheat germ and various organ tissue such as thymus, spleen, pancreas and liver obtained from calves or testes obtained from bulls. For purposes of economics and ease in preparation, the substance preferably contains at least 0.5% nucleoprotein, by weight. As used here, the term nucleoprotein material and nucleoprotein containing material will mean a material containing at least 0.5% nucleoprotein.

The salts are added in an amount sufficient to treat the amount of nucleoprotein present. This may be an amount sufficient to form a viscous sol. The amount of salts needed varies somewhat depending on the particular salt or salts employed. In this regard, as little as 0.04% by weight of sodium dodecyl sulfate has been found to form a viscous sol from a 0.1% by weight dispersion of nucleoprotein material, the weight percentages being based on the total weight of the aqueous phase. In contrast approximately 6.0% by weight of sodium chloride is needed to form a viscous sol from the same milt solids dispersion. The nucleoprotein material will normally be treated with salt in an amount of at least 10% by weight based on the weight of the nucleoprotein material, generally, at least 40% or more. Of course, excess of the salt or salts can be employed.

If the sol or treated nucleoprotein material is added in only small amounts to the waste containing aqueous medium, flocculation will take place without further treatment. However, if large amounts of sol or nucleoprotein material are added to the aqueous medium, the aqueous medium may become a stable viscous sol and flocculation will not take place. Acidification in this case will result in good flocculation. The pH is preferably about pH 6 or lower, although flocculation in many instances will take place even at a pH above pH 6. The acidification may be produced by adding any acid, typically, hydrochloric acid, sulfuric acid, phosphoric acid and the like, to the aqueous medium.

The acid may be added to the nucleoprotein sol prior to adding the sol to the waste containing aqueous medium. The acid may be added to the waste containing aqueous medium prior to adding the sol to the aqueous medium. Alternatively, the acid may be added to the mixture of the nucleoprotein sol and the aqueous medium. In each instance the acidification improves the flocculation. The following examples illustrate the process.

Example 1: An industrial waste from a soybean processing plant was collected. The waste contained 11.0% total solids and appeared to be a stable suspension. To 100 ml of the waste was added an equal volume of a viscous sol prepared from chicken blood erythrocyte material and sodium dodecyl sulfate. The erythrocyte material was prepared by treating chicken blood with heparin (0.2% by weight to prevent blood coagulation). After centrifugation at 40°F and 2,000 rpm for 20 minutes, 102.4 ml of the blood was separated into the supernatant (plasma) and a precipitate of 49.0 ml (mainly erthrocytes). The precipitate, composed of 69.8% moisture and 30.4% solids, was dispersed in five volumes of ethanol and stirred continuously at room temperature for 30 minutes.

After filtration, air drying and grinding, a dark powdery product was obtained. The viscous sol was prepared from 1,000 mg of the dark powdery product, 93 ml of water and 7 ml of 10% by weight aqueous sodium dodecyl sulfate. After mixing the sol with the waste, about 1.5 ml of 6N HCl was added dropwise to the resulting mixture. The precipitate or flocculate that formed was readily removed by filtration. Analysis of the filtrate showed that over 84% of the suspended solids had been removed.

Example 2: Industrial waste of the type described in Example 1 was treated with a viscous sol prepared from fish milt. Frozen, fresh salmon milt was thawed and blended in a Waring Blendor. The blended milt was dispersed in 10 volumes of 95% ethanol and brought to boiling. The solids were filtered and washed twice with 10 volumes of 95% ethanol, filtered, air dried and ground. The product was creamy white. A viscous sol was prepared by dispersing 100 mg of the product in 93 ml of water. Then 7 ml of a 10% by weight aqueous solution of sodium dodecyl sulfate was added with stirring. To 100 ml of the waste was added an equal volume of the viscous sol. After mixing, 6N HCl was added dropwise to the resulting sol with stirring. The flocculate that formed was readily removed by filtration. Analysis of the filtrate showed that 91.8% of the waste solids had been removed.

Example 3: Approximately 500 grams of clay containing soil was suspended in 3,500 ml of tap water. After standing at room temperature for 4 hours, the muddy supernatant was collected by decantation. To 100 ml of the supernatant was added 400 mg of freshly ground salmon milt and 7 ml of a 10% by weight aqueous solution of sodium dodecyl sulfate. Stirring was continued for a few minutes until a thick, viscous sol was obtained. Approximately 0.25 ml of 6N HCl was added in a dropwise manner with stirring. The addition of the acid brought about a complete loss in viscosity and produced floc formation. The suspended clay particles were visibly enmeshed by the floc. The floc was readily removed by filtration.

Example 4: Frozen, fresh salmon milt was thawed and blended in a Waring Blendor. The milt was then dispersed in 10 volumes of 95% ethanol and brought to boiling. The solids were filtered and washed twice with 10 volumes of 95% ethanol, filtered, air dried and ground. An aqueous dispersion containing 0.1% by weight of the dried milt product was prepared. 6% by weight NaCl was added. A thick viscous sol was obtained which was suitable for waste treatment according to the process.

Example 5: The process was used to remove fine taconite tailings which were suspended in water. The taconite suspension was waste obtained from Reserve Mining Co. of Silver Bay, Wis. The waste contained 1 to 2% suspended solid matter. A sol was prepared which contained, by weight, 0.7% of the creamy white product of Example 2, 0.7% sodium dodecyl sulfate and 99.2% water. The taconite suspension was divided into 10 samples. Sample 5-A was 10 ml of the suspension. Samples 5-B through 5-I were 10, 20, 30, 40, 50, 60, 70, 80 and 100 ml respectively of the suspension. Sample 5-J was the control suspension and was not treated. Samples 5-A through 5-I each received 1 ml of the sol and after mixing 1 drop of 6N hydrochloric acid was added. The nucleoprotein material was present in Sample 5-I at a level of about 10 parts per million. Flocculation was observed in Samples A through I, and after standing the floc settled out leaving the samples clear. The control, Sample J, remained a cloudy suspension.

Recycle Technique

A process described by *P.D. Kostenbader; U.S. Patent 3,738,932; June 12, 1973; assigned to Bethlehem Steel Corporation* relates to a method for treating acid water, such as, AMD (acid mine drainage) and diluted waste pickle liquor, which contain metallic values, such as ferrous and ferric iron and compounds of aluminum, calcium, magnesium, zinc, manganese, silicon and the like which compounds can be sulfates and/or chlorides, with a high calcium lime-water slurry mixed with a portion of the sludge formed in the method. The acidity of the water is neutralized and a substantial portion, if not all, of the metallic values are precipitated as solids in a reactor and settle out as a densified sludge in a separating container. In the method, a specified portion of the sludge is recycled and mixed with high calcium lime-water slurry for specified reaction times and within a specified pH range.

The process is diagrammatically shown in Figure 6.5. An alkali, for example, high calcium lime, magnesia, dolomitic lime and sodium hydroxide and the like, is mixed with water to form a slurry. High calcium limestone is a general term for stone containing calcium carbonate and little (2 to 4%) magnesium carbonate. The high calcium lime-water slurry, which for practical considerations can be 1 to 10% high calcium lime, flows to a first reactor. A portion of the sludge settled in the final step of the method, is passed to the first reactor.

It has been found that it is necessary to recycle a sludge containing not less than about 20 lb of solids for each pound of solids precipitated from the acid water. It is preferred to recycle a sludge containing about 20 lb to about 30 lb of solids for each pound of solids precipitated. The high calcium lime-water slurry and recycled sludge are mixed for a time, at least 1 second, to achieve the results of the process. The chemical reactions which occur between the high calcium lime-water slurry and recycled sludge in the first reactor are very rapid.

FIGURE 6.5: TREATMENT OF ACID MINE DRAINAGE

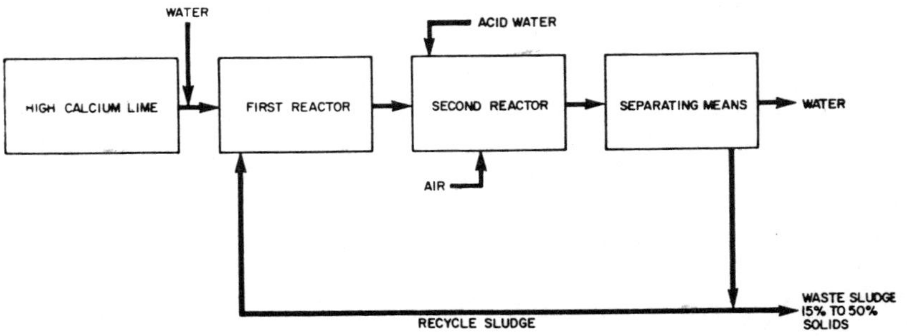

Source: P.D. Kostenbader; U.S. Patent 3,738,932; June 12, 1973

The high calcium lime-water slurry and recycled sludge mix flows to a second reactor. The acid water also flows into the second reactor. The high calcium lime-water slurry and recycled sludge mix react (in a neutralization-oxidation step) with the acid water in the presence of air, which is bubbled through the reacting mixes. The acidity of the water is neutralized, that is, the pH value is modified to not less than 7.0. The oxygen in the air oxidizes substantially all the ferrous iron to insoluble hydrated ferric oxide. The compounds of aluminum, calcium, magnesium, silicon and zinc are precipitated as insoluble hydroxides and sulfates. The time in the second reactor can be as little as 5 minutes and as long as necessary to obtain substantially complete oxidation and precipitation of metallic values as insoluble solids. The acidity of the water is neutralized, that is, the water has a pH of not less than 7.0 and may be as much as 9.0.

The high calcium lime-water slurry-recycled sludge-acid water mixture flows to a separating means, for example, a thickener or a settling pond. The precipitated solids settle out to form a sludge containing about 15% to about 50% solids. As noted previously, a portion of the sludge is recycled to the first reactor. The remaining sludge is disposed of as a waste product. The clarified neutralized water can be recycled in the system or can be disposed of in environmental surface water. In a specific example, AMD having a pH of 2.8 to 3.1 and having the following composition (in parts per million):

Fe^{+2}	225 to 350	CaO	250
Fe^{T}	300 to 350	MgO	50
Al_2O_3	200	SO_4	2,100

was treated for a period of about 8 hours according to the method. A throughput of 100 milliliters per minute was maintained in the apparatus. About 10 grams of hydrated lime (made from high calcium lime) was mixed with a liter of distilled water to make a lime slurry. An average of about 55 ml/min of recycled sludge containing 15% solids were mixed with the lime slurry. The retention time was varied from 0.1 to 20 minutes in the first reactor. The mixture flowed to the neutralization-oxidation step or second reactor. The retention time of the AMD and mixture in the second reactor was varied between 10 and 20 minutes.

The pH of the solution was within the range of 7.1 to 8.0. The AMD and mixture flowed to a thickener. About 1,000 ppm of solids were precipitated from the AMD. Settled sludge separated from the treated water contained about 50% solids. Sludge in excess of that required for recycle was drained from the bottom of the thickener and was discarded. The clarified neutralized water from the thickener had a pH within the range of 7.1 to 8.0 and was substantially free of dissolved iron. In another specific example, AMD having a pH of 2.8 to 2.9 and having the following composition (in parts per million):

Fe^{+2}	60 to 120	MgO	10 to 120
Fe^{T}	200 to 320	Mn	7 to 9
Al_2O_3	170 to 330	SiO_2	40 to 70
CaO	230 to 270	SO_4	1,900 to 2,200

was treated for 1 week according to the method. A throughput of 3 gal/min was maintained in the apparatus. From 0.1 to 1.0 lb of high calcium lime was mixed with 9 to 10 lb of water to make the high calcium lime slurry. About 1.5 gal/min of recycled sludge containing about 6.5% solids were mixed with the lime slurry in a retention time of about 20 minutes in the first reactor. The mixture was pumped to the neutralization-oxidation step or second reactor. The retention time of the AMD and mixture in the second reactor was 10 to 12 minutes.

The pH of the solution was within the range of 7.0 to 7.8. The AMD and mixture flowed to a thickener. About 1,054 ppm of sludge was precipitated from the AMD. The settled sludge separated from the water contained 15% solids. Sludge, in excess of that required for recycle, was drained from the bottom of the thickener and was discarded. The clarified neutralized water that overflowed from the thickener had a pH of 7.0 to 7.8 and had the following dissolved solids composition (in parts per million):

Fe^{T}	<1	Mn	8 (max)
Al_2O_3	2 to 8	SiO_2	10 (max)
CaO	850 to 1,040	SO_4	1,880 to 2,440
MgO	50 to 170		

Coke Layer as Coalescer

According to a process described by *K. Grob, J. Oertter and P. Wiesner; U.S. Patent 3,687,844; August 29, 1972; assigned to Metallgesellschaft AG, Germany* the aqueous refining phase from a solvent extraction step of a mineral wet processing process is purified of extraction agent residues by passing the aqueous refining phase through a granulated layer of hydrophobic material having acid surface action into a sedimentation chamber and withdrawing the thus purified refining phase from below. It has been found that the surface structure of a coalescer is decisive for the suitability of a filling material and that this surface must have a fissured sharp edge structure and furthermore must be hydrophobic and have a certain acidity. A material having these properties, cokes, resulting from the decomposition distillation of coal, tar, pitch or mineral oil residues, has proved suitable in several cases in which an oily dispersion phase is to be separated from an aqueous continuous phase.

The coke particles preferably have a granulation diameter of 2 to 20 mm. They are arranged in a shaft-like container as short and tall columns for comparison. The refining phase to be purified is passed upward through the columns and the surface load is 10 to 30 cubic meters of refining phase per square meter cross section of flow per hour.

The coke particles can be obtained, for example, by screening from the sifted crushed oven coke of a metallurgical coking plant. Such coking gives the coke particles an acid surface action. The coke particles can, however, also be produced by briquetting or pelletizing of fine coke or fine coal with bituminous binding agents such as tar, pitch or the like and coking of the briquettes or pellets. The coking temperature should not be much more than 800°C so as to avoid a graphitization of the particle surfaces which would destroy the acid surface function resulting from the coking. Suitable initial substances for the production of such coke particles are coal, lignite, peat or coke obtained therefrom, earth oil bitumen, petroleum coke and the like.

For filling the shaft-like container use is preferably made of particles of a small range of granulation so as to keep the flow resistance of the coke layer as small as possible. In such a coke layer, the direction of flow of the liquid is changed often and suddenly so that, more particularly with high velocities within the range of turbulent flow, there will be a good contact between the liquid and the coke particles. The weakly acid and hydrophobic surface with only a very small affinity to the acid aqueous solution, moistens itself with the droplets of the solvent dispersed in the liquid so that they agglomerate to large drops on the coke surface and can form a discretely coherent phase.

The refining phase flowing from the shaft-like container above the coke layer is led to a clarification container in which, for example, by means of a baffle weir, a runoff zone is delimited, from which the liquid rising from the bottom of the container can be siphoned off.

The clarification container may be united with the shaft-like container into a single constructional unit, for example, in such a way that an overflow trough surrounding the shaft at the upper end is subdivided in a suitable manner and is provided with diversion means for the purified refining product and the separated solvent. Since the clarification container often has to serve as an intermediate store, for example, when the ore leaching is carried out intermittently, use is made of clarification basins which may be at some distance from the shaft-like container of the coke layer. In these clarification basins there is in the intervals of time during which no purified material is withdrawn, a very small flow which considerably promotes the separation of enlarged drops of solvent.

Example 1: So as to be able to ascertain and compare the efficiency of different granulated materials for the purification of the acid solvent containing refined product from solutions of ore leaching, as test apparatus, a double-walled heatable glass tube with an inner diameter of 50 mm and a length of 1,000 mm was placed vertically and filled with the granulated material to be tested up to a height of layer of 800 mm.

As liquid to be treated, in all tests served refined material, of a uranium extraction, still containing residues of the solvent, prepurified in known manner in coalescers and separators. The solvent was a mixture of 93% by weight of kerosene, 2% by weight of isodecanol and 5% by weight of a tertiary amine, each with 8 C atoms in the alkyl groups. The degree of purity of the refined material treated according to the process was ascertained by measuring the base content remaining therein.

This liquid to be treated was passed through the layer of granulated material with a velocity of 30 l/hr which corresponds to a surface load of 5 m^3/m^2 cross section of flow per hour. The liquid treated was collected in high vessel and was left for about 10 minutes for clarification. Granular fillings of different granulation of gravel or ceramic material proved to be ineffective. The amine content of the liquid treated in this manner was unchanged as compared with the nontreated liquid.

Screened oven coke of a granulation filling of 10 to 15 mm had an amine content of 35 parts per million in the nontreated liquid reduced to 3 ppm in the treated liquid. The comparison tests were carried out at a temperature of 35°C. With the coke filling a temperature increase to 50°C resulted in an acceleration of the separation, which allowed an increase of the surface load to 12 $m^3/m^2/hr$.

Example 2: A large scale test with a coke layer of a height of 1 meter and a diameter of 1 meter was carried out with a surface load which varied between 20 and 30 $m^3/m^2/hr$ at a temperature of 35°C in the liquid to be treated. The coke particles had a granulation of between 5 and 10 mm. Here also, the amine content dropped from 35 ppm in the nontreated solution to 3 ppm in the treated solution.

Recovery of Vanadium and Chromium from Titanium Tetrachloride Residue

R.F. Mitchell and G.A. Carlson; U.S. Patent 3,655,344; April 11, 1972; assigned to PPG Industries, Inc. have discovered that the vanadium and chromium values present in titanium tetrachloride drier residue can be insolubilized by water leaching followed by raising the pH of the resulting leach solution to at least 4, but less than 12. Preferably, the pH of the leach solution is raised to from 6 to 8 and thereafter maintained at at least 4 but less than 12.

Titanium tetrachloride is commonly produced by reacting, at high temperatures, a titanium-bearing material, such as ilmenite, rutile or titanium-containing slag or like titaniferous material, with chlorine and a reducing agent such as carbon or other carbonaceous material in a static bed or fluosolids operation. The vaporous titanium tetrachloride removed with the reaction product mixture is condensed to form a crude titanium tetrachloride condensate. Since there are a number of elements in the raw material chlorinated, in addition to titanium, which react with chlorine to form volatile chlorides and which carry over with the titanium tetrachloride vapors, the titanium tetrachloride condensate, sometimes referred to as crude titanium tetrachloride, is impure. Some of the aforementioned volatile chlorides, other than titanium tetrachloride, found in the crude condensate include, for example, the chlorides of silicon, aluminum, niobium, tungsten, iron, manganese, magnesium, calcium, zirconium, vanadium and chromium.

One such chloride, which because of the proximity of its boiling point to titanium tetrachloride cannot be separated by fractional distillation, is vanadium chloride. The vanadium can be in the form of vanadium tetrachloride and/or vanadium oxychloride. Vanadium chloride is present in substantially all crude titanium tetrachloride condensate and comprises a major portion of the impurities in the condensate and, hence, is chiefly responsible for the yellowish color of crude titanium tetrachloride. By way of example, a typical crude titanium tetrachloride can comprise from 0.10 to 0.25% vanadium, based on the weight of the titanium tetrachloride.

In addition to the products of the chlorination reaction, solids including ore and carbonaceous reductant, are also carried over into the primary condenser. Following condensation, these solids are separated from the condensate by, for example, vaporization or settling, and these solids removed to sludge storage. The following examples illustrate the process.

Example 1: 50 gram portions of a titanium tetrachloride drier residue composite containing 8.5 weight percent vanadium and 1.7 weight percent chromium were leached with various amounts of tap water for 15 to 30 minutes and the resulting slurries filtered. About 75% of the residue was water-soluble. Analyses of the dark green filtrates for vanadium and chromium were performed. The results are tabulated in Table 1.

TABLE 1: WATER LEACHING OF DRIER RESIDUE

Run #	Water Leach (ml)	Time (min)	Percent Water-Soluble V	Percent Water-Soluble Cr
1	200	15	90.7	91.6
2	200	30	90.0	92.5
3	300	30	95.0	95.1
4	400	30	95.1	95.2
5	500	30	96.1	95.7
6	500	30	95.9	94.9
7	750	30	97.5	97.0
8	400*	30	97.0	96.5

*Filter cake washed with 350 ml additional H_2O to give a total filtrate of 750 ml.

Portions of the dark green filtrate from Runs 4, 5 and 7 in Table 1 were neutralized with 15 molar sodium hydroxide or 15 molar ammonium hydroxide which were added dropwise until the pH of the filtrate was raised from <1 to between 7 and 8. In each case, a colorless filtrate and voluminous, gelatinous residue were obtained after vacuum filtration. Results of the analyses of the filtrates for vanadium and chromium following vacuum filtration are tabulated in Table 2.

TABLE 2: BASIC TREATMENT AND FILTRATION OF LEACH SOLUTION FROM DRIER RESIDUE WATER LEACHING

Run # (from Table 1)	Original Leach (ml) H_2O	Base Added	Filtrate (ppm) V	Filtrate (ppm) Cr
4	400	NH_4OH	<1	<0.6
4	400	NaOH	<1	<0.6
5	500	NH_4OH	<1	<1
5	500	NaOH	<1	<1
7	750	NH_4OH	<1	<1
7	750	NaOH	<1	<1

The data of Table 1 show that both the vanadium and chromium values contained in the titanium tetrachloride drier residue are readily soluble regardless of whether 300 or 750 milliliters of tap water are used. The basic treatment of the aqueous slurry of titanium tetrachloride drier residue almost completely precipitates (greater than 99.9%) the vanadium and chromium values present in the liquid phase of the slurry (Table 2).

Example 2: 50 gram portions of the titanium tetrachloride drier residue composite of Example 1 were stirred in 500 ml portions of tap water. While monitoring the stirred slurry on a pH meter, various amounts of about 19 molar sodium hydroxide, 15 molar ammonium hydroxide or cell liquor were added to different pH levels. Each mixture was stirred for 30 minutes and filtered. The filtrates were analyzed for vanadium and chromium. The results of such analyses are summarized in Table 3.

TABLE 3: FILTRATE ANALYSES AT VARIOUS pH LEVELS

Base Used Type	Base Used Ml	pH Slurry	pH Filtrate	Filtrate Analysis (ppm) V	Filtrate Analysis (ppm) Cr
15 M NH_4OH	43	7.6	-	<0.5	<0.1
15 M NH_4OH	125	11.0	10.5	<0.5	<0.1
19 M NaOH	34.7	8	6.3	<0.5	0.3
19 M NaOH	36.3	10	9	<0.5	0.1
19 M NaOH	37.9	12	11.3	0.6	<0.1
19 M NaOH	41.5	13.5	12.9	5.9	0.1
19 M NaOH	73	14	14	540	0.1

(continued)

TABLE 3: (continued)

Base Used: Type	Base Used: Ml	pH: Slurry	pH: Filtrate	Filtrate Analysis (ppm): V	Filtrate Analysis (ppm): Cr
19 M NaOH	96	>14	>14	1,250	<1.0
Cell liquor*	170	7.5	5.8	0.5	1.0
Cell liquor*	419	14	14	673	0.9

*About 12 weight percent NaOH.

The data of Table 3 show that excess alkaline agent should be avoided since at pH values greater than 12 vanadium, and to a much lesser degree chromium, begins to redissolve. However, so long as the slurry pH was maintained equal to or less than 12 and above 4, the amount of vanadium and chromium in solution were each less than 0.6 ppm. Caustic treatment to a pH of 14 results in the solubilization of about 5% of the vanadium. Further addition of caustic caused further solubilization of the vanadium.

Example 3: 43 ml of either 15 molar sodium hydroxide or 15 molar ammonium hydroxide were mixed with either 500 or 750 ml of tap water and added to 50 gram portions of the titanium tetrachloride drier residue composite of Example 1. Each slurry, which had a resulting pH of from 7 to 8, was stirred for 30 minutes and filtered. In each case, the filtrate contained less than 1 ppm vanadium and chromium. This example shows that a one-step treatment by adding the required alkaline agent and tap water to the drier residue is as effective in rendering the vanadium and chromium values present in the residue insoluble as did the procedure of water leaching followed by basic precipitation illustrated in Example 1.

Example 4: The filter cakes of Example 3 were reslurried in 500 ml of water and acidified with concentrated hydrochloric acid to the pH levels shown in Table 4. The slurries were filtered and the filtrates analyzed for vanadium and chromium. The results of such analyses are summarized in Table 4.

TABLE 4: ACIDIFICATION OF BASIC-TREATED DRIER RESIDUE SLURRIES

Base	Slurry, pH	Filtrate, pH	Filtrate Analysis (ppm): V	Filtrate Analysis (ppm): Cr
NaOH	6.9	5.4	0.2	0.2
	4	4	131	9.3
NH_4OH	4.4	4.2	0.3	0.5
	3	3	48	4

The data of Table 4 show that acidification below a pH of 4 should be avoided as the vanadium, and to a lesser degree, the chromium, begin to redissolve especially in the sodium hydroxide medium.

SAND

Tar Sands

According to a process described by *W.J. Lang and D.A. Hentz; U.S. Patent 3,723,310; March 27, 1973; assigned to International Minerals & Chemical Corporation* finely-divided clay and hydrocarbons are flocculated and settled from an aqueous suspension by adding acrylic acid or a water-dispersible anionic acrylic acid-derived polymer, and allowing resulting flocculated solids to settle. Suspended silica may also be flocculated and settled by adding a water-dispersible cationic or nonionic polymer or an inorganic salt as a secondary flocculant.

This process is especially useful for flocculating and settling solids from aqueous suspensions such as those produced by utilizing the hot water process for recovering oil from tar sands. Tar sands, which are also sometimes referred to as oil sands and bituminous sands, are found in various areas throughout the world. One of the most extensive deposits of tar sands occurs in the Athabasca district of the province of Alberta, Canada.

The tar sands of Athabasca district typically contain from 6 to 20% by weight of oil, from 1 to 10% by weight of water, and from 70 to 90% by weight of inorganic solids. The oil is very viscous and has a specific gravity varying from 1.0 to 1.05. Fine grain coarse sand having a particle size greater than 45 microns and less than 2,000 microns constitutes the major portion of the inorganic solids. The remaining inorganic solid material, which is from 10 to 30% by weight of the solid inorganic content of the tar sands, has a particle size of less than 45 microns and is referred to as fines. The fines contain clay, such as chlorite, kaolinite, illite and montmorillonite, and silt including some very small particles of sand.

One of the best known methods for recovering the oil from tar sands such as found in the Athabasca district is known as the hot water process. This process comprises adding caustic soda and hot water to the tar sands to form a slurry which is introduced into a separation cell where an elevated temperature, e.g., 150° to 190°F, is maintained. The aerated oil rises as a froth which is drawn off, and the bulk of the coarse solids content of the tar sands settles as tailings at the bottom of the cell.

A middlings stream consisting essentially of the aforementioned fines with some oil is also drawn from the cell. This middlings stream is generally subjected to a secondary extraction operation, such as in scavenger cells, to recover its oil content. In these cells the middlings solids, consisting mainly of the fine grain quartz sand having a particle size greater than about 45 microns, settle as tailings and are thus separated from the oil. The oil comes off as a froth which can be added to the primary froth collected from the main separation cell. The tailings streams from the primary and secondary extraction operations are combined and constitute slimes which can be flocculated in accordance with this process. The tar sands of the Athabasca district and the hot water process which is currently used for recovering the oil content therefrom are described in further detail in the September 25, 1967 issue of *Oil Week,* published by Maclean-Hunter Publishing Co., Ltd. of Calgary, Alberta, Canada.

The slimes which are produced as tailings from the hot water process are complex oil-containing slimes having a suspended solid content which will not settle within any practical time and cannot be flocculated by utilizing any of the conventional flocculation methods. It was previously necessary to store these slimes since they could not be disposed of without creating a nuisance and pollution problems. As illustrative of the magnitude of the problem, the Great Canadian Oil Sands Limited is reported to produce, by its use of the hot water process for extracting oil from tar sands obtained from the Athabasca district, about 80,000 tons per day of a sand tailings slurry. The slurry typically contains in suspension 0.5 to 1.7% clay, 0.1 to 0.3% sand, and 0.2 to 0.6% oil after the coarse silt and sand fractions have settled out of the water. It is necessary to store these large volumes of slimes in tailings storage ponds because the Canadian Government prohibits the disposal of this material into nearby rivers.

It has been found that the clay and sand may be removed from the aqueous suspensions by utilizing acrylic acid or a water-soluble anionic acrylic acid-derived polymer in combination with a conventional coagulating agent as a secondary flocculant, i.e., a cationic or nonionic polymer or an inorganic salt, and preferably a nonionic polymer. The addition of the acrylic acid or anionic acrylic acid-derived polymer to the suspension initiates selective flocculation of some of the slime components, e.g., the clay and hydrocarbons in intimate association therewith, but will not produce a clarified supernatant liquid.

The use of a nonionic polyelectrolyte coagulant as the secondary flocculant and gentle agitation will cause all of the suspended material to coagulate into tough rubbery flocs which settle rapidly. The agglomerated material can then be easily separated from the bulk of the supernatant liquid by any conventional method such as decantation, filtration, centrifugation, and the like. The remaining liquid is clear and can thus be disposed of in nearby rivers and streams or recycled for processing water.

The preferred acrylic acid-derived polymers are graft copolymers prepared by graft copolymerizing acrylic acid onto a minor amount of a substrate which is a water-soluble polyhydroxyl-containing polymeric material. The substrates which are coreacted with the acrylic acid to produce graft copolymers which are useful are water-soluble polyhydroxyl-containing polymeric compounds. In general, these polymeric compounds may be characterized as having a molecular weight greater than about 350 and containing at least 5% free hydroxyl groups. The polymeric compounds may contain up to 30% and even higher free hydroxyl groups.

The useful polymeric substrates include water-soluble polysaccharides and polyhydroxyl-containing derivatives thereof. Although disaccharides and trisaccharides may be used, the preferred polymeric compounds are the tetra-, penta- and higher polysaccharides and oligosaccharides, i.e., saccharides containing at least four linked monosaccharide molecules. The most preferred substrates include derivatives of cellulose and the biochemically-synthesized heteropolysaccharides. The following examples are illustrative of the process.

Example 1: This example illustrates the effectiveness of the two-stage treatment of this process for flocculating and settling finely-divided material from an aqueous suspension. The suspension which was utilized in this experiment was a sample of the complex oil-clay-sand slimes, containing about 2.0% by weight of suspended solids, obtained from the tailings storage pond from the hot water process employed by the Great Canadian Oil Sands Limited. The primary flocculant was a graft copolymer produced by copolymerizing about 100 parts acrylic acid and about 1 part of carboxymethyl cellulose.

An aqueous solution containing 0.5% of the graft copolymer was added to a 100 ml sample of the slime in a graduated cylinder at the rate of about 1.5 lb of polymer per ton of slime solids. The graduated cylinder was inverted several times to allow mixing. It was noted that there was rapid flocculation of a portion of the suspended solid content of the slime. A 0.05% aqueous solution of a nonionic polymer sold as Polyox was then added to the liquid in the cylinder which was again inverted to allow mixing. A sufficient quantity of the aqueous solution was added to add the polymer at the rate of 2 lb of polymer per ton of original slime solids. It was noted that this combination of polymers resulted in the rapid coagulation and settling of all of the suspended material to leave a clear supernatant liquid.

Example 2: An attempt was made to repeat the experiment of Example 1 except that an anionic acrylamide polymer sold as AP-30 was added to the liquid, in the cylinder in 1 lb increments up to 10 lb per ton of suspended solids. The addition of this anionic polymer did not result in the flocculation of any of the solid content of the suspension.

Example 3: Another attempt was made to carry out the experiment outlined in Example 1 using a guar derivative anionic polymer sold as F-3 in 1 lb increments from 1 up to 10 lb per ton of suspended solids. It was again observed that the use of an anionic polymer other than an acrylic acid-derived polymer as hereinbefore defined was without effect.

Example 4: A series of experiments are conducted to demonstrate the usefulness of various materials as secondary flocculants. In these experiments the primary flocculant is a graft copolymer produced as hereinbefore described by utilizing methyl cellulose as the substrate. When alum, an acrylamide polymer sold as Separan NP-10, and thorium nitrate are each added to separate quantities of the suspension to which the graft copolymer is previously added, there is flocculation and settling of the solid contents of the suspensions.

Oily Sand

A process described by *J.A. Garcia; U.S. Patent 3,723,309; March 27, 1973; assigned to Esso Production Research Company* is directed to a method and system for cleansing oily solid material, such as oily sand, in which improved washing, rinsing and recirculation concepts permit disposal of clean solid material without contaminating the environment.

The cleaning system comprises, briefly, a flotation compartment; a cyclone for washing solid material contained in a mixture of oil and water and for separating the washed solid material from the liquid mixture; means for pumping a slurry of the solid material oil and water from the flotation compartment into the cyclone, the action of the cyclone aerating and causing gas bubbles to form in the liquid phase; means for recirculating the separated liquid phase into the flotation compartment, the gas bubbles aiding in separating the oil from the water in the flotation compartment; a rinse compartment; means for discharging the washed and separated solid material from the cyclone into the rinse compartment; and an overflow opening between the rinse and flotation compartments for permitting the liquid in the rinse compartment to overflow into the flotation compartment.

As shown in Figure 6.6a a sand separator system, indicated at **10** separates sand from oil production liquids introduced into the system through conduit **11**. The separated oil and water are sent to an oil and water separator through conduit **12** and the separated sand, indicated at **13**, is flushed from separator system **10** by water contained in a tank **14** and pumped through a conduit **15** by means of a pump **16**. The flushed water and oily sand is delivered to the sand washing system, indicated at **17**, through a conduit **18**. The sand washing system **17** is shown in detail in Figures 6.6b through 6.6d. Separated water and oil from system **17** is returned to water tank **14** through conduit **19**. Water is added to the system through conduit **20**. Clean sand is removed, as indicated at **21**, and removal of the separated oil and water is indicated at **22**.

Referring to Figures 6.6b through 6.6d, a series of chambers or compartments and cyclone or vortex liquid-solid separators are shown. A receiver-flotation compartment **30** has a common wall **31** with a rinse-flotation compartment **32**, which in turn has a common wall **33** with another rinse-flotation compartment **34**. Each of the compartments **30, 32** and **34** are formed with sloping bottoms indicated at **35a, 35b** and **35c**, which are V-shaped, as indicated in Figure 6.6c and terminate at one end in a sand-collection sump **36a, 36b** and **36c**. An opening **37** in wall **31** is connected to an overflow barrier or trough **38** and an opening **39** in wall **33** is connected to a similar barrier or trough **40** positioned above trough **38**.

Another wall member **41** of compartment **30** forms a common wall with a sand-foam separator compartment **42**. The wall member has an opening **43** to which is connected a trough **44** positioned below trough **38**. A vortex or cyclone **50** is mounted above compartment **30** and a sand delivery tube **51** extends from cyclone **50** into compartment **30**. Another similar cyclone **52** is mounted above compartment **32** and a sand delivery tube **53** extends from cyclone **52** into compartment **32**.

An additional cyclone **54** is mounted above compartment **34**. As shown in Figure 6.6b, each compartment **30, 32, 34** and **42** has a gas tight top or cover. Gas tight covers on these compartments permit the use of other gas in the cyclones (such as the nonoxidizing) as well as air. However, when air is to be used in the cyclones then covers for these compartments are not needed.

Perforated pipes **60a** and **60b** are connected to a source of water, not shown, through a conduit **20**. A conduit **65** containing a pump **66** connects through **66a** to cyclone **52**. Another conduit **67** containing a pump **68** connects through **36b** and cyclone **54**. A conduit **69** connects a source of sand cleansing agent, not shown, to either or both conduits **65** and **67** through a conduit **70** and **71**.

A conduit **75** connects trough **36c** to a pivotal liquid level control pipe **76**, located in a sand discharge compartment **77**, through a jet pump **78** and a conduit **79**. The pumping liquid to operate jet pump **78** is delivered to the pump through a conduit **80** which is connected to a conduit **20** and a source of water supply. A valve **81** controls operation of jet pump **78**.

FIGURE 6.6: SCHEMATIC OF FLOTATION-CYCLONE SEPARATION PROCESS FOR CLEANING OILY SOLID MATERIAL

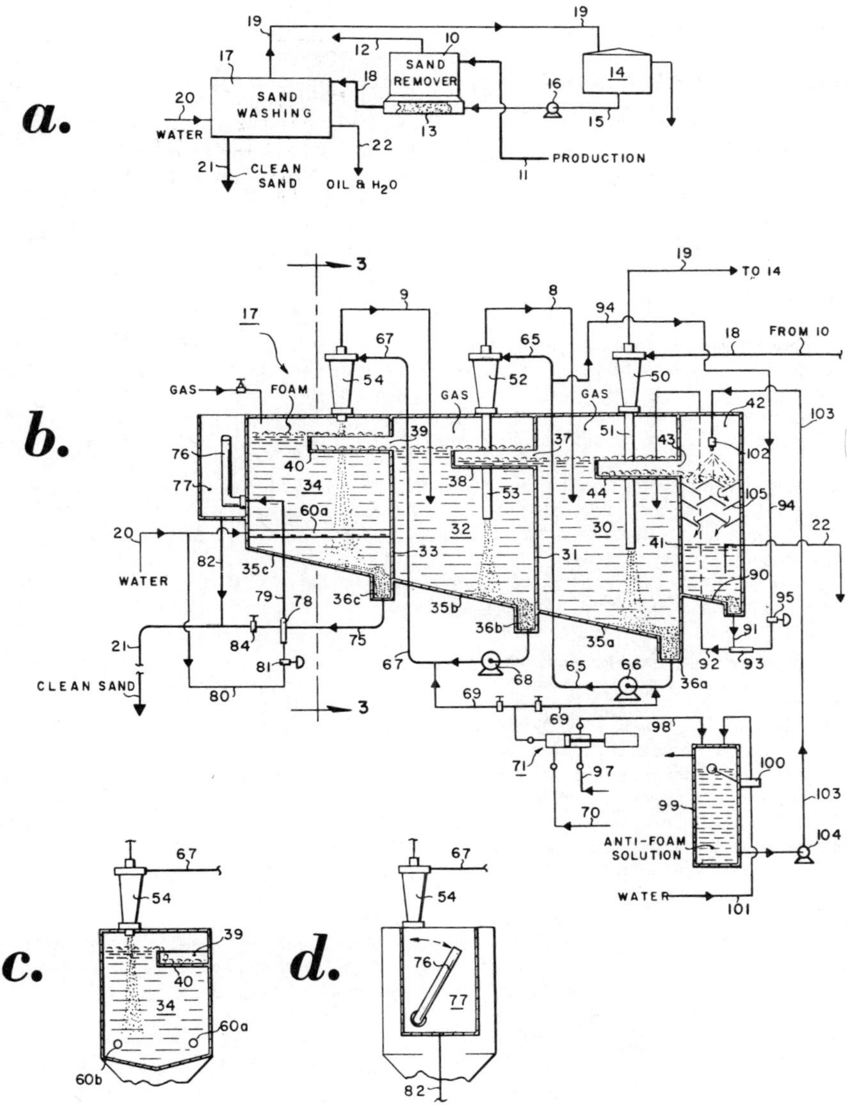

(a) Sand Washing System
(b) Detail View of Washing System
(c) View Along Line 3–3 of Figure 6.6b
(d) End Sectional View of System

Source: J.A. Garcia; U.S. Patent 3,723,309; March 27, 1973

A conduit **82** is connected to the bottom of compartment **77** to discharge clean sand collected in compartment **77** to a discharge conduit **21** which is also connected to conduit **75** through jet pump **78**. A normally closed valve **84** is located in conduit **75**.

Sand-foam separator compartment **42** also has a sloped bottom **90**. A sand discharge conduit **91** connects the bottom of compartment **42** to a conduit **92** through a jet pump **93**. Conduit **92** discharges into compartment **30** and the pumping fluid to operate jet pump **93** is supplied by bleeding the liquid slurry in conduit **65** through a conduit **94** which contains a dump valve **95**. The separated oil and liquid in chamber **42** is discharged to a sump system, not shown, through conduit **22**, which extends below the surface of the liquid in chamber **42**. An antifoam agent is pumped by pump **71** from a source, not shown, through conduits **97** and **98** into a storage vessel **99** provided with a liquid level control **100**. Dilution water is supplied to vessel **99** through a conduit **101** and the antifoam agent is delivered to a spray device **102** positioned in the upper end of chamber **42** through a conduit **103** by a pump **104**. A plurality of flat, tilted baffles or trays **105** are arranged in chamber **42** to permit fallout and separation of the sand.

In operation of the system, a mixture or slurry of water, oil and sand is delivered through conduit **18** to cyclone **50** in which oily sand is separated from the feed slurry. The oil and water separated in cyclone **50** are removed through conduit **19** and returned to tank **14** for reprocessing. Oily sand separated in cyclone **50** falls into compartment **30** through sand delivery tube **51**. The liquid level in compartment **30** is controlled by trough **44**.

Oily foam on the surface of the liquid in compartment **30** flows into compartment **42** where the sand is separated from the foam by flowing the foam over trays **105** and spraying the foam with a defoaming agent pumped by pump **104** from receptacle **99** through conduit **103**. The sand separated from the foam in compartment **42** is returned to compartment **30** through conduit **92** by a pumping liquid slurry bled from conduit **65** and fed through jet pump **93**. The sand which falls into compartment **30** settles to the bottom thereof. The sand is then pumped in the foam of a slurry from sump **36a** to cyclone **52** through conduit **65** by pump **66**.

A cleansing agent may be added to the sand slurry in conduit **65** through conduit **69**. In cyclone **52** the sand is washed by combined action of the cleansing agent and shearing action caused by impact of the sand on the wall of the cyclone and separated from the liquid phase by combined centrifugal and centripetal forces. The action of the liquids in the cyclone causes air, supplied from the atmosphere above the liquid level in compartments **30**, **32** and **34** when gas tight covers on these compartments are not used or from an air source (indicated as gas source) when gas tight covers are used, to be sucked up the center of the vortex formed by the swirling liquid mixture in the cyclone and beaten into the liquid to form small air bubbles in the liquid.

The oil attaches to the surface area provided by these small air bubbles. The oil, water and cleansing agent separated in cyclone **52** are recirculated to compartment **30** through conduit **8**. The bubbles in the liquid rise to the surface in flotation compartment **30** to aid in the separation of the oil and water. The sand separated in cyclone **52** falls into compartment **32** through tube **53**. The liquid level in compartment **32** is controlled by trough **38** which is located above trough **44** so that excess liquid in compartment **32** overflows into compartment **30** over trough **38** through opening **37**.

The sand settles to the bottom of compartment **32** and is pumped, as a slurry from sump **36b** through conduit **67** into cyclone **54** by means of pump **68**. A cleansing agent may be added to the sand slurry in conduit **67** through conduit **69**. The aerated liquid phase separated in cyclone **54** is recirculated to compartment **32** through conduit **9**. The sand separated in cyclone **54** falls into compartment **34**. The liquid level of the liquid in compartment **34** being controlled by the height of trough **40** is above trough **38** overflows into compartment **32** through opening **39**. As the sand settles to the bottom of compartment **34**, it is rinsed by water through perforated conduits **60a** and **60b** prior to being flushed periodically by a timer from compartment **34** by a jet pump **78** connected to the sand slurry discharge conduit **75**. The clean sand is pumped into compartment **77** through tube **76**. Tube **76** pivots, as indicated in Figure 6.6d, to adjust the liquid level of the liquid in compartment **34**. The clean sand and water slurry in compartment **77** is discharged through conduits **82** and **21**. Valve **84** might be used to discharge the sand slurry from compartment **34** in the event jet pump **78** is not used.

In this system the oil as well as the sand cleansing agent are washed off the sand prior to disposal of the sand. More liquid is introduced into the system than actually leaves with the sand. Although a cyclone **50** is used to provide initial separation of the oily sand and water and oil mixture, other types of separation might be used for this purpose. In addition, rinse compartment **34** may be omitted or additional rinse compartments might be added in particular applications of the system and process.

Also, rinse compartment **32** and flotation compartment **30** might, if desired, be provided with water rinse piping such as is employed in rinse compartment **34**. Further, although the use of a cleansing agent in the washing process is preferred, and in some cases may be essential depending upon the difficulty in separation of the oil from the oil coated surface, the cleansing agent may be omitted, if desired, in certain applications.

REFINERY OPERATIONS AND OTHER CHEMICAL PROCESSES

REFINERY OPERATIONS

Injection of Oxygen, Air and Catalyst

D.W. Wheeler; U.S. Patent 3,725,264; April 3, 1973; assigned to Signal Oil and Gas Co. describes a method for removing residual oil and dissolved sulfides from oil production wastewaters prior to disposal of the wastewater into the ocean. Primary oil production produces crude oil with as much as 20 to 50% water. In water flooding operations, commonly used as a secondary recovery method for oil production, oil is produced by injecting water into an oil bearing sand to push the oil and water in the sand formation towards adjacent wells where it is pumped to the surface. The oil thus produced is an oil-water mixture containing about 60 to 70% water. The oil in such production mixture is normally separated from the water by physical separation means, such as settling tanks.

The water from which the oil in the mixture is separated, still contains on the order of 50 to 500 ppm oil and must be subjected to further cleanup operations to remove such residual oil therefrom prior to discharge into adjacent bodies of water. The discharge of this wastewater into the ocean is regulated by governmental water quality control agencies which establish quality requirements for the wastewater discharged into the ocean in order to protect the beneficial uses of the ocean. The State of California, for example, prohibits discharge of such wastewaters into the Pacific Ocean unless such water contains less than 20 ppm oil and grease.

The wastewater cleanup operation employed conventionally is a flotation process by which gas is injected into the water and the mixture pressurized by pumping the oil-water mixture from a surge tank into a flotation cell where the pressure is released, so that gas bubbles through the water and attaches to the suspended oil globules or particles which reduces the effective specific gravity of the particles and causes the oil to move upward where it can be skimmed off in the flotation cell.

Another requirement for wastewater discharge into the ocean, is a limitation of the amount of dissolved sulfide, which is usually present as hydrogen sulfide either un-ionized or ionized depending upon the pH of the water. The federal Environmental Protective Agency currently requires that water discharged into the ocean, contains less than 0.5 ppm of ionized, dissolved sulfide. Hydrogen sulfide has been removed from water by several methods, e.g., air or gas stripping, chlorination, precipitation and air oxidation. Gas or air stripping physically removes the hydrogen sulfide from the water and transfers it to the off-gas and ultimately to the atmosphere, which is objectionable from an air pollution standpoint. The stripping process also tends to raise the pH of the wastewater which increases the scaling tendency of the water. The removal of hydrogen sulfide by chlorination is applicable from a practical standpoint, only to water of low sulfide content. Where the amount of hydrogen sulfide to be removed is substantial, the cost of the chlorine gas becomes excessive. Precipitation methods, for example, with zinc chloride, are also limited to wastewaters having relatively small sulfide content due to high treating costs. Air oxidation processes for removing hydrogen sulfide from wastewaters on an economically feasible basis, require an alkaline solution, since air oxidation of hydrogen sulfide in neutral or acidic wastewaters is normally too slow to be economically feasible.

An air oxidation process for purifying secondary recovery wastewaters using a soluble nickel catalyst is disclosed in U.S. Patent 3,576,738 to J. Duffy. The process described in this patent has been found satisfactory for treatment of wastewater of moderate sulfide content, that is, below about 40 ppm, however, when the sulfide content is above about 40 ppm sulfide, this process was found to be inadequate for reducing the sulfide content of wastewaters to acceptable levels, even with higher catalyst amounts.

A process to remove sulfide in a flotation wastewater cleanup process by injecting an oxygen enriched gas into the wastewater which with an oxidation catalyst effects rapid oxidation of the hydrogen sulfide has been found. Air is also added to the wastewater during processing, to facilitate removal of the sulfur, residual oil, and solid particles in a flotation cell separation. It has been found that in addition to effecting an efficient sulfide separation, the process enhances solids removal effected in the flotation cell without employing additional chemicals, e.g., reverse emulsion breakers, emulsion breakers, or polyelectrolytes.

Referring to Figure 7.1, oil production from water flooding operation is produced from a well and pumped through line **10** to settling tank **12**, commonly called a settling or free water knockout tank. In the settling tank **12**, the gas products are released and sent to gas processing operations and the oil, with approximately 20 to 30% water therein, is transmitted through line **14** to a heated wash tank **16**. In tank **16** the partially dewatered oil is heated and the oil-water mixture separated by settling. The water content of the separated oil is reduced to about 1 to 3%. Wash tanks such as tank **16**, or heater treaters, as gas fired tanks are sometimes called, contain means for heating the oil-water mixture to about 100° to 210°F depending upon the type of crude oil being separated. For example, an 18 gravity crude oil is heated to about 140°F but generally heavier crudes are heated to higher temperatures.

The oil separated in wash tank **16**, is pumped to shipping tanks or a pipeline for transporting to storage or refinery processing. The wastewater removed in the settling tank **12** is combined with the separated wastewater from wash tank **16** in line **18** and additional oil removed therefrom in skim tank **20**. The skimmed water withdrawn from skim tank **20** still has approximately 50 to 100 ppm residual oil which must be removed prior to discharging into the ocean. This water is passed through line **22** to a container vessel or surge tank **24** from which it is pumped by pump **26** into retention tank **28** from which it is introduced into flotation cell **30**. Each of the processing and apparatus items heretofore described are conventional and well-known oil-water separation techniques. In accordance with the process, oxygen is injected into the wastewater at the suction side of the pump **26** to effect a rapid dissolution of oxygen into the pressurized wastewater.

It was found that when wastewaters with more than about 40 ppm of sulfide are treated, the oxygen requirements increased and when air is used as the oxidizing gas it becomes necessary to inject so much air into the wastewater in order to satisfy the oxygen requirements, that the excess air creates turbulence which reduces the efficiency of the flotation cell. Also, the use of air rather than oxygen, necessitates the use of a larger retention tank at higher sulfide concentrations, in order to allow the additional retention time needed to get oxygen (from the air) sufficient to oxidize the sulfide, into solution. A wastewater treating process using air rather than oxygen is also less efficient since air injection causes more nitrogen than oxygen to be dissolved in the wastewater.

FIGURE 7.1: RESIDUAL OIL AND SULFIDE REMOVAL PROCESS

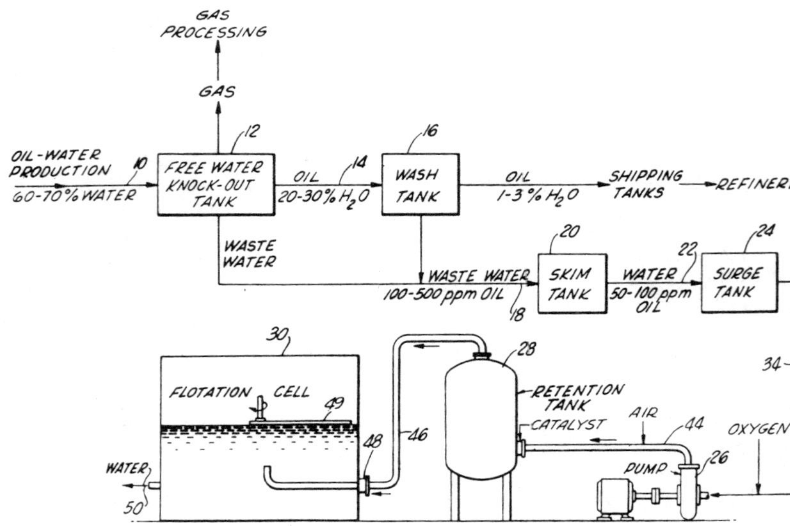

Source: D.W. Wheeler; U.S. Patent 3,725,264; April 3, 1973

An important feature of the process is the addition of oxygen to the wastewater prior to the injection of air into the wastewater. This sequence causes the oxygen to go into solution before nitrogen is dissolved in the wastewater. The oxidation of sulfides in the wastewater is enhanced by the use of higher oxygen concentrations in solution per unit volume of gas added to the wastewater, and as previously stated, this higher oxygen concentration is accomplished by the injection of oxygen into the wastewater prior to the introduction of air. When oxygen is introduced into the wastewater in relatively pure form for oxidation purposes, dissolved air is still needed in order to accomplish a good separation of solids and oil in the flotation cell.

The use of air alone, however, when treating water of higher sulfide concentrations with larger amounts of air, results in flotation cell turbulence. Additionally, since air is only about one-fifth oxygen, a larger retention tank is needed when air is used in order to provide the additional retention time necessary to get the air into solution for oxidizing the sulfide. Air is needed in the flotation cell, since the flotation cell separation is accomplished by the release of dissolved air which effectively floats sulfur, oil particles, and solids to the surface in the flotation cell where these materials may be skimmed off.

The catalyst is preferably added to the wastewater after the injection of oxygen and air to avoid absorption of the catalyst by the oil, but not so soon as to initiate the reaction substantially upstream from the retention tank which would result in the formation of elemental sulfur which would clog the pipe between the pump and the retention tank. If the catalyst is absorbed by the oil, the sulfide removal will be reduced significantly.

It has been found advantageous to add the oxidation catalyst to the wastewater at the retention tank, since the catalyst is effective to initiate the reaction which almost immediately results in a substantial amount of sulfur precipitating in small colloidal particles. These particles would tend to deposit on the pipe wall between the pump and the retention tank or on the internal surfaces of the pump should the catalyst be added to the pump discharge or suction respectively.

The aerated wastewater is pumped into the retention tank **28** from line **34** and under the pump pressure, flows through line **46** into the flotation cell **30** through back pressure valve **48** which is positioned at the flotation cell entry. The flotation cell **30** is maintained at atmospheric pressure, and as the wastewater which is saturated with gas under pressure enters the flotation cell, the reduction in pressure causes the gas to be released from solution in the form of small bubbles which attach to oil particles, free sulfur, and residual solids and carry them to the surface of the water in the flotation cell. The materials may then be removed from the water surface by mechanical skimming means, e.g., a rotary skimming arm **49**. Clean wastewater is removed from the flotation cell through line **50**. The type and the amount of catalyst is substantially the same as described in U.S. Patent 3,576,738, i.e., about 0.1 to 1.0 ppm nickel. The following examples illustrate the process.

Example 1: In the treating system shown in Figure 7.1, the wastewater from the surge tank contained 55 ppm dissolved sulfide and 20 ppm oil. This wastewater was treated by injection of about 0.5 scfm oxygen per 10,000 barrels of wastewater, injected on the suction side of the pump and about 1.0 scfm air per 10,000 barrels of wastewater injected on the discharge side of the pump. The catalyst addition was about 0.08 pound per day (as nickel) which is equivalent to 0.24 ppm. Nickel was added as nickel sulfate hexahydrate. The retention time in the retention tank which is an average of retention time of the water moving through the tank, was about 2.8 minutes.

The water recovered from the flotation cell contained less than 0.1 ppm sulfide and about 10 ppm oil. In the run comparable to the one described above where the air was injected into the wastewater ahead of the oxygen, the sulfide reduction was about 55 to 5 ppm. This reduction was not acceptable to the EPA which requires sulfide content of less than 0.5 ppm.

Example 2: In a system similar to that shown in Figure 7.1, 18,200 barrels per day of wastewater were pumped into a 35 barrel retention tank at a pressure of 42 psig. The retention time was about 2.8 minutes in the retention tank. The catalyst was injected into the wastewater on the suction side of the pump as nickel sulfate hexahydrate. The sulfide and oil removal data are tabularized in the table below.

Into retention tank S=(p.p.m.)	Out retention tank S=(p.p.m.)	Out float cell S=(p.p.m.)	Injected at pump suction (s.c.f.m.)	Injected ahead of retention tank (pump DSCH) (s.c.f.m.)	Pressure (p.s.i.g.)	Ni++ (p.p.m.)
65	1.0	<0.5	1.0	1.2	42	.7
65	3.0	1.5	0.5	2.0	42	.7
65	1.0	<0.5	0.5	2.0	50	.7
65	<0.5	<0.5	1.0	1.0	50	.7
65	1.0	<0.5	1.0	1.0	42	.35
65	<0.5	<0.5	1.0	1.5	42	.35

Subterranean Injection

J.F. Tate; U.S. Patent 3,817,859; June 18, 1974; assigned to Texaco Inc. describes a process of disposing of certain process effluent waste streams by injecting them into subterranean formation. The method of inhibiting the formation of precipitates involves lowering the pH of the mixed streams and optionally removing any organic phase created thereby before injection.

A major reason for failure of many deep well industrial waste disposal wells is plugging of the formation face with solids present in the injected stream or formed due to incompatibility of the stream with formation material or connate water. Such solids may come from four sources: (1) suspended solids and emulsions such as clay from river water used in an industrial process present in waste stream as a result of poor surface treatment, for instance, coagulation and sedimentation, (2) solids formed by the mixing of two or more solids-free streams by combination of an ion from one with an ion from the other to form a precipitate, (3) solids formed by mixing of the stream(s) with formation (connate) water at the formation face and (4) solids formed by precipitation when certain streams contact and react with the formation rock itself. Suspended solids and emulsion, (1) above, may be removed by more effective surface treatment using techniques well-known in the art. This process is directed to solving the plugging problems posed by the last three situations mentioned above.

While it is difficult to classify streams according to chemical types, in order to detail the operation of the process, it is necessary to make certain arbitrary classifications. Refineries, to cite a typical example, may have at least two general types of waste streams which could be disposed of by injection into deep formations. These streams will be arbitrarily labeled Group 1 streams and Group 2 streams.

Group 1 streams may contain some or all of sulfides, cresylates and soluble iron salts. The separate Group 1 streams have pH values in the alkaline range. The various streams classified here in Group 1 also may contain other ingredients, but the ones mentioned are adequate to illustrate the problems encountered. It must be reiterated that there may be several streams from a given refinery classified in Group 1, each of which may have a different composition from the other Group 1 streams, and each of which may contain one or more of the constituents mentioned above in various proportions.

Group 2 streams are composed of alkaline earth metal ions such as calcium, barium, and magnesium as well as iron. Some of these streams also have pH values in the alkaline range. The most definite difference between Group 2 and Group 1 streams is that Group 2 streams have no organic constituents in significant quantities. Again, as in Group 1, each Group 2 stream may differ from other Group 2 streams and contain one or more constituents in varying proportions.

Formation of precipitates which will plug the injection wells may arise in a variety of ways. (1) Mixing Group 1 streams with other Group 1 streams. Here iron sulfide may precipitate and plug the formation. (2) Mixing Group 2 streams with other Group 2 streams. Here iron hydroxide may precipitate out. (3) Contacting the injection horizon with Group 1 streams may cause reaction between the formation water containing alkaline earth metal ions and the cresylates forming insoluble alkaline earth metal cresylates. (4) Contacting the calcareous formation itself with Group 1 streams may cause precipitation of calcium or barium cresylates. (5) Mixing Group 1 and Group 2 streams will cause iron sulfide, alkaline earth metal cresylates and/or iron hydroxide to form when contact with the formation and formation water is attained upon injection.

It has been found that all of these problems may be solved by lowering the pH of the streams to the acidic range before mixing the streams as they are mixed or after mixing. The precipitates which would normally form in an alkaline medium will not form or will dissolve in an acidic medium, with the exception of the freed cresylic acids which will oil out as a layer over the aqueous acidic medium. These cresylic acids may be removed by skimming or other means known in the art. The pH should be lowered to about 5 to 7, but in any event sufficiently low to prevent precipitation or to provide near complete dissolution of existing precipitate and, if desired, to insure oiling out of any cresylic acids present.

It is especially preferred to lower the pH to about 5. The pH may be lowered by use of an acid, preferably the mineral acid hydrochloric (HCl). Though sulfuric acid (H_2SO_4) may be used, there is a strong likelihood that gypsum ($CaSO_4 \cdot 2H_2O$) would precipitate and plug the formation. In one example of the process, the streams are lowered to a pH value in the acidic range before the streams reach the subterranean formation. This will dissolve the precipitates already formed and inhibit the formation of any further precipitates. This may be carried out by lowering the pH of each individual stream before mixing or by lowering the pH of the streams after mixing commences and before the mixed streams reach the formation face.

Injectivity tests of effluent waste streams from an industrial plant were made to determine whether and how extensively precipitation occurs upon mixing these various streams together and with representative formation waters. The injectivity measurements were carried out with a device used routinely in oil field practice to determine plugging tendency of a liquid stream toward a formation face. The rate of flow of the stream across a millipore filter (0.45 micron pore size) at a constant head pressure was measured. As suspended matter accumulates upon the filter face, thus plugging the pores, the flow rate decreases (incremental time for passage of incremental volume increases), simulating formation face plugging. A measurement using distilled water was made for comparison purposes.

Plugging tendency of each separate stream and their mixtures via millipore injectivity tests are shown in Tables 1 and 2. In Table 3 the results of mixing the effluent streams and their mixtures with representative formation waters from the Salem Field are depicted.

TABLE 1: MILLIPORE INJECTIVITY TESTS ON GROUP 1 STREAMS, 20 psi

	Volume, ml.									
	100	200	300	400	500	600	700	800	900	1,000
Stream, seconds:										
A	9	17	26	35	44	55	66	78	90	103
B	32	53	66	79	91	104	115	127	139	151
C	9	17	26	36	50	64	81	99	118	137
D	42	104	162	210	264	325	400	500	608	725
E	19	42	68	99	128	156	180	208	232	254
F	18	44	60	74	88	Millipore disintegrated				
Mixture of above, seconds	10	18	35	79	215	498	1,290	1,840		
Distilled water, seconds	10	19	29	39	50	62	75	88	102	117

TABLE 2: MILLIPORE INJECTIVITY TESTS ON GROUP 2 STREAMS, 20 psi

	Volume, ml.									
	100	200	300	400	500	600	700	800	900	1,000
Stream, seconds:										
P_1	120	540	1,380							
G	17	65	115	245	347	460	585	720	870	1,035
H	9	17	26	35	44	54	65	77	90	105
J	10	19	28	37	46	55	65	74	83	93
P_2	720	Millipore plugged								
K	15	38	70	112	160	219	285	355	437	527
L	10	18	27	35	44	53	62	72	81	91
Mixture of above,[1] seconds	35	173	290	380	585	825	1,080	1,340	1,620	
Distilled water, seconds	10	19	29	39	50	62	75	88	102	117

[1] Excluding P and P_2.

TABLE 3: COMPATIBILITY OF EFFLUENT STREAMS AND THEIR MIXTURES WITH REPRESENTATIVE FORMATION WATERS

	Stream plus—	
	Warsaw water	Salem water
Stream:		
A	No precipitate	No precipitate.
B	Heavy white precipitate	Heavy white precipitate.
C	Heavy precipitate	Milky solution.
D	Heavy orange precipitate	Heavy orange precipitate.
E	do	Do.
F	Heavy white precipitate	Heavy white precipitate.
Mixture of above [1]	Heavy precipitate	Heavy precipitate.
G	Not run	No precipitate.
H	do	Do.
J	do	Do.
K	do	Not run.
L	do	No precipitate.
Mixture of above [1]	do	Do.
Proportion of two mixtures (1:1) [1]	do	Heavy precipitate.

[1] After millipore filtration.

All individual streams except P_1, G, and P_2 have good individual injectivities. The extremely poor injectivities of P_1 and P_2 are attributable to oil carry-over and emulsion formation in the case of P_1, and inadequate sedimentation of suspended solids from river water in the case of P_2. The suspended or emulsified matter in these streams must be more efficiently removed by chemical and mechanical pretreatment filtration, coagulation, sedimentation, etc., before their injection may be contemplated. The injectivity of stream G was acceptable even though not as good as most of the other streams.

The separate mixing of both the Group 1 streams with other Group 1 streams and the Group 2 streams with other Group 2 streams results in diminished injectivity in both cases.

In the case of Group 1, combination of dissolved iron in A stream effluent water, and sulfide in B and F streams, probably produced a precipitate of iron sulfide. In addition, insoluble calcium (and possibly barium) cresylates, from combination of cresylates in C stream and E stream with calcium (and barium) in A stream, were formed. Further, precipitation of lead sulfide, by combination of dissolved lead in D stream with sulfide from above cited sources is suspected.

Injectivity of the Group 2 mixture was diminished by precipitation of an orange-brown colloidal solid, much like iron hydroxide in appearance. If the G stream is omitted from the mixture, no precipitation occurs after standing overnight. The iron therefore appears to be furnished by this stream. The hydroxide ion in this mixture is furnished by the highly alkaline L stream.

It is believed that both these stream mixtures would eventually plug a formation (unless highly fractured), even if they are compatible with connate formation water. When the pH of these two separate stream mixtures is lowered to about 5, the formed precipitates dissolve in each case, as evidenced by ½ hour centrifugation of the acidified stream mixtures. The sulfides are acid-soluble and the calcium (or barium) cresylates dissolve, allowing the free cresylic acid to oil out on the surface of the aqueous mixture from which it may be removed by skimming or other conventional means. Exposure of the acidified mixtures to an Illinois limestone core (calcium carbonate) for one hour produced no precipitate as evidenced by 15 minutes centrifugation of the exposed solution or dissolution of the core as evidenced by evolution of carbon dioxide gas.

It is concluded that pretreatment of both the mixed streams with acid to produce a pH of 5 would allow injection of a solids-free solution which would be compatible with a limestone formation, assuming the solutions are compatible with the connate water itself. Exposure of the Group 1 mixture to a representative formation water produced a heavy precipitate (Table 3). This precipitation is due to the reaction of the basic cresylates with calcium (and barium) in the formation water. The addition of formation water to the Group 2 mixture resulted in no precipitation since neither the Group 2 mixture nor the formation water contained ions which when mixed would precipitate.

Reduction of pH to about 5 solubilized the precipitates formed freeing the cresylic acid which oiled out as a separate phase when the Group 1 mixture was exposed to the representative formation water. It may be concluded therefore that acidification can render each separate stream compatible with the connate water. A three component mixture formed by adding the two millipore filtered effluent stream mixtures (Groups 1 and 2) to formation water 1:1:1 proportion by volume resulted in copious precipitation. This precipitate probably consists of a mixture of insoluble sulfides, hydrous iron oxides and the insoluble calcium (and barium) cresylates.

Reduction of the pH of this triple mixture to 5 resulted in complete dissolution of the precipitate and oiling out of the cresylic acid as a surface layer. Exposure of an Illinois limestone core to the acidified (pH about 5) solution resulted in no precipitation nor evolution of gas (carbon dioxide) over an hour's exposure. It may be concluded, thereof, that lowering of pH can produce a solids-free solution when both Groups 1 and 2 are mixed with a representative connate water and limestone formation.

Mixtures with Hydrocarbon Fuel and Partial Oxidation

J.P. Tassoney and R.M. Dille; U.S. Patent 3,725,270; April 3, 1973; assigned to Texaco Inc. describe a continuous pollution abatement process for sulfide-bearing wastewater, e.g., sour water, treated sour water, and other refinery wastewaters. A stream of wastewater is introduced into the refractory-lined reaction zone of a free-flow noncatalytic synthesis gas generator in admixture with a hydrocarbonaceous fuel. By the partial oxidation reaction at an autogenous temperature in the range of about 1500 to 3000°F and a pressure in the range of about 1 to 250 atmospheres, the aforesaid feed to the gas generator is reacted with an oxygen-rich gas to produce a hot effluent stream of synthesis gas substantially comprising a mixture of H_2, CO, CO_2 and H_2O and containing a relatively minor amount of H_2S, COS, A, N_2, CH_4 and particulate carbon. The H_2S in the effluent synthesis gas may be recovered; also purified water may be separated from the effluent gas and used elsewhere in the process or discarded to the sewer.

In the second example of the process the hot effluent synthesis gas from the gas generator is cooled and scrubbed with sulfide-bearing wastewater. Particulate carbon is thereby removed from the effluent synthesis gas by the sour wastewater. A wastewater particulate carbon dispersion is produced. The particulate carbon may adsorb any oil and phenol in the wastewater. Any ammonia in the wastewater may be used for pH control. For example, the ammonia may react with any formic acid produced in the quench zone. Simultaneously, the effluent synthesis gas strips the hydrogen sulfide from the sour quench water. The hydrogen sulfide going overhead with the effluent synthesis may be recovered downstream and converted into by-product sulfur.

Sulfide-bearing aqueous waste liquors include by definition sour water, treated sour water, and other refinery wastewaters. Such wastewaters are essentially dilute solution of water having a pH of about 7 or above and which contain small amounts of H_2S, NH_3, hydrocarbons and other organic compounds. The following example illustrates the process. All quantities are on an hourly basis.

Example: The effluent gas stream from a hydrocracking unit in a petroleum refinery is washed with water. The resulting wastewater, also known as sour water, contains the following impurities in parts per million by weight: 4,800 ppm sulfide, 40 ppm sulfite, 1.0 ppm thiosulfate, 3,360 ppm NH_3, 904 ppm oil, and 1,315 ppm phenol. The pH of the wastewater is 9.0. On an hourly basis, about 892 pounds of wastewater are heated to a temperature of about 730°F and mixed with about 219 pounds of fuel oil at a temperature of 225°F to form a dispersion. The fuel oil has the following properties: API gravity 16°; viscosity, (SSF/122°F), 119; and heat of combustion 18,357 Btu/lb. The ultimate analysis was (weight percent): C 84.96, H 11.54, S 2.05, N 0.50, O 1.12, and ash 0.03.

By means of an annulus type burner, the dispersion of wastewater, and fuel oil is introduced into the closed compact refractory lined reaction zone of a vertical free-flow synthesis gas generator in the absence of packing or catalyst. About 238.9 pounds of substantially pure oxygen (greater than 95% O_2 by volume) at a temperature of about 94°F are passed through the burner and into the reaction zone of the gas generator. In the reaction zone, the oxygen is intimately mixed with the dispersion of wastewater and fuel oil and reaction by partial oxidation then takes place at an autogeneous temperature of about 1800°F and at a pressure of about 1,600 psig.

The hot gaseous effluent stream from the reaction zone of the synthesis gas generator is quickly cooled below the reaction temperature, to about 575°F by direct quenching in 5,723 pounds of water at a temperature of about 475°F. The quench water is contained in a closed quench vessel. The turbulent condition in the quench chamber, caused by large volumes of effluent synthesis gas bubbling up through boiling quench water helps to scrub substantially all of the particulate solids from the effluent gas. About 6,025 pounds of a dispersion comprising about 1.5 weight percent of particulate carbon in quench water is removed from the bottom of the quench tank and sent to a carbon-recovery zone, as previously described. Clear water reclaimed in the carbon-recovery zone is recycled to the quench tank.

Vaporized quench water or steam passes out of quench vessel along with the effluent synthesis gas. When required, additional gas scrubbing may be effected by means of a conventional gas-liquid contact apparatus, for example, a spray tower venturi or jet scrubber, bubble plate contactor, packed column, or in a combination of such equipment. About 9,556 scfh of effluent synthesis gas leaves the cooling and scrubbing zone having the following analysis (dry basis-mol percent): H_2 37.73, CO 12.86, CO_2 32.73, H_2S 1.03, COS 0.05, CH_4 14.94, A 0.12, and N 0.50. CO_2, H_2S and other acid gas constituents may be removed by conventional refrigeration techniques, or by a combination of refrigeration and physical or chemical absorption with solvents, such as n-methyl pyrrolidone or an alkanolamine solution. The purified effluent stream of synthesis gas may be used as a source of feed gas for the synthesis of hydrocarbons or oxygen containing organic compounds. The H_2S and COS-containing solvent may be regenerated by flashing and by stripping with nitrogen. The H_2S and COS are then converted into by-product sulfur by a suitable process.

Oxidation and Carbon Monoxide Reduction

P. Urban and R.H. Rosenwald; U.S. Patent 3,672,835; June 27, 1972; assigned to Universal Oil Products Company describe a process for the conversion of a hydrocarbon charge stock containing sulfurous and nitrogenous compounds where an aqueous waste stream containing substantial quantities of NH_3 and H_2S (typically present as NH_4HS) is produced by contacting the effluent from the conversion zone with a water stream. This wastewater stream is treated by the process to recover elemental sulfur and to produce a treated water stream suitable for recycle to the water-contacting step of the hydrocarbon conversion process in order to remove additional quantities of NH_3 and H_2S, to abate a substantial pollution problem and to minimize make-up water requirements.

The first step of the treating method involves catalytically treating the aqueous waste stream with oxygen at oxidizing conditions effective to produce an effluent stream containing NH_4OH, $(NH_4)_2S_2O_3$ and elemental sulfur or ammonium polysulfide. Sulfur is in the second step, separated from the effluent stream from the first step to produce a water stream containing $(NH_4)_2S_2O_3$. The final step involves treating the aqueous stream from the second step with carbon monoxide at reduction conditions selected to form a substantially thiosulfate-free treated water stream.

Referring to Figure 7.2, a light gas oil enters the combination process through line **1**. This light gas oil is commingled with a cycle stock at the junction of line **11** and line **1** and with a recycle hydrogen stream at the junction of line **7** with line **1**. The resulting mixture is then heated via a suitable heating means to the desired conversion temperature and then passed into hydrocarbon conversion zone **2**.

FIGURE 7.2: TREATMENT OF WASTE STREAM FROM HYDROCARBON CONVERSION UNIT

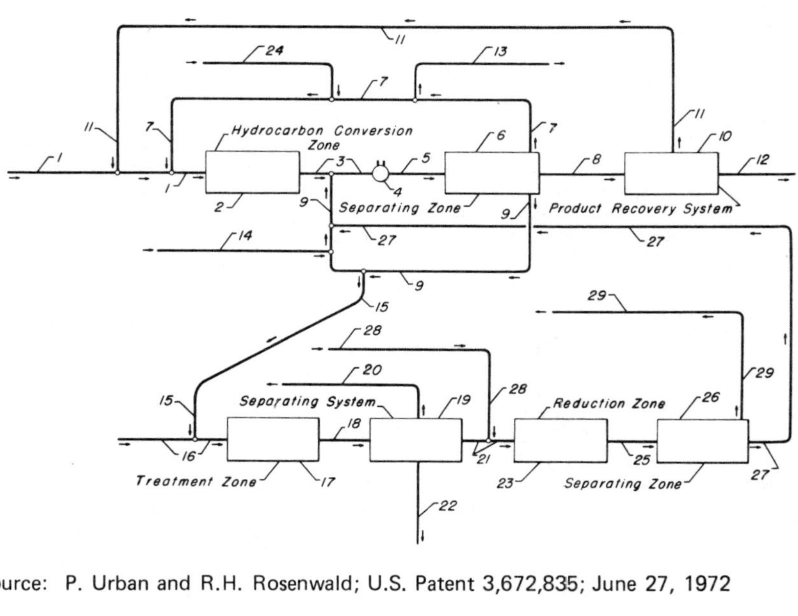

Source: P. Urban and R.H. Rosenwald; U.S. Patent 3,672,835; June 27, 1972

An analysis of the light gas oil shows it to have the following properties: API gravity at 60°F of 25°, an initial boiling point of 421°F, a 50% boiling point of 518°F, and an end boiling point of 663°F, a sulfur content of 2.21 weight percent, and a nitrogen content of 126 ppm by weight. Hydrogen is supplied via line **7** at a rate corresponding to a hydrogen circulation rate of 10,000 standard cubic feet of hydrogen per barrel of oil charged to hydrocarbon conversion zone **2**. The cycle stock which is being recycled via line **11** is a portion of the 400+ fraction of the product stream which is separated in product recovery system **10** as will be hereinafter explained.

The catalyst utilized in zone **2** comprises nickel sulfide combined with a carrier material containing silica and alumina in a weight ratio of about 3 parts silica per part of alumina. The nickel sulfide is present in amounts sufficient to provide about 5 weight percent nickel in the final catalyst. The catalyst is maintained within zone **2** as a fixed bed of ⅛" x ⅙" cylindrical pills. The conditions utilized in zone **2** are hydrocracking conditions which include a pressure of about 1,500 psig, a conversion temperature of about 600°F, and a liquid hourly space velocity of about 2.0 hr^{-1} based on combined feed.

An effluent stream is then withdrawn from zone **2** via line **3** and commingled with a water stream, in a water-contacting step, at the junction of line **9** with line **3**. The resulting mixture is passed into cooling means **4** where it is cooled to a temperature of about 100°F. The cooled mixture is then passed via line **5** into separating zone **6** which is maintained at a temperature of about 100°F and a pressure of about 1,450 psig. The amount of water injected into line **3** via line **9** is about 5 gallons of water per 100 gallons of oil. As explained previously, the reason for adding the water on the influent side of cooling means **4** is to insure that it does not become clogged with sulfide salts.

In a separating zone **6**, a three-phase system is formed. The gaseous phase comprises hydrogen, hydrogen sulfide and a minor amount of light ends. The oil phase contains a relatively large amount of dissolved H_2S. The water phase contains about 5 weight percent ammonium hydrosulfide with a slight excess of ammonia. The hydrogen-rich gaseous phase is withdrawn via line **7** and a portion of it (about 20% by volume) is vented from the system via line **13** in order to prevent buildup of excessive amounts of H_2S in this stream. The remainder of the hydrogen stream is passed via line **7** through compressive means, not shown, and is commingled with additional make-up hydrogen entering the process via line **24** and passed back to hydrocarbon conversion zone **2**. The oil phase from separating zone **6** is withdrawn via line **8** and passed to product recovery system **10**.

In this case, product recovery system **10** comprises a low pressure separating zone and a suitable train of fractionating means. In the low pressure separating zone, the oil stream is maintained at a pressure of about 100 psig and a temperature of about 100°F in order to strip out dissolved H_2S from this oil stream. The resulting stripped oil stream is fractionated to recover a gasoline boiling range product stream and a cycle oil comprising the portion of the product stream boiling above 400°F. The gasoline product stream is recovered via line **12** and the cycle oil is recycled to hydrocarbon conversion zone **2** via line **11**.

Returning to the aqueous phase formed in separating zone **6**, it is withdrawn via line **9** and continuously recycled back to line **3**. Additional make-up water is injected through line **14** during start-up of the process. It is a feature of the process that the requirement for make-up water is minimized since water is a product of the oxidation step. A drag stream is withdrawn from the water stream flowing through line **9** at the junction of line **9** and line **15**. This drag stream is passed via line **15** to line **16** where it is commingled with an air stream and the resulting mixture is passed into treatment zone **17**. This drag stream is the input water stream to the process.

Treatment zone **17** contains a fixed bed of a solid catalyst comprising cobalt phthalocyanine mono-sulfonate combined with an activated carbon carrier material in an amount such that the catalyst contains 0.5% by weight phthalocyanine. The activated carbon granules used as the carrier material are in a size of 30 to 40 mesh. The ammonium hydrosulfide is present in the input water stream withdrawn from separating zone **6** in an amount of about 5% by weight. This stream is charged to treatment zone **17** at a liquid hourly space velocity of about 1 hr^{-1}. The amount of air which is also charged to treatment zone **17** via line **16** is sufficient to react about 0.7 atom of oxygen per atom of sulfide contained in the water stream. This is an amount less than the stoichiometric amount necessary to convert the sulfide to sulfur, and consequently ammonium polysulfide is formed within treatment zone **17**. The conditions utilized in this zone are a temperature of 95°F and a pressure of 50 psig. Because of side reactions, a minor amount of the sulfide contained in the input water stream is oxidized to higher oxides of sulfur, principally $(NH_4)_2S_2O_3$.

Depending somewhat upon the life of the catalyst utilized within zone **17**, about 1 to 10% of the ammonium hydrosulfide will be ozidized to $(NH_4)_2S_2O_3$. Accordingly, the effluent stream withdrawn from zone **17** via line **18** contains ammonium polysulfide, NH_4OH, $(NH_4)_2S_2O_3$, and a minor amount of nitrogen gas. The stream is passed to separating system **19** which in this case comprises: a gas separator, a polysulfide decomposition zone and a sulfur recovery zone. In the gas separator, the minor amount of nitrogen gas contained in the effluent stream from treatment zone **17** is vented from the system. In the polysulfide decomposition zone, the liquid effluent stream from the gas separator is heated to a temperature of about 280°F and passed into a distillation column where an overhead stream containing NH_3, a minor amount of H_2S, and H_2O is recovered via line **20**.

A bottom stream comprising liquid elemental sulfur is withdrawn via line **22** and a side stream comprising an aqueous solution of $(NH_4)_2S_2O_3$ and NH_4OH is withdrawn via line **21**. In this case, the sulfur recovery zone is the bottom of the distillation column. The resulting elemental sulfur-free aqueous stream containing a minor amount of ammonium thiosulfate is withdrawn from separating system **19** via line **21**, commingled with a carbon monoxide stream at the junction of line **28** with line **21**, and the resulting mixture is passed into reduction zone **23**.

The amount of carbon monoxide commingled with this aqueous stream is sufficient to provide a mol ratio of 5.6 mols of carbon monoxide per mol of $(NH_4)_2S_2O_3$. An analysis of the aqueous stream containing a minor amount of $(NH_4)_2S_2O_3$ withdrawn from separating system **19** via line **21** shows it to contain about 0.2% by weight sulfur as $(NH_4)_2S_2O_3$, and about 4 mols of NH_4OH per mol of $(NH_4)_2S_2O_3$.

Reduction zone **23** contains a reduction catalyst comprising cobalt sulfide combined with an activated carbon carrier material. The catalyst is utilized in a particle size of about 12 to 20 mesh and contains 1% by weight cobalt on an elemental basis. The reduction catalyst is supported in reduction zone **23** as a fixed bed and the mixture of carbon monoxide and the thiosulfate-containing aqueous stream are passed in downflow fashion over the catalyst. The conditions utilized in reduction zone **23** are: a temperature of 200°C, a pressure of 500 psig, and a liquid hourly space velocity of 1 hr^{-1}.

An effluent stream is withdrawn from reduction zone **23** via line **25** and passed to separating zone **26** where a gaseous phase separates from a liquid aqueous phase. The gaseous phase comprises unreacted CO, CO_2, H_2S, NH_3 and H_2O. It is withdrawn from separating zone **26** via line **29** and vented from the system. The aqueous phase formed in separating zone **26** is withdrawn via line **27** and recycled via line **27** and line **9** back to the water-contacting step of the hydrocracking process. An analysis of the treated water stream flowing through line **27** shows that 99% of the ammonium thiosulfate charged to zone **23** is converted at a selectivity for sulfide of about 98%. Accordingly, this stream is substantially free of thiosulfate.

Hydrogen Sulfide Treatment of Thiosulfate Solution

P. Urban; U.S. Patent 3,773,662; November 20, 1973; assigned to Universal Oil Products Company describes a four-step method for treating an aqueous stream containing a water-soluble thiosulfate compound where the thiosulfate compound is first reacted with H_2S under conditions selected to result in a relatively rich polysulfide compound, the resulting polysulfide compound is at least partially decomposed by rapid cooling under a positive H_2S partial pressure, sulfur is then recovered from the resulting cooled solution without substantially changing the pressure of the effluent stream, and finally H_2S is separated from the resulting cooled and substantially elemental sulfur-free effluent stream at a reduced pressure to form a treated water stream of greatly reduced thiosulfate content.

In one specific aspect the process involves a method for the treatment of an aqueous stream containing sodium thiosulfate to produce elemental sulfur and treated aqueous stream, which has a greatly reduced total sulfur content relative to the input thiosulfate-containing aqueous stream, by a procedure which involves the production of an effluent stream containing a relatively rich sodium polysulfide compound and unreacted H_2S by a reaction between H_2S and the input solution and the subsequent decomposition of same to yield elemental sulfur by means of a cooling step and a sulfur separation step both of which are conducted under a positive H_2S partial pressure.

As part of the price that has to be paid for a modern industrial society, large quantities of aqueous solutions of thiosulfate compounds are currently produced. Aqueous solutions containing ammonium thiosulfate are an undesired side product of many economically significant industrial processes in the chemical, petroleum, natural gas, paper and steel industries. For instance, in the petroleum industry an aqueous solution containing ammonium thiosulfate is produced as a drag stream from sulfur recovery systems that employ an oxidation method to recover sulfur from ammonium hydrosulfide solutions, which are commonly available as side streams from such typical refinery processes as hydrorefining, hydrocracking, catalytic cracking, etc.

Another source of these thiosulfate-containing streams are processes for natural gas sweetening, coal gas purification, town gas purification, and the like processes where hydrogen sulfide is scrubbed from a gaseous mixture containing same, and thereafter oxidized to elemental sulfur in a regeneration step. The following examples illustrate the process.

Example 1: In order to demonstrate an importance of the presence of H_2S during the sulfur separation step of the process, two experiments were performed in which the effect of H_2S partial pressure on the recovery of elemental sulfur from a freshly decomposed polysulfide solution was studied. In the first experiment, a simulated partially decomposed polysulfide solution was prepared by mixing 80 grams of finely divided elemental sulfur with 260 cc of an aqueous solution containing 0.016% by weight sodium sulfide along with an equivalent amount of sodium bicarbonate. The resulting solution was heated in an appropriate vessel to 125°C under 30 psig of nitrogen for about one-half hour. A liquid sulfur phase was then withdrawn from the vessel and found to contain 43 grams of elemental sulfur.

This result indicated that about 37 grams of the elemental sulfur originally charged to the vessel was hydrolyzed to sodium polysulfide. That is, about 46% of the elemental sulfur present in the simulated effluent from the polysulfide decomposition step was lost by back hydrolysis in the sulfur separation step when the elemental sulfur is just simply separated from the treated solution.

The second experiment was conducted according to this process and essentially involved the addition of 70 psi of H_2S to the separation vessel. The experiment repeated above was then repeated with this modification and it was found that 72 grams of the elemental sulfur was recovered. Thus the effect of performing this sulfur separation step under an H_2S partial pressure was to increase the yield of elemental sulfur by a factor of about 1.7 or by 67% relative to the control run.

Example 2: A water stream, containing 18.2% by weight sodium thiosulfate and 0.6% by weight sodium hydroxide, is heated by a conventional heating means to the temperature of about 350°F. The pressure of the resulting heated stream is then raised by a conventional pumping means to the value of about 790 psig. The resulting heated and high pressure aqueous stream is then admixed with a gas stream containing 91 mol percent H_2S,

4.8 mol percent CO_2 and 4.2 mol percent H_2O. The temperature of the gas stream is about 100°F and its pressure is about 780 psig. The resulting mixture is then passed into the top of a conventional, vertically positioned reduction zone containing a fixed bed of a reduction catalyst comprising a catalytically effective amount of cobalt sulfide combined with an activated carbon carrier material. More particularly, the catalyst comprised about 2.3% by weight cobalt as cobalt sulfide combined with 10 to 30 mesh particles of activated carbon. The inlet temperature and pressure of the reduction zone are fixed at about 350°F and about 780 psig, respectively. In addition, the flow rate of the aqueous solution is adjusted to achieve a liquid hourly space velocity of about 1.5 hr^{-1} based on the volume of the catalyst bed.

An effluent stream is then withdrawn from the bottom region of the reduction zone and found to be at a temperature of about 392°F and at a pressure of about 770 psig. The positive differential temperature across the reduction zone is of course caused by the exothermic reaction taking place therein; likewise, the negative differential pressure is caused by the pressure drop associated with the internals of the reduction zone. The relative rates of circulation of the liquid stream and the gas stream through the reduction zone are adjusted to result in the mol ratio of H_2S to sodium thiosulfate entering the reduction zone per unit time of about 5.6:1.

Analysis of the total effluent stream from the bottom of the reduction zone indicates that it contains, 0.6% by weight unreacted H_2S, 0.04% by weight CO_2, 1.06% by weight sodium thiosulfate, 1.6% by weight sodium bicarbonate and 20.9% by weight sodium polysulfide. The total effluent stream from the reduction zone is then quickly passed to a conventional cooling zone where its temperature is dropped to about 250°F while its pressure is maintained constant at about 770 psig. The flow rate of this effluent stream and the distance between the reduction zone and the cooling zone are adjusted so that the cooling of the effluent occurs within 5 minutes after the effluent leaves the bottom of the reduction zone.

This cooling step corresponds to a reduction in the temperature of the effluent stream of about 142°F and is effective to decompose the polysulfide solution and release substantial quantities of elemental sulfur. It is to be noted that this cooling step occurs under a partial pressure of H_2S. The resulting cooled effluent stream, now containing elemental sulfur, is then passed into a conventional sulfur separating zone where a liquid sulfur phase is allowed to separate from the cooled effluent stream. The separating zone is maintained under a pressure of 770 psig and a positive partial pressure of H_2S.

A stream of substantially pure, liquid elemental sulfur is then withdrawn from the lower region of this sulfur separating zone. A supernatant liquid also withdrawn from the upper region of this sulfur separating zone and passed through a suitable pressure reduction means to a gas-liquid separating zone maintained at a temperature of about 250°F and a pressure of 50 psig. A gas phase is allowed to separate from the liquid phase in this last separating zone and the gas phase is withdrawn as an overhead stream and the liquid phase comprises the bottoms stream.

The overhead gas stream comprises primarily H_2S admixed with some H_2O and CO_2, and it is withdrawn from the separating zone, admixed with additional make-up H_2S and passed back through suitable compressing means to the reduction zone in order to provide the gas stream used therein. The liquid phase is likewise withdrawn from this last separating zone and it comprises the treated water product stream of this process. An analysis of the treated water stream indicates it contains about 1% by weight sodium thiosulfate. Calculations based on the disappearance of the thiosulfate indicate that about 93% of the thiosulfate present in the input water stream charged to the reduction zone is eliminated in this particular case.

Conversion of Hydrogen Sulfide to Sulfites

A process described by *W.E. Luetzelschwab; U.S. Patent 3,821,110; June 28, 1974; assigned to Marathon Oil Company* removes unacceptable pollutants comprising H_2S, and light hydrocarbons from a refinery sulfidic water stream and discharges them as acceptable sulfite or bisulfite salts, (or sulfates or bisulfates, upon oxidation). H_2S is stripped from the sour water, then burned to produce SO_2 which is absorbed into a portion of the H_2S-stripped sulfidic water in an SO_2 absorption zone to form an aqueous bisulfite stream, which in turn, is recycled back to mix with the incoming sulfidic water stream. The aqueous bisulfite stream from the SO_2 absorption zone can contain sulfites, bisulfites, sulfates, or bisulfates, according to pH and degree of oxidation.

While the process is adaptable to any stream containing water, H_2S and NH_3, and/or alkali metal and/or any other cation forming water-soluble sulfide, its preferred commercial use will be in the purification of sour water streams which commonly result from the refining of sulfur-containing crude oils. Such streams are contaminated with both H_2S emitted from, e.g., the crude distillation step of the refining process and with NH_3, produced by, e.g., the hydrogenation step of the refining process, or added to fractionator overheads to neutralize the H_2S to inhibit corrosion.

Continuous Two-Stage Processing of Sludge

D.E. McCoy, W.N. Gilmer and E.C. Johnston; U.S. Patent 3,692,668; September 19, 1972; assigned to Texaco Inc. describe a continuous two-stage process for deoiling and dewatering refinery sludges in which, in the first stage, sludges mixed with a diluent oil, such as recycle oil recovered in the process, are heated at an elevated temperature and at a pressure which is about atmospheric or slightly above to effect separation of the sludges into a water vapor phase, and an oil-solids slurry phase. Separation of most of the oil from the oil-solids slurry phase is accomplished preferably by mechanical means, such as by centrifugation, after which the concentrated sludge is heated at an elevated temperature and at substantially atmospheric pressure or below to effect removal of the remaining oil by vaporization thus yielding dry, oil-free, free-flowing solids.

Petroleum based refinery sludges are widely encountered waste by-products of the petroleum refining industry. Such sludges are typically stable emulsions of oil, water and solid particulates. The solids originate as sands and clays contained in crude oil, stray catalyst particles, coke fines and as precipitates, scale and rust derived from various refinery processes.

Since petroleum sludges are produced continuously in refineries, they pose a constant disposal problem. For example, large amounts of sludges accumulate on the bottoms of settling ponds used to remove suspended matter from refinery water effluent streams. Eventually the pond bottoms must be dredged in order to maintain effluent quality and large volumes of sludges are generated for disposal. API separator bottoms and oil-soaked sand and soil resulting from oil spills also contribute to the sludge disposal problem of refineries. The nature of the sludge compositions has made it difficult over the years to economically deoil and dewater the sludges for pollution-free disposal.

A typical petroleum sludge contains about 35% solids, about 38% water, with the balance being oil. The high content of the oil and its value as a cracking stock make it desirable to recover as much of the oil present in the sludges as possible for removal and recycling to the refinery cracking unit.

In the process, the refinery sludges which are highly viscous and abrasive due to heavy oils and solid particles present and are not readily pumped through restricted areas such as heat exchanger tubes are mixed and diluted with oil such as recycle oil derived from the process or any hydrocarbon oil fraction derived, for example, from refinery operations and preferably having a boiling point of not less than about 300°F to provide a sludge mixture of improved pumpability and heat transfer. Generally from about 0.25 to 5.0 volumes or more and, preferably, about 0.5 to 2.0 volumes of the hydrocarbon oil are mixed with each volume of the sludge. It has been found that when the sludge mixture is passed through a first heating zone maintained at a temperature of about 220° to 300°F and at a pressure of about 15 to 75 psia and, preferably, at a temperature of about 220° to 270°F at a pressure of about 15 to 40 psia, it is possible to obtain an excellent separation of the mixture into a water vapor phase and an oil-sludge slurry phase. From the heating zone the heated mixture is passed to a suitable separator vessel which serves to separate the water vapor phase (i.e., steam) from the remaining oil-solids slurry phase. Steam withdrawn from the separator is condensed to yield water of reduced chemical oxygen demand while the oil-solids slurry phase is sent to the primary oil recovery section of the process.

In the primary oil recovery section a substantial part of the oil, i.e., about 50 to 90% by weight of the oil is separated from the oil-solids slurry phase by, for example, mechanical means such as through the use of a centrifuge optionally in conjunction with a filter. Advantageously, a portion of the recovered oil is recycled for mixing with the untreated sludges while the remainder is routed to the refinery for processing. The concentrated sludge from the primary oil recovery section still contains appreciable amount of absorbed oil and is, therefore, routed to the secondary oil recovery section of the process.

In the secondary oil recovery section the concentrated sludge is passed through a second heating zone maintained at a temperature of about 400° to 800°F at a pressure of about 1 to 20 psia and, preferably, at a temperature of about 550° to 750°F and at a pressure of about 5 to 15 psia, such as a rotary kiln where the remainder of the oil contained in the solids is vaporized. The oil vapors which are removed from the kiln by means of a vacuum system, by eluting with a flow of hot inert gas or by any other convenient method are subsequently condensed, for example, by exchanging heat in the feed-effluent heat exchanger.

Although it is possible to employ a part of the oil recovered from this process as fuel to heat the fired heater and kiln, it is preferred to utilize the recovered oil as catalytic cracking feedstock. The dry solids produced in the second heating zone are useful for a variety of purposes and, for example, are readily disposable in a sanitary landfill.

High Pressure Thermal Treatment of Sludge

According to a process described by *H.V. Hess and E.L. Cole; U.S. Patent 3,716,474; February 13, 1973; assigned to Texaco Inc.* oil-water emulsion type sludges, in particular refinery pit sludges are thermally treated at a temperature of at least 750°F under pressure of at least 3,000 psi to give a gas phase containing saturated hydrocarbons, suitable as fuel or cracking stock, a usable oil fraction, a water phase and a useful silt-coke phase.

In the process, coke deposits on the silt particles which are suspended in the feed as it passes through the heater and then into an insulated hold vessel. A large amount of sulfur is given off (as H_2S in the produced gas) indicative of the destruction of sulfur-containing emulsion formers in the sludge. Further, a large amount of CO_2 is given off in the gas indicative of decarboxylation reactions which can be related to destruction of the biological component. After sufficient time has elapsed to carry out the thermal treatment, and the system is finally cooled, it separates into a gas phase, oil phase, water phase, and silt and coke phase which has settled to the bottom of the water phase.

Waste Gas Scrubbing

According to a process described by *I.S. Shah; U.S. Patent 3,655,338; April 11, 1972; assigned to Chemical Construction Corporation* magnesium sulfite slurry, derived from the scrubbing of a waste gas stream for sulfur dioxide removal with a circulating aqueous magnesium oxide-sulfite slurry, is processed to recover a solids cake rich in magnesium sulfite which may be subsequently calcined. The aqueous magnesium sulfite slurry is passed through a thickener so as to decant clear liquor, and the thickened slurry is heated to convert magnesium sulfite hexahydrate to the trihydrate form. The slurry is then passed through centrifugal separation means under controlled conditions of feed rate, initial total solids content and speed in revolutions per minute of the centrifuge, so as to produce a residual wet solids cake with a moisture content of less than 5% by weight and an aqueous solution phase containing less than 1% residual solids by weight.

The principal advantage of the process is that minimum moisture content cake is produced by the centrifugal separation, and therefore, the heat requirement of the subsequent dryer and calciner are reduced. In addition, the handling characteristics of the cake are improved and change at low moisture content. In addition, minimum solids remain in the mother liquor from the centrifugal filtration.

The system was developed in a laboratory semiworks scale investigation. In a general study of operating parameters, the feed to the centrifuge was an aqueous slurry containing 5 to 25% solids content by weight, at a pH of 7 to 9. Analysis of the slurry solids phase indicated a content of about 90% hydrated magnesium sulfite, 5% hydrated magnesium sulfate, and 5% magnesium oxide. The liquor phase contained about 0.4% sulfur dioxide as magnesium sulfite and 1.0% sulfur dioxide as magnesium sulfate. The centrifuged cake solids contained about 85% hydrated magnesium sulfite, 5% hydrated magnesium sulfate and 5% magnesium oxide, balance surface water.

The mother liquor from the centrifuge contained in the range of 0.3 to 0.7% residual solids, and the liquor analyzed 0.3 to 0.5% sulfur dioxide as magnesium sulfite and 1.0 to 2.0% sulfur dioxide as magnesium sulfate in the liquor phase, and 50 to 70% hydrated magnesium sulfite, 1 to 5% hydrated magnesium sulfate and 20 to 50% magnesium oxide in the solid phase. It is important to attain minimum moisture content cake, as this affects the Btu requirements of the dryer and changes the handling characteristics of the cake, and to attain minimum solids in the mother liquor. Following are test results obtained by varying operating parameters.

Effect of Feed Rate

Feed Rate, liters/minute	% Solids in Feed Slurry	Bowl, rpm	% Moisture in Cake
7.57	13.3	5,000	25.0
11.0	13.3	5,000	9.0
14.8	13.3	5,000	7.7
16.7	20.0	6,000	6.9
28.8	20.0	6,000	6.5

Effect of Bowl, rpm

Feed Rate kg/minute	Bowl, rpm	% Moisture in Cake	% Solids Recovery
3.53	6,000	0.5	99
4.27	5,000	5.5	98
11.9	6,000	5.5	98
11.2	5,000	8.5	97
17.3	6,000	3.0	97
18.2	5,000	5.7	96

In the above table tests were conducted with 5.23% solids in the feed slurry. As the centrifuge bowl rpm decreases from 6,000 to 5,000, percent moisture increases, and percent solids recovery decreases. Hence, to attain less than 5% moisture, and maximum solids recovery, the bowl rpm should be greater than 5,000 under these conditions.

Effect of % Solids in Feed

Feed Rate kg/minute	% Solids in Feed Slurry	Bowl, rpm	% Moisture in Cake
17.7	20.0	6,000	6.9
17.2	5.23	6,000	3.0
17.2	13.3	5,000	7.7
18.1	5.23	5,000	5.7

The higher the precent solids in the feed slurry, the higher the moisture content. To obtain required moisture content, with increase in percent solids, the bowl rpm should be increased. With lower percent solids, the bowl rpm could be decreased. Thus, with 13% solids or higher in the feed slurry, the bowl rpm should be higher than 6,000, however, with 5 to 13% solids in the feed slurry, the bowl rpm should be 5,000 or higher. In an actual industrial installation, the bleed slurry drawn off from the circulating aqueous scrubbing slurry and containing solid particles of magnesium sulfite together with magnesium sulfate and magnesium oxide is passed to a thickener or settler, from which clear liquor and thickened slurry are drawn off. The thickened slurry is heated to a temperature above 40°C to convert magnesium sulfite hexahydrate to the trihydrate form, and the resulting slurry is passed to the centrifuge.

In a large scale installation, the bowl rpm range would be at least 1,500, or in the range of 1,500 to 3,600 rpm or up to 6,000 rpm as a maximum. The bleed slurry stream derived from waste gas scrubbing contains solids in a total proportion less than about 15% by weight, and generally in the range of 1 to 5% total solids. The aqueous slurry is passed through a thickener to produce a thickened slurry containing 5 to 25% solids by weight, and the thickened slurry is heated to a temperature above 40°C and preferably in the range of 50° to 80°C, to convert magnesium sulfite hexahydrate to trihydrate form.

The slurry is fed to the centrifuge at an initial total solids content in the range of 5 to 25% by weight, consisting principally of magnesium sulfite together with minor proportions of magnesium sulfate and magnesium oxide. A wet solids cake is produced with a surface moisture content of less than 5% by weight, and preferably in the range of 0.5 to 3% by weight. The aqueous solution phase discharged from the centrifuge contains less than 1% residual solids by weight, and typically contains in the range of 0.3 to 0.7% residual solids by weight.

FERTILIZER

Multistage Liming Process

C.C. Legal, Jr.; U.S. Patent 3,725,265; April 3, 1973; assigned to W.R. Grace & Company describes a multistage liming process which purifies wastewater by controlling the pH of the wastewater, e.g., as from a phosphate fertilizer plant, and thereby permits selective precipitation of impurities in the wastewater and their recovery as useful by-products of the process.

In one method lime is added in a multistage process to wastewater, more particularly wastewater containing phosphorus and fluorine pollution and silica, such as may be found in the wastewater of fertilizer plants. The pH of wastewater which is the result of fertilizer plants is generally quite acidic and it typically is 1 to 3. Lime is added directly to the wastewater until the pH is about 5. As soon as the liming process begins, fluorine, mainly as calcium fluoride, begins to precipitate. When the pH reaches about 5 to 5.3, the liming is stopped and the precipitate which has been formed is separated by any convenient separation procedure. After the separation, the filtrate is again treated with lime until a pH of about 6.5 to 8.5 is attained. This second liming step causes a second precipitate to form which is rich in P_2O_5 in the form of available or citrate soluble calcium phosphates.

The second precipitate, hereafter referred to as calcium phosphate, contains a mixture of phosphate including dicalcium phosphate and tricalcium phosphate. The ratio of fluorine to phosphorus in this second precipitate is generally in the range of 1:15 to 45. This is quite similar to Curacao rock which is a common source of fertilizer grade phosphate and is also used directly as a source of phosphate for chicken feed supplements.

In another method lime is added in the multistage process to the wastewater, more particularly wastewater containing phosphorus and fluorine pollution and silica, such as may be found in the wastewater of fertilizer plants. However, the first stage liming is stopped when the pH of the water reaches 5.0, preferably when the pH is in the range of 3.5 to 4.5, and more preferably at 3.5. At a pH of 3.5 about 21% of the fluorine remains in the wastewater, about 90% of the P_2O_5 still remains, and all the silica is still in the wastewater. Obviously, this means that there is less P_2O_5 contaminant in the first stage precipitate when the precipitate is removed at a pH of 3.5 rather than at a pH of 5.0. When the wastewater reaches the pH of 3.5 to 5.0, the precipitate which is formed is separated from the wastewater by any convenient separation procedure. After the separation the filtrate is again treated with lime until a pH of about 6.5 to 8.5 is attained. This second liming step causes a second precipitate to form which is rich in P_2O_5 in the form of available or citrate soluble calcium phosphates.

Of course, an intermediate step is also possible if the original precipitate is removed when the pH of the wastewater is 3.5. Since less fluorine has precipitated when the pH is 3.5 than when it is 5.0, there is more fluorine present to contaminate the by-product P_2O_5. Thus removal of all precipitate at about 5.0 will leave a substantially fluorine-free wastewater from which P_2O_5 can be extracted. The following examples illustrate the process.

Example 1: 132.1 grams of an approximately 5% CaO solution prepared by suspending and dissolving 47 grams CaO in 953 grams of deionized water (providing in effect, a solution and/or suspension of calcium hydroxide) were added to 2,000 grams of wastewater having a pH of 2.0, and the analysis given at the end of the example. The solution was stirred for 30 minutes, after which time the pH had risen to 5.0 and a precipitate had formed. The precipitate was collected and found to weigh 58.6 grams when wet and 13.2 grams when dry. The weight of the filtrate was 2,063 grams. An analysis performed on both the solid and filtrate gave the following results:

	Percentage of-			
First Stage Liming	P_2O_5	F	CaO	SiO_2
Solid	15.13	28.15	49.01	0.04
Filtrate	0.1584	0.0047	0.1302	0.0954

Based upon the percentages in the analyses given above, it can be calculated that the composition of the first precipitate (solid) is approximately as follows: 57.7% CaF_2, with most of the remainder being dicalcium phosphate, along with minor amounts of $CaSO_4$ and phosphates of iron and alumina, and a trace of silica. This is also the approximate qualitative composition of the first precipitate in each of the following examples.

24.3 grams of the approximately 5% CaO solution was then added to 1,000 grams of the filtrate to form the second precipitate. After stirring for 30 minutes the pH had risen to 7.10. After filtration the weight of the filtrate was 987 grams. The weight of the wet precipitate was 27.9 grams and after drying the precipitate weighed 4.63 grams. An analysis of the filtrate and precipitate and the analysis of the original wastewater are given below:

	Percentage of-			
Second Stage Liming	P_2O_5	F	CaO	SiO_2
Solids	35.00	0.98	42.01	11.10
Filtrate (pH 7.1)	0.0046	0.0003	0.0504	0.0444
Untreated wastewater	0.2810	0.1830	0.1358	0.0725

Example 2: 198.6 grams of a 5% CaO solution prepared as in Example 1 were added to 2,000 grams of wastewater having a pH of 1.95 and the analysis given below. The solution was stirred for 35 minutes, after which time the pH had risen to 5.1 and a precipitate had formed. The precipitate was collected and found to weigh 80.15 grams when wet and 20.0 grams when dry. The weight of the filtrate was 2,108.3 grams. An analysis performed on both the solid and filtrate gave the following results:

	Percentage of-			
First Stage Liming	P_2O_5	F	CaO	SiO_2
Solid	17.23	23.65	48.82	0.05
Filtrate	0.2051	0.0027	0.1282	0.0950

31.9 grams of the 5% CaO solution was then added to 1,000 grams of the filtrate. After stirring for 30 minutes the pH had risen to 7.05. After filtration the weight of the filtrate was 997.8 grams. The weight of the wet precipitate was 29.20 grams, and after drying the precipitate weighed 5.85 grams. An analysis of the filtrate and precipitate is given below:

	Percentage of			
Second Stage Liming	P_2O_5	F	CaO	SiO_2
Solid	31.73	0.68	38.92	10.92
Filtrate	0.0076	0.0005	0.0448	0.0454
Untreated wastewater	0.3970	0.2440	0.151	0.0770

Example 3: The following examples illustrate the minimization of P_2O_5 present in the first precipitate. 372.0 grams of 5% CaO solution prepared by suspending and dissolving 47 grams CaO in 953 grams of deionized water were added to 3,750 grams of wastewater having a pH of 1.90 and the analysis given at the end of the example. After 25 minutes the pH of the mixture had risen to 3.50 and a precipitate had formed. The precipitate was recovered and found to weigh 142.95 grams when wet. 2,250 grams of the raw pond water was then added to the wet solid, heated to a boil, and stirred for 15 minutes. Filtering was again used to obtain a second precipitate which weighed 125.02 grams when wet. 100.02 grams of this wet precipitate was dried in an oven at 105°C and 34.96 grams of dry solids with the following analysis was recovered:

	Percentage of			
	P_2O_5	F	SiO_2	CaO
Dry solids	6.80	36.00	0.44	42.00
Raw pond water	0.4168	0.3370	0.0967	0.0510

To 2,000 grams of the first stage filtrate was added 69.77 grams of the 5% CaO solution prepared heretofore. After mixing 15 minutes the pH had risen to 7.0 and a precipitate had formed. No. 1 filter paper was then used to recover the 98.38 grams of wet precipitate by filtration. The wet precipitate was then dried in an oven at 105°C and 15.30 grams of dry solids were recovered. The dry solids and filtrate had the following analysis:

	Percentage of			
	P_2O_5	F	CaO	SiO_2
Dry solids	34.30	0.56	38.68	2.70
Filtrate	0.0116	0.0003	0.0250	0.0960

Example 4: 20 grams of the dried solids of the first stage liming of Example 3 were then treated with hot tap water as follows:

(a) slurried with 1,000 ml hot tap water and filtered (the filtrate had a pH of 3.0);
(b) the precipitate from (a) was then washed in a Büchner funnel with 1,000 ml hot tap water and again filtered (the filtrate had a pH of 3.5);
(c) the washed precipitate from (b) was slurried with 1,000 ml hot tap water and filtered (the filtrate had a pH of 3.5);
(d) the precipitate from (c) was slurried in 1,000 ml hot tap water and filtered slightly raising the pH of the filtrate;
(e) the precipitate from (d) was then slurried with 1,000 ml hot tap water twice and filtered after each slurrying (the filtrate's pH was then 5.9);
(f) the precipitate from (e) was then washed twice in a Büchner funnel with 500 ml of hot tap water and filtered after each washing (the pH of the filtrate was then 6.7);
(g) the wet precipitate was dried in an oven at 105°C. 16 grams of dried solids having the following analysis were recovered:

	Percentage of			
	P_2O_5	F	SiO_2	CaO
Dried solids	5.10	38.95	Nil	56.7

Phosphate Plants

G.F. Palm; U.S. Patent 3,699,212; October 17, 1972 describes systems for use and handling of water in phosphate plants to reduce air and stream pollution, to reduce losses, and to reduce costs. This process provides treatment systems which eliminate air pollution from fluorides evolved from the acid water ponds, and eliminate stream pollution due to leakage or seepage from the acid water ponds.

Two general systems are described, one in which waters of the acid water pond and of the gypsum disposal pond are separately pH-adjusted and then mixed for reuse in the plant, with or without supplementary pH adjustment of the mixed waters, and the other in which the waters of the acid water pond and of the gypsum disposal pond are separately pH-adjusted, at distinct pH levels, and separately reused in the plant, with or without supplementary pH adjustment of the separate water streams. The latter system is more efficient and permits more nearly ideal uses of the waters and recoveries of the valuable materials contained in the waters.

Conversion of Phosphorus Values to Calcium Phosphate Dihydrate

R.B. Hudson, R.E. Mesmer and R.A. Rose; U.S. Patent 3,650,686; March 21, 1972; assigned to Monsanto Company describe a process for the recovery of phosphorus values from an effluent stream containing dilute soluble phosphate values by adding lime to form and precipitate basic calcium phosphate having a general formula $Ca_{10-x}(PO_4)_6(OH)_{2-x}\cdot 2.5XH_2O$ and thereafter draining the excess water to obtain a sludge. The improvement comprises slurrying the sludge with a substantially equivalent amount (on P_2O_5 basis) of phosphoric acid to form crystalline calcium phosphated dihydrate whereby a compound readily filterable and easily removed from the effluent is obtained. The resulting calcium phosphate dihydrate after conventional dewatering process is ready for use as a conventional feed supplement.

Example 1 (Comparative): A series of experiments were carried out to demonstrate the nature of the basic calcium phosphates and several process modifications. Lime was added to the phosphate-bearing water as a freshly prepared slurry of $Ca(OH)_2$ at 13% solids. Phosphate-bearing water was a solution of a mixture of sodium orthophosphates and polyphosphates such that the pH was 7 and 90% of the total P_2O_5 was present as orthophosphate at concentrations as indicated in Table 1 below. Three different methods of contacting the phosphate-containing water with lime slurry are given.

(1) Single-Stage Contact — 800 ml of phosphate-bearing water was placed in a 2,000 ml beaker equipped with a variable speed stirrer having a chord-shaped agitator 10 cm long by 2.5 cm at mid-chord. Lime slurry was added rapidly and mixed with the water at an agitator speed of 100 rpm. This was followed by agitation at 50 rpm for 10 minutes to flocculate the precipitate. After flocculation the precipitate was allowed to settle for 10 minutes before any sample was withdrawn.

(2) Sludge Recycle Simulation of a Commercial Reactor-Clarifier — This batchwise simulation was done in the same equipment as the single-stage procedure. The single-stage procedure was followed and after 10 minutes settling, the supernatant liquid layer was decanted. A fresh charge of lime slurry was then added to the sludge and mixed at 100 rpm for 2 minutes. A second 800 ml charge of phosphate-bearing water was then added and the 5-minute mixing, 10-minute flocculation, and 10-minute settling sequence was repeated. The cycle could be repeated indefinitely if sludge is withdrawn periodically but only 2 cycles were used for this work except where noted otherwise.

(3) Two-Stage Parallel Reaction — This scheme used the equipment previously described to simulate a two-stage continuous reactor system. All of the lime slurry is mixed rapidly for 5 minutes with half of the phosphate-bearing water. Initial production of a lime-rich slurry reduced deactivation of the lime by surface coating and furnished nucleating centers. The remaining half of the water is then added, mixed and flocculated for 30 minutes before settling. The two-stage and sludge recycle procedures yielded somewhat better P_2O_5 removal

than the single-stage procedure. Contrary to expectation from the trade literature, little difference in solids settling characteristics was found that could be attributed to the type of process. This is probably due to the relatively high phosphate concentration of the water.

TABLE 1: COMPARISON OF LIME-TREATMENT PROCESSES

Process	Phosphate Concentration, mg/l as P Initial	Residual	% Removal	Solids Content of Slurry After 24 Hours Settling, percent
Single-stage	1,800	41	98	7.5
Two-stage	3,600	7.7	99+	11.8
Sludge recycle	1,800	4.5	99+	8.0
	3,600	2.2	99+	10.5

Ca/P mol ratio is 2.0.
Total contact time is ca 30 minutes.
Pebble quicklime slaked at 90°C.

TABLE 2: EFFECT OF INITIAL PHOSPHATE CONCENTRATION ON SETTLING AND DEWATERING

Initial Phosphate Concentration, mg/l as P	Solids Content of Settled Slurry, (weight/volume) After 24 Hours	After 1 Week
450	2%	No data
900	4%	No data
1,800	8%	19%
3,600	10%	23%
5,400	12%	25%
22,500	16%	33%

Ca/P mol ratio is 2.0.
Two-stage procedure.
Percent solids (w/v) is (gram solids x 100)/(ml slurry).

This illustrates the effect of process and concentration on basic calcium phosphate.

Example 2 (Comparative): Supernatant water was decanted from a portion of the basic calcium phosphate slurry from the two-stage reaction as shown in Example 1 after 24 hours settling. The settled slurry was 11.8% solids. One hundred ml of this slurry was filtered on a 11 cm vacuum filter with suction at 30 mm Hg absolute. Suction was supplied for 15 minutes until the filter cake was dewatered and shrinkage cracks developed. The resulting wet cake contained 73% water as determined by drying loss at 150°C. The resulting dried cake was examined by x-ray diffraction and found to be largely amorphous or very poorly crystalline. Very weak diffraction patterns for hydroxyapatite and $Ca(OH)_2$ were the only crystalline components that could be detected. This illustrates the poor crystallinity and difficulty in dewatering basic calcium phosphates.

Example 3: A quantity of 24 hour settled sludge at 10.5% solids obtained from the experiments described in Table 1 was poured slowly into reagent grade phosphoric acid (54% P_2O_5) at 40°C with vigorous agitation until the pH reached 6.5. The sludge was added over a 15 minute period. The slurry was stirred for an additional 15 minutes during which time small additions of slurry were added to bring the final pH to 7.0. After only 10 minutes settling without agitation, the slurry was 18% solids (w/v).

The supernatant liquid was decanted and 100 ml of slurry was filtered and dried as described in Example 2. Filtration was complete in 5 minutes and the water content of the wet cake was only 42%. X-ray diffraction of the dried material showed it to be strongly crystalline and essentially all $CaHOP_4{\cdot}2H_2O$. This material was more than 30% plus 100 mesh and 50% plus 270 mesh in particle size. The filtrate contained 1,100 mg/l soluble P.

Example 4: The experiment described in Example 3 was repeated with the exception that the reaction temperature was held at 70°C. In this case, the filtration required 8 minutes to complete, the wet cake contained 53% water and the dried product was anhydrous $CaHPO_4$ rather than the dihydrate. These examples show the great improvement in dewatering that can be obtained by converting the amorphous basic calcium phosphates to crystalline dicalcium phosphate.

Ammonium Nitrate

I.R. Higgins; U.S. Patent 3,775,088; November 27, 1973; assigned to Small Business Investment Company of New York, Inc. describes a continuous process for treating a fertilizer plant waste aqueous stream containing ammonium, hardness metal and nitrate values to recover an ammonium nitrate fertilizer and a demineralized and substantially nitrate-free aqueous stream for recycle.

The process includes initially removing the ammonium ion and hardness metal ion values from the fertilizer plant waste aqueous stream with a cation exchange resin bed and then contacting the essentially ammonium and hardness metal ion-free aqueous stream with an anion exchange resin bed to recover a second ammonium nitrate fertilizer stream and a demineralized and nitrate-free aqueous stream.

The apparatus shown in Figure 7.3 comprises a cation exchange column loop system **10** in combination with an anion exchange column loop system **13** for recovering valuable fertilizer values from a fertilizer plant waste stream while simultaneously producing from the same waste stream process water suitable for recycle within the total system, thus providing an ideal overall operation from both an ecological and economical viewpoint.

FIGURE 7.3: TREATMENT OF FERTILIZER PLANT WASTE STREAMS

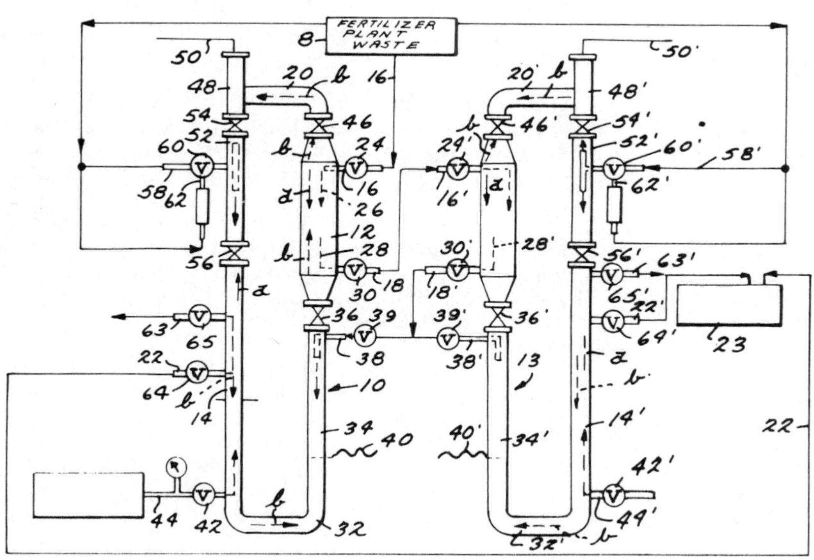

Source: I.R. Higgins; U.S. Patent 3,775,088; November 27, 1973

The continuous cation exchange column loop system **10** has two operating ion exchange sections, i.e., a loading section **12** and a stripping section **14**. The direction of liquid flow is indicated by the solid arrows **a** and the direction of resin flow is indicated by the open arrows **b**. The loading section **12** is positioned adjacent the upper portion of one of the legs of the column loop and fertilizer plant waste enters through conduit **16** adjacent the upper portion of the loading section **12**. This loading section contains a strong acid type cation exchange resin in the acid or salt form, and as the fertilizer plant waste stream flows downwardly, the hydrogen ions of the ion exchange resin are exchanged into solution and the ammonium and metal ions such as calcium and magnesium are adsorbed on the resin. Thus, a softened product water is withdrawn from the lower portion or region of the loading section **12** through conduit **18** and is delivered to the anion exchange column loop system **13**.

Thus, the ammonium and metal ion rich resin now in the loading section **12** is shifted upwardly around the upper loop end **20** and is passed downwardly to the stripping section **14**. In the stripping section a nitric acid regenerating solution is passed upwardly through the bed of loaded resin, and the ammonium and metal ions loaded thereon are exchanged into solution with adsorption of hydrogen ions from the regenerating solution. Thus, an ammonium nitrate containing stream is removed from the stripping section via conduit **22** and is delivered to ammonium nitrate product recovery station **23**.

The aqueous stream leaving the loading section **12** of the cation exchange column loop system **10** via conduit **18** is characterized in that it has substantially reduced hardness and ammonia values. The stream is delivered, directly or indirectly via storage vessels, not shown, to the continuous anion exchange column loop system **13** which has two operating anion exchange sections, i.e., a loading section **12'** and a stripping section **14'**. Again, the direction of liquid flow is indicated by the solid arrow **a** and the direction of resin flow is indicated by the open arrow **b**. As in the cation exchange column loop system **10**, the aqueous stream in these sections of the anion exchange column loop system generally flows countercurrently to the movement of the resin.

The softened and reduced ammonia value aqueous stream from cation exchange column loop **10** is introduced into the loading section **12'** via conduit **16'**. The loading section **12'** contains a weak base anion exchange resin and as the softened aqueous stream flows therethrough, the hydroxyl ions are exchanged into solution and the nitrate anions of the softened aqueous stream are absorbed on the resin.

Thus, a substantially contaminating nitrate anion and cation free aqueous stream, i.e., a demineralized stream is taken off from the loading section **12'** via conduit **18'**, the demineralized aqueous stream advantageously being recycled within the overall operation as a rinse medium for both the regenerated cation and anion resin via conduits **38** and **38'**.

The contaminating nitrate anion rich resin now in the loading zone **12'** is shifted to the stripping section **14'**. In stripping section **14'**, a regenerating agent, an aqueous ammonia solution, is passed upwardly through the resin and the nitrate anions are exchanged into solution with absorption of hydroxyl ions. Thus, an ammonium nitrate containing stream is removed from the stripping section **14'** via conduit **22'** and is delivered to ammonium nitrate product recovery station **23**.

This system can conveniently be operated to treat a fertilizer plant waste stream typically containing the following constituents: 340 ppm NH_3; 4.8 ppm Mg; 60 ppm Ca; 1,240 ppm NO_3; 53 ppm Cl; 72 ppm SO_4; 15 ppm SiO_2; 27 ppm (average) urea. The cation exchange column loop **10** is operated for a 13.5 hour period during which time 4,792 gallons of fertilizer plant waste feed were treated. This yielded a decationized water containing less than 5 ppm NH_3 and an ammonium nitrate (AN) product of 91 gallons with an average AN concentration of 0.81 pounds per gallon.

The following operating conditions prevailed: 2.25 minutes run time; 10 seconds pulse time; 13 inches pulse length; 1.33 cu ft/hr resin rate; 6.15 gal/min feed rate; 31.4 gpm/ft^2 flow. The resin was Dowex HCR-W, having a size ranging from 16 to 100 standard mesh; the regenerating agent was aqueous nitric acid having a concentration in the range of 15 to 35% by weight.

The anion exchange column loop **13** was operated for a total of 16.5 hours to treat 4,300 gallons of decationized stream from the cation exchange column loop **10** to yield a demineralized, essentially nitrate free water containing less than 5 ppm NH_3 and an ammonium nitrate (AN) product of 111 gallons with an AN concentration of 0.495 lb/gal. The following operating conditions prevailed: 2.25 minutes run time; 10 seconds pulse time; 14 inches pulse length; 1,443 cu ft/hr resin rate; 4.5 gal/min feed rate; 23 gpm/ft^2 flow. The resin was IRA-93 having a size ranging from 16 to 100 standard mesh, weak base. The regenerant was a 4 to 25% by weight aqueous ammonia solution.

The ammonium nitrate fertilizer solution produced from both the cation and anion exchange columns has a concentration in the range of 15 to 25% solids by weight and in the product tank receiving the same, the excess acid contained therein is neutralized with ammonium hydroxide. Prior to introducing the fertilizer plant waste aqueous stream into the cation exchange column loop **10**, preferably this stream is passed through one or more filters to remove particulate matter and urea.

Further, prior to recycling the demineralized aqueous stream from the anion exchange column loop to both the cation exchange column loop and the anion exchange column loop as cation resin rinse medium and anion resin rinse medium, respectively, which demineralized aqueous stream can also be recycled to the principal ammonium nitrate production system for use in the cooling towers, this demineralized aqueous stream can be introduced into a second bed of an anion exchange resin bed, fixed or otherwise, and preferably a strong base anion exchange resin to remove silica values therefrom. Thereafter, this second bed of anion exchange resin can be regenerated using sodium hydroxide as the regenerating agent with the effluent being directed to a waste stream.

Additionally, the demineralized and essentially silica free aqueous stream produced in the second anion exchange resin bed can be recycled to the main fertilizer plant producing ammonium nitrate as boiler feed water or cooling tower water. The diameter of the loading zone as well as the diameter of the stripping section of each column loop can be varied. Generally, the diameter of the loading zone will be at least about 12 inches in order to achieve an economical as well as efficient treatment of the fertilizer plant waste aqueous stream. The columns, preferably, are constructed of mild steel and the sections thereof can be belled to 12 inch diameter butterfly valves and 12 inch elbows.

PAINT

Electrophoresis Painting

A process described by *R. Eisenmann; U.S. Patent 3,772,190; November 13, 1973; assigned to Otto Durr KG, Germany* relates to separating particles of paint, enamel or the like from the wash water of electrophoresis painting apparatus by mixing the polluted water with air so as to produce a foam, for example, a paint foam, and by subjecting the foamy mixture to a separating process in which the paint or other undesired substance is removed from the water.

Referring to Figure 7.4, the first stage **I** of the process is carried out in a mixing apparatus **10**, preferably in the form of a circulating container, while the second stage **II** of the process is likewise carried out in a mixing apparatus **11** which preferably also consists of a circulating container. The numeral **13** generally indicates the separating and calming unit in the form of four adjacent containers **14, 15, 16** and **17** the upper edges of which have different levels so that a part of the contents and especially the foam of the first container **14** which has the highest liquid level **14a** can flow over an overflow **18** into the next container **15** which has a medium liquid level **15a**, and from the latter over an overflow **19** into the next container **16** which has the lowest liquid level **16a**.

The upper part of this last-mentioned container **16** is provided with a partition **20** from which a ramplike wall **20a** ascends over which the foam is finally discharged from the apparatus. Due to this partition **20**, a foamless chamber **21** is formed in the container **16** from which the partly purified water may flow over an overflow **22** into the next container **17**. The impure mixture of water and paint or enamel is first supplied through a line **23** to the mixer **10** of the first stage, preferably a circulating container, into the lower end of which the water which has previously been partly purified in the container **15** is supplied through a line **24** by means of a pump **25** so as to form a driving jet which has an injector effect.

The mixing container **10** is further supplied through a line **28** with the coagulant from a container **27** which may be provided, for example, with a stirrer **26**, and through a line **29, 30** in which check valves are provided; compressed air is also passed into the container for producing the foam. Due to the injector effect, the mixture of water, paint or enamel, coagulant and compressed air is thus circulated frequently in the mixing container **10** and the ingredients of this mixture are thus thoroughly mixed with each other and whipped up to a foam.

FIGURE 7.4: PURIFICATION OF WASH WATER FROM ELECTROPHORESIS PAINTING APPARATUS

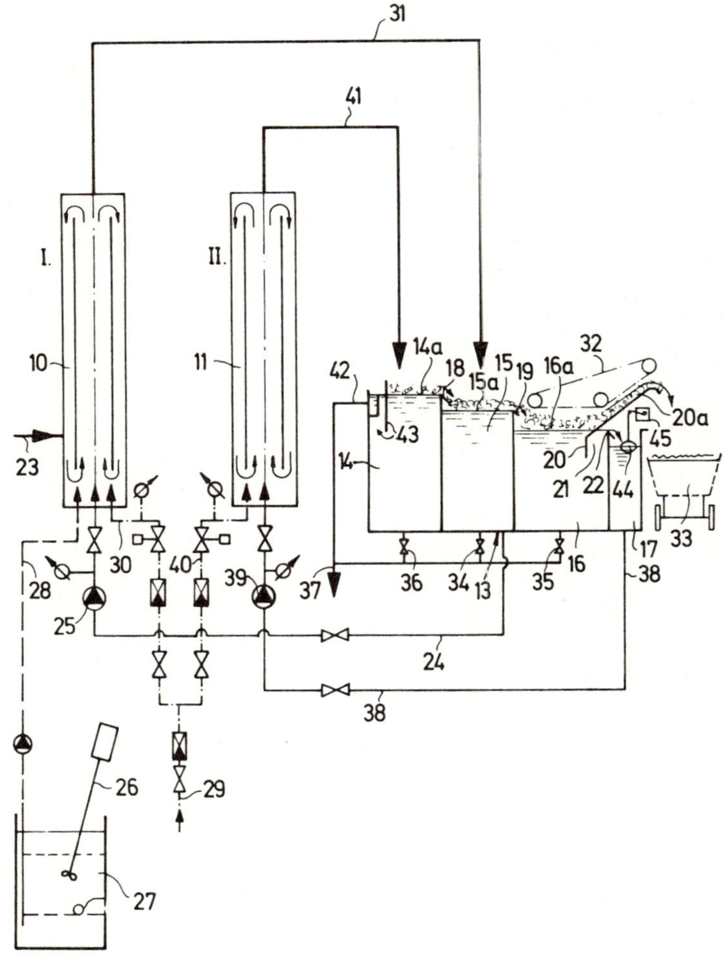

Source: R. Eisenmann; U.S. Patent 3,772,190; November 13, 1973

The foamy mixture is then conducted from the container **10** through a line **31** to the container **15** in which the first separation of this mixture takes place. The foamy mixture of water and paint which is then deposited on the surface of the water in the container **15** will then flow over the overflow **19** into the next container **16** in which a further separation of water and foam takes place. The foam may then be wiped or stripped off the surface **16a** of the water in the container **16**, for example, by means of a conveyor belt which is provided with wipers which push the foam over the ramp **20a** from which the foam will then drop into a lorry or other vehicle **33**.

For emptying the containers **15** and **16** and also the container **14**, these containers are provided with discharge valves **34**, **35** and **36** which are connected to a common discharge line **37**. Since a purifying process as described above is often not sufficient for attaining a degree of purity of the water which is satisfactory or as high as desirable, this process provides an additional purifying stage which is carried out as follows. After being separated from the paint foam, e.g., in the container **16**, and after passing through the chamber **21** into the last container **17**, the water which still contains particles of paint or enamel is conducted under pressure, preferably by means of a pump **39**, through a line **38** to the mentioned mixing apparatus **11** of the second stage **II**.

Thus, the water which is supplied from the container **44** through the line **38** is thoroughly mixed with compressed air in the apparatus **11** by being circulated numerous times therein. The compressed air may for this purpose be supplied to the mixing apparatus **11** from the line **29** through a branch line **40** which is controlled by valves or similar means. The forced circulation of the mixture of water and compressed air in the mixing apparatus **11** is again produced by an injector effect similarly as in the apparatus **10**. In this case it is, however, usually not necessary to supply the apparatus **11** with a coagulant, although if desired this may also be done, for example, through a line from the container **27**.

The foamy mixture of water and paint or enamel which is formed in the second stage is conducted through a line **41** to the container **14** in which another separation of the foam from the water is carried out. Since the surface of the liquid in this container **14** has a lever **14a** which is higher than the level **15a** of the liquid in the following container **15**, the foam can flow over the overflow **18** into this container **15** and can then pass together with the foam formed in this container over the overflow **19** into the following container **16** in which all of the foam is separated from the liquid.

From the upper part of the container **14** in which the water has already reached a high degree of purity this water passes over an overflow **42** which is located at a slightly lower level than the overflow **18** and is screened by a suitable screen **43** from the foam which floats on the surface **14a** of the liquid in the container **14**. From the overflow **42** the purified water finally passes out of the apparatus through the discharge line **37**.

In order to insure that the amount of the partly purified water which passes back into the circulating apparatus through the line **38** equals as accurately as possible the amount of paint-polluted water which is supplied to the apparatus through the line **23**, a float element **44** is provided in the container **17**. This float element **44** is connected to a switch **45** and is adapted to control the flow of water through the line **38** in accordance with the level of the water in the container **17**, for example, by varying the operation of the pump **39**.

In related work *R. Eisenmann; U.S. Patent 3,764,013; October 9, 1973; assigned to Otto Durr KG, Germany* describes a method and apparatus for separating paint or the like from water, especially for purifying the water before passing it into a lake or river, by initimately mixing the paint-contaminated water with a coagulant and compressed air so as to form a foamy mixture, then allowing the foam to float quickly to the surface of the water, wiping the foam off the surface and filtering the water before discharging it from the apparatus. The apparatus is preferably designed as a circular unit requiring relatively little ground space in which the various separating containers are of an annular shape surrounding each other about a common axis, while the coagulator is mounted above the central cylindrical space within the innermost annular container and extends likewise coaxially thereto.

Electrodeposition and Ultrafiltration

L.R. Le Bras, R.M. Christenson and R.R. Zwack; U.S. Patent 3,749,657; July 31, 1973; assigned to PPG Industries, Inc. have found that electrodeposition rinse water may be treated by a selective separation process comprising a separation through a physical barrier such as a membrane or filter in an ultrafiltration process; thereby removing or concentrating the solids which may either be disposed of or returned directly or indirectly to the electrodeposition bath.

The resultant filtrate or effluent, in turn, may be returned directly to the electrodeposition bath, if desired, or more usually may be employed as a rinsing agent, either as such or in admixture with or in combination with conventional rinsing media such as deionized water, tap water, or any other source of water usable as rinse water. The rinsing may be conducted in such a way that at least a portion of the rinse water containing solids is returned to the electrodeposition bath. If desired, the filtrate or effluent may be treated as waste and disposed of in an appropriate manner.

It is noted that the rinse water containing dragout is frequently deficient with respect to solubilizing agent. This can be explained by the fact that the resin being electrodeposited migrates toward the electrode being coated. When the resin is coated, it gives up the solubilizer counter-ion which migrates toward the counter-electrode. There is frequently an area created immediately surrounding the electrode being coated which is richer in resin and poorer in solubilizing counter-ion than the remainder of the electrodeposition bath. Thus, the dragout frequently tends to be solubilizing agent deficient.

Before treating the dragout-containing rinse water by the process described, it is desirable when the rinse water is in an unsatisfactory pH condition or in an unstable condition due to a deficiency in solubilizing agent to add sufficient solubilizing agent, e.g., in the case of a polyacid resin a base such as an amine or alkali metal hydroxide such as potassium hydroxide, in order to adjust the pH of the rinse water to the point where the solubilized resin is in a stable form, that is, so that it does not precipitate or agglomerate, especially when concentrated.

It is usually desirable to add the solubilizing agent which is present in the electrodeposition bath, especially if the concentrated solids are to be returned to the electrodeposition bath, although any solubilizing agent may be employed. Generally, in the case of polyacid resins, it is preferred to process rinse water having a pH of at least 6 and usually at least 7.5. Obviously, the optimum pH depends on the nature of the resin and optimum pH of the electrodeposition bath. The following examples illustrate the process.

Example 1: The vehicle resin in this example is a maleinized tall oil fatty acid-adipic acid ester of a styrene-allyl alcohol copolymer of 1,100 molecular weight and 5 hydroxyl functionality comprising 39.8% of the copolymer, 52.9% tall oil fatty acids, 1.3% adipic acid, and 6.0% maleic anhydride as a 90% solution in 4-methoxy-4-methyl pentanone-2 having an intrinsic viscosity of 31,500 centipoises and an acid value of 41. The electrodepositable material had the following composition:

	Percent		
Non-volatiles			100
Vehicle non-volatiles		86.05	
20 percent maleinized linseed oil	5.05		
Vehicle resin	85.44		
Alkylether of methylolated phenol (Methylon 75108)	8.38		
Cresylic acid	0.88		
Surfactant (Witco 912)	0.25		
Pigment		13.95	
Lead silicate	4.15		
Manganese dioxide	8.03		
Anthracite coal	83.40		
Strontium chromate	4.15		
Solubilizing amine	¼ diethyl/triethylamine		

The composition was reduced to 12% solids with deionized water. The solvent comprised 2.6% of the total composition. Steel parts electrocoated from this bath were rinsed with deionized water and the water containing dragout was collected. This rinse water was subjected to a batch selective filtration utilizing a Diaflow Ultrafilter (PM 30). The filter passed ultrafiltrate and retained the coating solids. However, the pigment tended to flocculate and settle when subjected to agitation. An additional portion of rinse water was treated with triethylamine (0.5 part per 400 parts of rinse water), the pH rising to 10.1 and the conductivity at 75°F rising from 330 to 600 mmhos/cm. This rinse water was subjected to ultrafiltration without precipitation. The following properties were noted.

	pH	Conductivity 75° F., mmhos/cm.	Solids
Base adjusted rinse	10.1	660	0.6
Concentrate	9.9	1,560	4.5
Filtrate	10.1	570	0

The filtrate was utilized to rinse additional electrocoated steel panels. No discernible difference in appearance was noted compared to deionized water. The concentrate, a stable dispersion, was returned to the electrodeposition bath. No discernible difference was noted in the steel panels coated from the bath after addition as compared to panels coated before addition.

Example 2: The vehicle resin in this example is a maleinized tall oil fatty acid-adipic acid ester of a styrene-allyl alcohol copolymer of 1,100 molecular weight and 5 hydroxyl functionality (Shell X-450) comprising 39.7% X-450, 52.9% tall oil fatty acids, 1.3% adipic acid and 6.1% maleic anhydride as a 90% solids solution in 4-methoxy-4-methylpentanone-2 having a viscosity of 36,700 centipoises and an acid value of 38.2. The electrodeposition primer had a composition:

	Percent		
Non-volatiles			100
Vehicle non-volatiles		87.35	
Allylethyl of methylolated phenol	7.63		
20% maleinized linseed oil (grinding vehicle)	5.82		
Vehicle resin (above)	85.10		
Surfactant (combination nonionic surfactant oil-soluble sulfonate—Witco 912)	0.60		
Cresylic acid	9.85		
Pigment		12.65	
Red iron oxide	17.05		
Anthracite coal	57.65		
Bentone 11 (a montmorillonite clay modified with the trimethyl-octyl ammonium ion and containing 0.65% nitrogen)	1.75		
Strontium chromate	3.30		
Lead silicate	14.52		
Manganese dioxide	5.78		
Organic solvent	4-methoxy-4-methylpentanone-2 in 20/80 ratio to vehicle resin (above).		
Amine	1/4 diethyl/triethylamine.		

The composition was diluted to 11% solids with deionized water. Phosphatized steel panels were coated at 250 volts for 2 minutes (75°F). These panels were rinsed with deionized water and the dragout-containing rinse was collected. Triethylamine was added to raise the pH of the rinse to 9.0 and the rinse water ultrafiltered as in Example 1.

The filtrate was utilized to rinse additional electrocoated steel panels. No discernible difference in appearance was noted compared to deionized water rinse. The concentrate, a stable dispersion, was returned to the electrodeposition bath. No discernible difference was noted in the steel panels coated from the bath after addition as compared to panels coated before addition.

Example 3: The vehicle resin of this example is an acrylic resin comprising 55% butyl acrylate, 25% styrene, 15% methacrylic acid, and 5% hydroxyethyl methacrylate. The polymer was utilized as a 74% solids solution in butyl Cellosolve (the monobutyl ether of ethylene glycol). The resin had a viscosity of 128,000 centipoises and an acid value of 68.2. The electrodeposition bath had the following composition in parts by weight: 223 parts vehicle resin (above); 42 parts ethoxymethoxymethyl melamine (XM1116); 24 parts diisopropanolamine; and, 35 parts deionized water. The bath contained 5.5% solids, with a pH of 8.8 and had a specific conductivity at 77°F of 490 mmhos. Tin-free steel panels were coated from this bath and the dragout material was collected by rinsing with deionized water and collecting the rinse water. The panels were coated at a bath temperature of 80°F at 300 volts for 5 seconds.

The dragout-containing rinse water containing 0.52% solids was subjected to ultrafiltration as described in Example 1. A total of 10,000 parts of rinse water were treated. The rinse water was ultrafiltered to a point where the solids of the retentate was 5%. The retentate was returned to the electrodeposition bath and panels were again coated. The reconstituted bath contained 5.2% solids with a pH of 8.6 and a specific conductivity at 77°F of 52 millimhos per centimeter.

Panels coated and baked under the same conditions as from the original bath showed no discernible difference in film thickness or properties. The ultrafiltrate was subsequently employed as a rinsing agent for electrodepositable panels and again so discernible difference was noted between panels rinsed with the deionized water or panels rinsed with the ultrafiltrate. Returning the dragout-containing ultrafiltrate rinse water to the electrodeposition bath affected neither bath stability nor thickness or properties of the subsequently coated panels.

Latex

According to a process described by *C.M. Taubman and A.M. Carabine; U.S. Patent 3,753,901; August 21, 1973; assigned to DeSoto Inc.* aqueous wastes, and particularly aqueous latex paint wastes, are flocculated using a polymer of benzyl C_1 to C_4 dialkyl acryloyl- or methacryloyl-oxyalkyl ammonium chloride. Preferred polymers are produced by aqueous solution polymerization at a temperature below about 90°C and a proportion of acrylamide or methacrylamide is desirably present in the polymer.

The latex wastewater is derived from the cleaning processes used in the production of latex paint, as when the equipment used is washed down at the end of a production run. This waste is characterized by very high levels of colloidal and suspended particles which cannot be discharged directly to sewer lines. Many commercial polyelectrolytes specifically recommended for flocculation of aqueous industrial wastes are either not effective at all, or are effective only at high dosage levels in separating the suspended solids from latex wastewater in the form of a mobile sludge.

The copolymers used in this process are approximately as efficient for flocculation as the homopolymers, but have the advantage of an extended optimum dosage range. The dosage range is the extent to which excess flocculant over that needed for complete flocculation can be tolerated before a tendency for redispersion is encountered. Thus, the copolymers are preferred, especially copolymers with acrylamide or methacrylamide, because these copolymers are less affected by dosage error. The following examples illustrate the process.

Example 1: The preparation of a preferred acrylamide-containing copolymer is as follows. To a four necked round bottom flask equipped with a stirrer, nitrogen sparge, thermoregulator for heating or cooling, monomer inlet and condenser, are charged 700 parts of deionized water and heat is applied to bring the water to 70°C. In a separate flask, 250 parts of acrylamide and 385 parts of a 65% aqueous solution of benzyl dimethyl methacryloyloxyethyl ammonium chloride are dissolved in 365 parts of deionized water to form a monomer premix.

A portion of the monomer premix (20%) is added to 70°C water together with 25.4 parts of free radical generating catalyst formed by dissolving 1.6 parts of potassium persulfate in 100 parts of deionized water. The remainder of the monomer premix and the balance of the catalyst solution is added in increments over a period of about an hour while maintaining 70°C. During the monomer addition an exotherm occurs requiring cooling. The peak temperature in this example was 88°C before air cooling stopped the temperature rise and cooled the mixture back to 70°C. Then, 0.5 part of potassium persulfate in 100 parts of water are added and 70°C is maintained for one hour and the solution product is cooled and discharged ready for use. It has a nonvolatile content of 27%.

Example 2: The treatment of aqueous latex paint waste is as follows. A typical aqueous latex paint waste is characterized by a substantial concentration of suspended solids (10,000 or more milligrams per liter); a substantial chemical oxygen demand (15,000 or more milligrams per liter); and a substantial total solids concentration (15,000 or more milligrams per liter). If this suspension is placed in the quiescent state, a very small reduction in pollutional constituents is obtained. This reduction is very small compared to the reduction required for discharge of the waste to surface water or to a municipal sewer system.

The suspended solids of this waste are, in general, nonbiodegradable pigments and resinous latex solids which represent a high inert solids loading problem to a municipal waste treatment plant as well as a serious sanitary sewer plugging problem. Adding to the municipal waste treatment problem is the high level of surfactants present in the waste. These surfactants present problems to a treatment plant which utilizes the activated sludge process (the most common installation). Poor oxygen transfer and nonsettling sludges are the main difficulties associated with surfactant-containing aqueous latex paint wastes. The surfactants also produce unsightly foaming.

An aqueous latex paint waste of the type described above (see Table 1 for further details) is treated by mixing the same with a small proportion of the aqueous polymer-containing solution prepared in Example 1 to provide a concentration of 150 milligrams per liter of the polyquaternary chloride polymer contained in the solution. The mixture of polyquaternary polymer and aqueous latex paint waste is deposited in a flocculating unit where it is gently agitated until the suspension and emulsion are broken up and a settlable floc is produced.

Flocculation is generally complete after 3 minutes of agitation with the agitator rotating at 100 rpm, followed by 3 minutes of agitation at 50 rpm. Desirably, agitation is stopped when flocculation is complete and the floc is allowed to settle, but continued agitation is not critical since continued stirring at 50 rpm for a total of 30 minutes instead of 3 minutes is not detrimental.

After settling is complete, it is found that more than 95% of the solids and more than 50% of the chemical oxygen demand have been eliminated from the supernatant liquid. The sludge is pumpable and can be removed in this manner for conventional concentration and disposal in a land fill. The clarified effluent can be used as wash water to minimize the volume of clarified effluent which must be disposed of.

The improvements obtained in accordance with the process will be more fully appreciated by comparing simple gravity settlement as reported in Table 1 with typical results obtained by the process. Listed in Table 1 below are the results obtained from averaging 15 typical batches of latex wastewater at a latex paint industrial plant.

TABLE 1

	Well mixed	Gravity settled without treatment for 30 minutes	Percent reduction
Chemical oxygen demand (milligrams/liter	26,000	21,300	18
Total solids (milligrams/liter)	30,500	22,000	27
Total suspended solids (milligrams/liter)	23,000	17,000	26
pH	7.2		
Color	Opaque and variable		

To further illustrate the improvement achieved in the process, there is presented Table 2 which shows the results obtained utilizing the product of Example 1 in comparison with a conventional commercial cationic flocculant which is recommended for the treatment of latex wastewater and which represents the best commercially available flocculant for this purpose. Referring more particularly to Table 2, it will be seen that the latex waste treated contained a higher proportion of suspended solids than the latex wastewater treated in Table 1. It will further be observed that the Example 1 product was used to provide a dosage of 600 milligrams of polymer per liter of wastewater being treated, this dosage representing an optimum dosage for producing a fast settling thick floc for the particular waste being treated.

At the same dosage, the commercial cationic polymer exhibited almost no flocculating activity. Accordingly, Table 2 reports the results of using 900 milligrams of the commercial polymer per liter of wastewater which was the dosage at which the commercial polymer exhibited its greatest fluocculating activity.

TABLE 2: TREATMENT OF LATEX WASTEWATER

Treatment	Dosage (mg./l.)	Chemical oxygen demand (mg./l.)	Total solids (mg./l.)	Total suspended solids (mg./l.)	Percent reduction			Proportion of waste forming clarified effluent after 30 minutes settling, percent
					Chemical oxygen demand	Total solids	Total suspended solids	
Raw		52,000	72,900	45,550				
Example 1 solution	600	5,580	2,440	54	89	96.7	99.9	90
Commercial cationic polymer	900	12,500	9,180	916	76	87	97.9	10

PHENOL

Phenol-Formaldehyde Resins

M. Adegeest; U.S. Patent 3,655,047; April 11, 1972; assigned to Corodex, NV, Netherlands has found that wastewater in which one or more phenols and formaldehyde have been dissolved may be purified by adding phenol(s) thereto in such quantities that the phenol:formaldehyde molar ratio will become from 1:1.02 to 1:1.12, and by adjusting the normality of the total acid from 0.015 to 0.10 N; heating the solution at a temperature of 80° to 85°C and subsequently removing the water by heating at a temperature that is at or above 5°C below the boiling point of the liquid mixture.

The composition of the wastewater is determined by the type of resin that was prepared. It will generally vary between the following limits: 3 to 7% by weight of phenols (normally ± 4%); 5 to 8% by weight of formaldehyde (normally ± 6%); 5 to 7% by weight of methanol (summer or winter formalin); and, small percentages of esters formed during the course of the reaction (BP ± 40°C); organic chlorides; methylal (BP 44°C) formed during the course of the reaction; and, formic acid. The HCl normality (if HCl is used as a catalyst) of wastewater to be purified is ± 0.015.

The phenol:formaldehyde molar ratio is of the greatest importance. The molar ratio, in the wastewater, is generally 1:4.18, i.e., the percentage of formaldehyde is too high. By adding phenol the molar ratio is brought from 1:4.18 to 1:1.02 to 1:1.12. The formaldehyde will now be bound by the phenol, as a result of which a thin, liquid resin mixture will be separated off, while the formation of very large resin molecules, i.e., a viscous resin is avoided.

Acid, preferably hydrogen chloride, is added in order to maintain the normality of the total acid between 0.015 and 0.10 N. Under such conditions a resin reaction is maintained that produces a nonadherent resin, so that the fouling of the apparatus is avoided. The fluid is subsequently heated at 80° to 85°C for 30 hours or several days, so that part of the methanol and other volatile compounds formed (in which the resin produced would readily dissolve) will escape. As a result, the ratio of separation of resin is considerably improved. The methanol and other volatile substances evolved may be recovered. The separation of resin may be accelerated by injecting very finely divided air. Very finely divided air may be provided in this instance, by injecting air into the liquid through acid-resistant, sintered, porous filters. The preliminary heating and evaporation of the volatile products is effected as near the surface of the liquid as possible, in order to prevent any turbulence from spreading through the entire quantity of liquid, and to give the formed resin the opportunity to settle in a zone which can be drained and is unheated. In this way further polymerization of the resin is prevented and the heat zone is kept substantially free from resin.

When the greater part of the liquid is free from methanol and other volatile substances, it is heated to at least 5°C below the boiling point of the liquid mixture, so as to evaporate the water. The temperature may not be higher than 116°C because at higher temperatures a gradual adherence of resite or C-stage resin occurs upon the heating surfaces, thus causing the rate of heat transfer to decrease within a short period of time and correspondingly causing evaporation to proceed at a slower rate. The evaporating surface of the source of heat has therefore to be such that at a steam temperature of below 116°C the liquid to be after-reacted and prepurified will keep on boiling.

The after-reaction thus completed will yield very small resin molecules having 2 to 6 nuclei, whereby it is observed that a four-nuclei molecule, for instance, is a molecule which contains 4 phenol and 3 formaldehyde units. Two-nuclei molecules, such as diphenyldihydroxymethane (in 6 isomers) may be observed as an emulsion of extremely fine crystals. The resin obtained according to the procedure described above may be further processed without presenting any difficulties. It may be used, inter alia, in preparing usual Novolac resin, though in this instance there is less formaldehyde present in the recovered resin condensation mixture than in the prior process.

Example: 624 liters (683 kg) of wastewater, remaining after the preparation of a phenol-formaldehyde resin which had been condensed in a basic medium, was admitted to the mixing tank together with 312 liters (341 kg) of wastewater resulting from the condensation of a phenol-formaldehyde resin in an acidic medium. 84 liters (90 kg) of water containing acid was mixed with these wastewater portions, this water-containing acid having been drawn through the continuous drain from the after-reactor of the evaporating plant. 109 kg of phenol was added to this mixture.

Before the addition of phenol, the phenol:formaldehyde molar ratio was 0.43 gram mol per liter to 1.8 gram mols per liter, i.e., 1:4.18. This is 40.46 grams of phenol and 54 grams of formaldehyde per liter. After the addition of phenol the phenol:formaldehyde ratio was 0.57 gram mol per liter to 0.61 gram mol per liter, i.e., 1:1.07. This is 53.64 grams of phenol and 18.38 grams of formaldehyde per liter. The normality of the total acid mounted to 0.015.

After the addition of phenol the mixture was held at a temperature of 80° to 85°C for 48 hours. The water was subsequently removed by being heated to boiling temperature. There was thus obtained 200 kg of a thin, liquid resin mixture which was separated off by settling. Inasmuch as 109 kg of phenol was added, there have thus been recovered 91 kg of resin product (i.e., ± 8.8% by weight) from the acid wastewater. Subsequently, another 84 liters of water containing acid was obtained from the continuous drain from the reaction-evaporation zone. This water is required for maintaining the acidity in this zone at a constant level, while it also serves to adjust the acidity in the mixing tank.

Phenolated Effluent

J.-P. Zumbrunn and F. Crommelynck; U.S. Patent 3,711,402; January 16, 1973; assigned to L'Air Liquide SA Pour L'Etude Et L'Exploitation Des Procedes Georges Claude, France describe a process for the complete purification of industrial effluent (industrial sewage, industrial residual water) by destruction of the phenols and hydroxylated aromatic compounds substituted on the core by electron donor groups, in solution. In this process, the phenols and other compounds are degraded by oxidation to products which have lost the aromatic character of the pollutant and are, consequently, very much more readily biodegradable. Phenolated residual water is found in the effluent of gasworks, coking plant refineries, chemical synthesis works (manufacture and conversion of phenols), works manufacturing plastics materials (phenoplasts) and, in general, works in which there are processed coal, tars and their derivatives, pesticides, dyestuffs, etc. The residual phenol concentrations vary widely, depending on the specific industry concerned. They may attain several grams per liter. It is known that the phenols are toxic to fish at levels as low as 0.1 mg/l. Furthermore, in water which is to be rendered drinkable by the addition of Javel water, a phenol content as low as 0.01 mg/l will suffice to impart an extremely disagreeable taste thereto, due to the formation of chlorophenols; it is thus necessary to process such water, in order to eliminate the phenols therefrom.

Previously, the processes for the purification of phenolated water have not been very numerous and, above all, they have not been totally effective. These known processes may be classified in accordance with two large categories. First of all, the processes involving physical recuperation or elimination, operation either by liquid-liquid extraction by means of a solvent, steam distillation, absorption on active charcoal or on an ion-exchanger resin, or by a foaming process whereby a surface-active agent is added to the water and the phenol accumulates in the foam.

Then, among the best known processes involving chemical and biological destruction, there may be mentioned treatment by activated sludges and bacterial beds. Depending on the nature of the biological bed employed, it is possible to purify effluents of initially 50 to 100 mg/l of phenol (such as may issue from an installation for physical recovery such as those enumerated hereinabove) and to expel after several hours water which still contains some mg/l of phenol. Some extremely elaborate plants make it possible to go down as far as 0.2 to 0.5 mg/l, but only by dint of extremely strict monitoring of the composition of the water to be treated (pH, concentration in respect of phenol and foreign bodies, poisons of the bacterial bed).

It has also been proposed to effect oxidation treatment by ozone or permanganate; these two products are extremely costly; the second results in the production of large quantities of sludges. Treatment by chlorine may be considered to be interesting, but it frequently produces chlorophenols and this almost always opposes the achievement of the desired aim, which is precisely to avoid the formation of such by-products. Treatment by means of catalyzed hydrogen peroxide is also known. Apart from the chemical oxidation processes, all the other processes are never quantitative and always leave a certain residual phenol content ranging between 0.2 and several mg/l.

A process has been found for the complete purification of industrial effluent polluted by phenolated impurities where the degradation of the pollutant is obtained by the action of an oxidizing reagent containing the HSO_5^- anion. The oxidizing reagent is a member of the group constituted by monopersulfuric acid and its salts. The oxidizing reagent is an aqueous solution containing HSO_5^- anion and containing furthermore NH_4^+, HSO_4^- ions and, optionally, alkaline cations. The oxidizing reagent is an aqueous solution containing the HSO_5^- anion and containing furthermore hydrogen peroxide and H^+, HSO_4^- and $SO_4^{=}$ ions.

According to the process, anion HSO_5^- may be utilized in various forms: (a) aqueous solutions in which the monoperoxysulfuric anion is associated with ammonium, potassium or sodium ions and HSO_4^- and H^+; (b) crystallized salt or a mixture of salts the active principle of which is monoperoxysulfate of potassium $HKSO_5$; and (c) aqueous solutions containing, in addition to the oxidizing anion, H^+ and HSO_4^- ions and molecules of H_2O_2, but free from metallic cation. These various oxidizing reagents may be obtained by the reaction of hydrogen peroxide on oleum, sulfuric acid, or sulfuric chlorohydrin, or by the hydrolysis of dipersulfates $S_2O_8^-$. The mechanism concerned in the reaction affording the destruction of the phenolated impurities is complex and depends on the composition of the oxidizing solution employed and on the proportion of reagents.

Monoperoxysulfuric acid permits the oxidation also of diphenols, polyphenols, substituted phenols, notably those carrying electron donor groups such as the diphenols (pyrocatechol, hydroquinone), resorcinol and the cresols which, with simple phenol, are the phenolated impurities most frequently encountered in aqueous industrial effluent; the reaction mechanisms are analogous to those described with reference to C_6H_5OH. This very thorough degradation involves the utilization of a relatively large proportion of oxidizing reagents relative to the phenol to be destroyed; however, this proportion depends also on the type of oxidizing reagent selected; in all cases, at least one mol of HSO_5^- per mol of phenol is necessary. The following examples illustrate the process. Examples 1 through 3 below correspond to tests effected with oxidizing solutions of type (a).

Example 1: To an aqueous solution containing 50 mg/l of phenol there is added an aqueous solution of peroxymonosulfuric anion obtained from the hydrolysis of ammonium dipersulfate for example, and containing the equivalent of 200 g/l of H_2SO_5. With a quantity of 37.4 ml of peracid solution per liter of effluent, i.e., a ratio HSO_5^-/C_6H_5OH equals 120, complete elimination of the phenol is obtained in 30 minutes at ambient temperature, such elimination is demonstrated by the amino antipyrine analysis method which gives a zero result on the treated solution.

The pH of the reaction is maintained at a value of 9 to 10 by the addition of an alkali. Most frequently, the alkaline product utilized is sodium hydroxide in an aqueous solution. If a ratio HSO_5^-/C_6H_5OH equals 90 is employed, the purification treatment is less rapid. In fact, 0.1 mg/l of phenol is revealed after a contact time of one hour.

Example 2: The purification is effected of an aqueous 50 mg/l phenol solution under the conditions described with reference to Example 1, but the quantity of oxidizing reagent is doubled, i.e., 74.8 ml of an aqueous solution equivalent to a 200 g/l monopersulfuric acid. The ration HSO_5^-/C_6H_5OH employed is then 240. Purification is total in five minutes, at ambient temperature.

Example 3: Under the conditions of Example 1, an aqueous 0.5 mg/l phenol solution is operated upon. The ratio HSO_5^-/C_6H_5OH equals 120 necessitates the addition of 0.38 ml of the aqueous solution equivalent to a 200 g/l H_2SO_5 monopersulfuric acid. Disappearance of the phenol is total in 90 minutes. Smaller ratios involve longer and more incomplete treatments, as compared with the amino antipyrine method. Thus, with a ratio of 90, 0.1 mg/l of phenol remains after four hours.

Example 4: There is treated, at pH 10, a solution of 25.8 ppm of phenol, by means of a solution of monopersulfuric acid analogous to the solution of type (c), containing 520 g/l of H_2SO_5, and having a molar ratio of HSO_5^-/C_6H_5OH equals 7. After 10 minutes of contact, the phenol concentration measured by the amino antipyrine method is zero. On restoring the mixture to an acid pH, a vigorous evolution of carbon dioxide gas is observed, showing that a portion of the organic carbon has been mineralized.

Example 5: By means of a monopersulfuric acid solution of type (c), containing 525 g/l of H_2SO_5, there is treated an aqueous refinery effluent containing 1.5 ppm of phenol and reducing products in quantities such that the oxygen chemical requirement is of the order of 100 to 200. Operation is effected at pH 10 with a ratio of reagent HSO_5^-/C_6H_5OH equals 100. The destruction of the phenol is followed by amino antipyrine dosing.

After 25 min there remains no more than 0.5 ppm of phenol. After 18 min there remains no more than 0.2 ppm of phenol. After 120 min there remains no more than 0.1 ppm of phenol. After 180 min there remains no more than $<$0.01 ppm of phenol.

Fiber Glass Operation

J.E. Etzel, C.H. Helbing and C.A. Justus; U.S. Patent 3,791,807; February 12, 1974; assigned to Certain-teed Products Corporation describe a process for the utilization of wastewaters typically encountered in the manufacture of fiber glass in which the wastewater is treated both mechanically and chemically to remove solid materials contained in the wastewater such as fiber glass, dust and other like materials. For many years fiber glass manufacturing operations have encountered a serious problem in the disposal of wastewaters which normally contain substantial quantities of phenolic resins. These phenolic resins present in wastewaters are normally utilized as a binder material in the manufacture of glass fiber insulation products. In a typical operation for the manufacture of fiber glass insulation material the fiber glass is deposited on moving chain conveyors from air streams. Water solutions of phenolic resins are sprayed on the glass fibers as they are collected on the chain conveyors in the form of a glass mat and the mat containing the binder is then transferred to ovens where the binder system is dried and cured. This process is described in U.S. Patent 3,215,585.

As a result of the spraying of resin binder on the fiber glass, considerable quantities of the resin used deposit on the chain conveyor collection system itself and unless the deposit is continuously removed, the conveyors become fouled and inoperable due to the drying and curing of the resin on the metal chain or wire comprising the conveyor. Caustic soda baths and water sprays have been used conventionally to clean the chain conveyor used in the manufacture of fiber glass and the water resulting from this cleaning operation results in a wastewater stream which contains considerable quantities of fiber glass and phenolic resin binder as well as caustic soda.

Waters of this character, as will be readily appreciated by those skilled in the art, cannot be disposed of by recourse to conventional sewer systems since they represent a pollution hazard to the streams and lakes in which sewer systems are typically discharged. Various systems have been proposed for solving this problem. Thus, in a paper presented at the Purdue Industrial Waste Conference on May 4, 1971, at Purdue University, West Lafayette, Indiana, one closed wastewater system was described by Angelbeck, Reed and Thomas.

Another system has been described for removal of impurities from fiber glass plant wastewaters. This system is described in an article titled "A Solution to the Phenolic Pollution Problem in Fiber Glass Plants: A Progress Report" by J.M. Baloga, F.B. Hutto, Jr. and E.I. Merrill published in *Chemical Engineering Progress Symposium Series, Water,* 1969, No. 97, Volume 65 at pages 124 through 127. While the two systems mentioned above are useful in reducing pollution hazards encountered in fiber glass manufacturing operations, further improvements can be made.

Thus in accordance with this process, wastewaters produced during the cleaning of chain conveyor systems are collected and screened to remove suspended solids. The screened water is then separated into a plurality of streams. One of the streams is recycled back to the spraying devices utilized to clean the chain devices while the second stream is discharged to a mixing tank in which the wastewater containing both usable and unusable phenolic resin binder is treated with a high molecular weight cationic polymer to separate unusable phenolic resins from this wastewater.

The water resulting from this polymer treatment which is a mixture of water containing usable dissolved phenolic resin binder and coagulated unusable phenolic resin solids is then filtered to remove the coagulated unusable phenolic resin from the wastewater. The water after filtration contains usable phenolic resin and is then recycled to a wastewater reclamation tank where it is utilized as makeup water for the manufacture of further quantities of phenolic resin binder solution for utilization as binder in the fiber glass insulation manufacturing process.

The treatment system is characteristically simple, compact, highly effective in recovering usable phenolic binder and completely closed so that all wastewaters are recirculated in the plant, thus requiring no water outfall from the wastewater plant streams formed in cleaning. In Figure 7.5a, glass fiber threads **1** are shown being drawn into the face of a forced gas flame emanating from a burner **2**. The fibers attenuated in the face of the flame are blown in short staple lengths onto a forming chain **3**. Prior to their deposition on the surface of the forming chain **3** a phenolic type resin binder is applied from spray head **4** to the fibers. The mat **5** formed on the forming chain **3** is passed over a roller **6** to a suitable oven. The forming chain bends around the roller **6**, passes over a series of rollers **7** and **8** and is admitted to tank **10** containing soft water. The chain **3** during its passage over the rollers to the tank **10** is sprayed with water from a spray **11**. The chain passes through tank **10** over rollers **9** and **12** and then passes over rollers **13, 14,15** and **16** and is subjected to high pressure water spray from spray device **18**. The chain then passes over rollers **17, 18** and **19** and is returned to the mat forming area for the collection of further mat **5**. All water introduced into the system onto the forming chain **3** via nip roll spray **11**, soft water bath **10** and the high pressure spray **18** is collected in a large floor reservoir or sump generally indicated as **20**. The reservoir is provided in the central portion thereof with a ditch or sump **22**. Located within the sump is pump **23**. All water collected in the ditch or sump area is pumped into a solids separation device **27** via line **24**.

Solids are removed at the face **28** of separator **27** and flow onto a conveyor **29** for removal. Liquid from the solids-liquid separator is removed via line **33** and is passed to an effluent tank **34**. The liquid material from the tank is removed in three lines **30, 31** and **32** which are connected to pumps **35** and **36** respectively. Line **32** feeds liquid from tank **34** to the liquid treatment tank **38** which is provided with a line **37** running from a tank **39** to supply chemicals to tank **38** to treat the water.

The treated water from tank **38** is fed to filter **41** via line **40** and the filtrate is removed via line **42** and is fed to a binder mixing tank **43**. Binder solutions from tank **43** are fed to spray **4** via line **44**. Pump **35** feeds the water from line **30** via line **46** to the high pressure spray system **18** and pump **36** feeds water from line **31** via line **47** to the spray system **11**. Makeup water to the system is supplied from water softener **50** via line **51** to the tank **10**.

In Figure 7.5b which is an enlarged view of the preferred solids separator system, the material pumped from the concentrated ditch or sump pool **22** by pump **23** is passed via line **24** to the solids separator **27**. In the preferred embodiment the separator is a sieve device having positioned on a generally vertical face thereof a screened element **28**. During the operation water is fed into a chamber **26** of the separator and overflows at lip **25**. The solids and liquid pass down the face of the screen **28** and solid materials are removed from the face of the screen and pass onto a dewatered solids collector **29**. The liquid passes through the screen into a reservoir **21** and from there is passed via pipe **33** to a screened effluent tank **34**. Thus three exit lines are preferably provided in the screened effluent tank **34**.

One effluent pipe **30** is connected to the high pressure spray system **18** via pump **35** and line **46** where it is utilized as wash water for the forming chain **3** in this portion of the process. A second feed line **31** is provided to feed water via pump **36** and line **47** to spray system **11** for utilization in cleaning the forming chain. A blow down line **32** is also provided in the screened effluent tank **34** which feeds the remainder of water to a coagulation tank **38** provided with a suitable mixer **49**. Also entering the tank **38** as shown more clearly in Figure 7.5c is a polymer supply line **37** from a supply tank **39**. In the tank a polymeric flocculant typically a cationic polymer is fed via line **37** and mixed thoroughly with the water from the blowdown line **32** of the screened effluent tank **34**.

The material flocculated and agitated in the mixing tank **38**, is then passed via line **40** to a paper or cloth media pressure filter **41**. In the filter the solid materials resulting from the flocculation occurring in the mixing tank **38** are removed and an essentially clean water stream is removed from the filter via line **42** and is passed to a binder mixing tank **43**. In the mixing tank **43** the necessary ingredients for preparing binder are added and mixed and the resulting binder solution is fed to the spray head **4** of the fiber glass forming system shown in Figure 7.5a via line **44**.

The overall system utilizes essentially all of the water which is used in the cleaning of the fiber glass forming chains **3** used in the manufacture of the fiber glass product. This water is used as recycle water for cleaning purposes and the water from the chemical treatment area is utilized, after it has been filtered to remove the unusable phenolic binder, as makeup water for further binder solution manufacture. Makeup water is added to the system in tank **10** preferably. Two particularly effective polymeric materials for utilization in the system are a high molecular weight polymeric coagulant, known under the trademark Aquafloc 418, and Primafloc C-7, a high molecular weight polyamine bisulfate polymeric flocculant. The mixing vessel in which flocculation occurs is regulated during operation to provide therein a pH in the range of between 7 and 11, preferably between 7.5 and 8.5.

The flocculated material recovered from tank **38** in line **40** must be filtered to remove the material coagulated in that mixing vessel. In the preferred case, high pressure filters utilizing filter paper or filter cloth and preferably horizontal plate type filters are used. Thus, filters may be used with or without filter aids being added to the system. By high pressures is meant pressures on the order of from 30 to 110 psig and typically in the range of between 40 to 60 psig. While pressure filtration forms the preferred embodiment, recourse to conventional atmospheric filters may be had where desired.

It is found in operating the system that the phenolic content of the water found in the sump **20** generally ranges on a weight basis between 400 and 800 parts per million as undistilled phenol. Effluent from the mixing tank **38** in line **40** is found to contain on the same basis between 150 to 300 parts per million phenol. This phenol is in usable form and thus the water is readily available for utilization in the binder mixing operation. Since the phenol content is quite low but usable, the binder mixing operation can be adjusted to take into consideration the small quantity of usable phenolic resin present in the water.

The process will now be described with reference to Figure 7.5a and a manufacturing operation in which five forming stations such as shown in Figure 7.5a are employed, each of the forming stations having a chain conveyor **3**, spray systems **11** and **18**, a tank **10**, feed lines **46, 47** and **51** and all of the water from each of the five stations being collected in the common sump **20** and central sump **22**.

The fiber glass fibers **1** in each of the five stations are drawn in front of a hot gas pressure blast from a gas burner **2** where they are attenuated into short fine diameter lengths and are collected in each station on a chain **3**. Fibers are sprayed in each station with a binder from one or more spray devices **4**, the binder used being that described in U.S. Patent 3,215,585. The mat **5** formed in each of the five stations passes to an oven associated with each station. The chain **3** of each station passes over rollers **6, 7** and **8** located in each station to a cleaning tank **10** in each station through which it is passed by rollers **9** and **12**, located in each tank **10**.

The chain **3** in each station then passes over rollers **13, 14, 15** and **16** in each station below a spray system **18** located in each station. The chain **3** in each station is then returned by rollers **17, 18** and **19** in each station to the mat forming area of each station for the collection of fibers **1** and preparation of mat **5** in each station. Each tank **10** of each of the five stations during the passage of the chain **3** thereto, has continuously fed thereto soft water at the rate of 3 gallons per minute. Water overflows the top of tank **10** in each of the stations and the total water from each of the tanks **10** in each station is collected in sump **20** which is used as a common collector for all five stations.

Chains **3** in each station have applied to them through the spray system **11** in each station, water at the rate of 36 gallons per minute. Also during their passage through each station each of the chains **3** have applied by high pressure spray system **18** water at the rate of 116 gallons per minute. The water from both spray systems **11** and **18** in each of the five forming stations are also collected in the common sump **20** and flow to the central ditch or reservoir **22** therein.

The contents of the ditch **22** are continuously pumped via pump **23** to the solids-liquid separator **27** at the rate of 775 gallons per minute. The solids-liquid separator **27** is comprised of Hydrasieve units. Four of the units are Model 552-60 and the fifth is Model 552-70. The 775 gallon flow of water is distributed to the five units on an approximate equal basis and water is removed from the solids-liquid separator system **27** at a rate of about 775 gallons per minute. This water is passed to the screened effluent tank **34** via line **33** from which it is removed in three lines **30, 31** and **32**. The water is passed to line **30** at a rate of 580 gallons per minute and is delivered to a pump **35**, one such pump being supplied for each forming station and the water in line **30** being equally divided among them. Each pump **35** in each station delivers water at the rate of 116 gallons per minute to a spray system **18** located in each of the five stations via a line **46** supplied in each station. The water in line **31** is passed at a rate of 180 gallons per minute to a pump **36**, one such pump being supplied for each forming station and the water in line **31** being equally divided to each pump **36**. Each pump **36** delivers water in a line **47** associated with each to a spray system **11** in each forming station at a rate of 36 gallons per minute.

Water from spray systems **11** and **18** clean the chain **3** before and after its immersion in tank **10** respectively. The overflow of water from each of the tanks **10** in each of the five forming stations is about 3 gallons per minute. The water in line **32** is passed at a rate of 15 gallons per minute to the mixing tank **38**. This tank is agitated by a motor driven mixer **49** and has added thereto a 1% solution of Primafloc C-7 to provide in tank **38** a concentration of Primafloc C-7 in the range of 30 to 100 parts per million. Water from tank **38** is passed to a filter system **41** operating at the rate of 15 gallons per minute.

The filter system is a horizontal plate filter operating at 40 psig. The water leaving the filter system **41** passes into tank **43** in line **42** at about 15 gallons per minute. The water in tank **43** is used to make up binder solution in accordance with U.S. Patent 3,215,585 and the binder solution is then fed via line **44** to spray **4** in each of the five forming stations.

FIGURE 7.5: WASTEWATER RECLAMATION IN FIBER GLASS OPERATION

a.

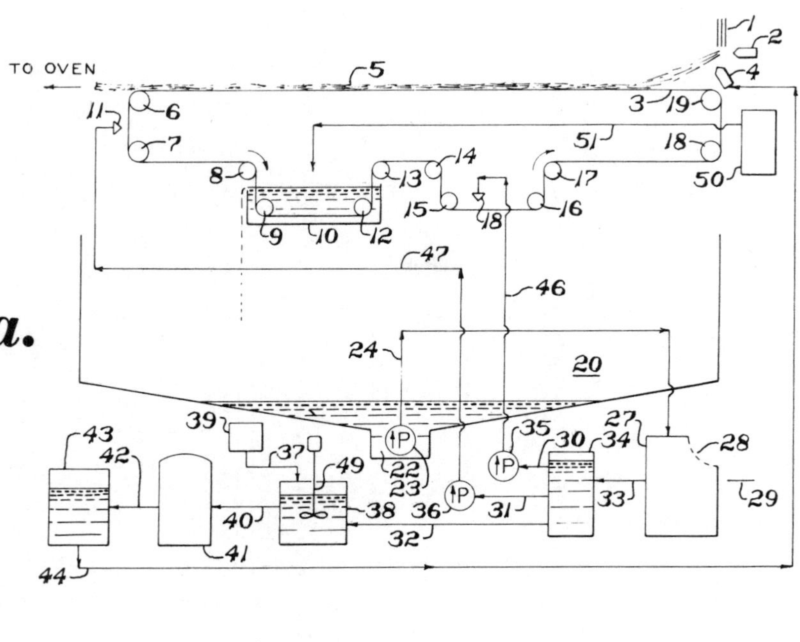

Fiber Glass and Wastewater Reclamation Operation

(continued)

FIGURE 7.5: (continued)

b.

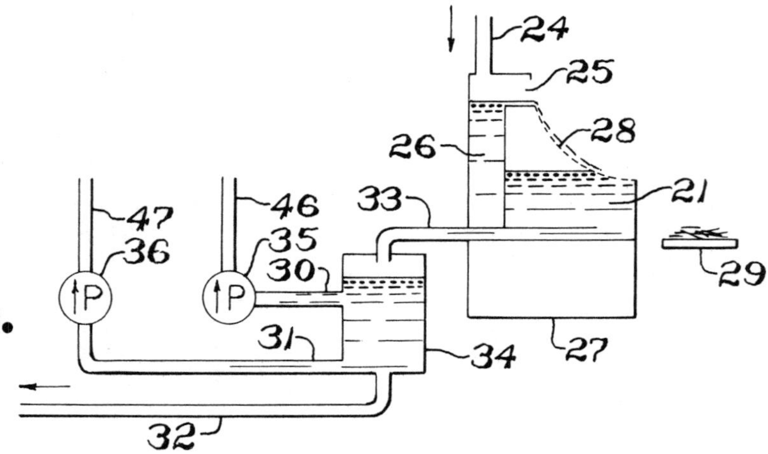

Enlarged View of Solids Dewatering System

c.

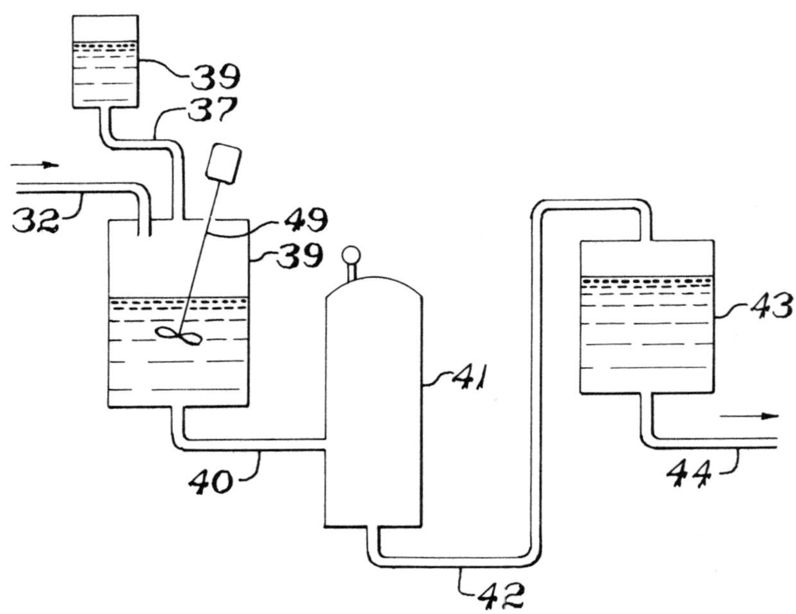

Enlarged View of Chemical Reclamation System

Source: J.E. Etzel, C.H. Helbing and C.A. Justus; U.S. Patent 3,791,807, February 12, 1974

Liquid Membrane Process

N.N. Li, R.P. Cahn and A.L. Shrier; U.S. Patent 3,779,907; December 18, 1973; assigned to Esso Research and Engineering Company have found that a dissolved species may be efficiently removed from an aqueous solution by contacting the aqueous solution with an emulsion, the emulsion comprising an exterior phase which is immiscible with the aqueous solution and permeable to the dissolved species, and an interior phase comprising a reactant which is capable of converting the dissolved species to a form in which it is no longer permeable. The dissolved species in its nonpermeable form is thus trapped in the interior phase and the concentration gradient of the permeable form of the compound between the aqueous solution and the interior phase is maintained. The concentration gradient acts to drive the dissolved compound efficiently through the exterior phase of the emulsion into the interior phase. By use of this technique small volumes of emulsion may be used to treat large volumes of aqueous solution efficiently. Thus the process is useful in water pollution control, where large bodies of water must be treated to remove small amounts of dissolved impurities, e.g., the dissolved species.

The various dissolved species which may be separated from their aqueous solutions include weakly basic compounds, e.g., NH_3, various amines, and other nitrogen-containing organic compounds. In general these are reacted in the interior phase with a strong acid, e.g., H_2SO_4, HCl, HNO_3, which neutralizes the basic compounds and converts them into an ionic, i.e., nonpermeable form. By maintaining a pH difference between the aqueous solution and the interior phase, the concentration of the permeable, unionized form of the compounds can be maintained high in the aqueous stream and low in the interior phase.

Weakly acidic compounds, e.g., CO_2, SO_2, acetic acid, H_2S, phenol, HCN, citric acid and many other organic acids are reacted in the interior phase with a strong base, e.g., NaOH, K_2CO_3, KOH, and are converted to ionic, i.e., nonpermeable forms. Various dissolved species may be precipitated in the interior phase, and thus converted to a nonpermeable form. Examples include the removal of H_2S by reaction with copper ions and the removal of HCN by reaction with silver ions, the ions being dissolved in the interior phase.

Ionic species may also be separated from their aqueous solutions by use of this process, the only proviso being that the ionic species be at least somewhat soluble in the water immiscible exterior phase of the emulsion. For example, ions of tin, iron and copper, being slightly soluble in aromatic and olefinic solvents, may be separated from aqueous solution by utilizing a water-in-oil emulsion comprising an aromatic or olefinic solvent as the exterior phase. The following example illustrates the process.

Example: In the following runs, phenol was removed from an aqueous solution. The solution was introduced into a separation zone which was provided with a mixer. The zone was maintained at a temperature of 25°C and a pressure of 1 atmosphere. The surfactant used was Span 80, a commercial name for sorbitan monooleate with a viscosity of 1,000 centipoises at 25°C, in an amount of 2% by weight. The mixture of surfactant, sodium hydroxide solution and surfactant solvent (a high molecular weight isoparaffin having a carbon number range of from 25 to 35) was emulsified and then contacted with the phenol-containing aqueous feed. After the emulsion and the aqueous solution were mixed at a rate of 100 rpm for about 50 minutes, the emulsion and the aqueous solution were separated and analyzed. The results and additional details of the experiments are included in the following table. The aqueous solution was analyzed by UV spectrography method for phenol concentration and by titration with acid for caustic concentration.

The table indicates that in Run 1 where no sodium hydroxide was added, a high phenol concentration remained after 43 minutes of agitation. On the other hand, in Run 2 where 0.4 wt percent of caustic was used, at the end of 19 minutes only 3 parts per million of phenol remained. In Run 3 where the caustic concentration was increased slightly to 0.5 wt percent, phenol was reduced from 1,000 to 33 parts per million in just 53 minutes. It should be noted that the Na content of the waste water rose very slowly, indicating very little permeation of Na ion from the inside of the emulsion to the outside wastewater stream.

Experimental Phenol Removal from Waste Streams

	Run Number 1	Run Number 2	Run Number 3
Aqueous Phenol Solution:			
Volume, cc	250	250	250
Phenol, ppm	1,000	200	1,000
Emulsion:			
Exterior phase	Isoparaffin	Isoparaffin	Isoparaffin
Volume of exterior phase, cc	200	200	200
Surfactant	2% Span 80	2% Span 80	2% Span 80
Interior aqueous phase, cc	100	96.8	97
Caustic concentration, wt percent	0	0.4	0.5

Run Data

Time (min)	Phenol in Aqueous Solution (ppm)	Percent Removed	Time (min)	Phenol in Aqueous Solution (ppm)*	Percent Removed	Time (min)	Phenol in Aqueous Solution (ppm)**	Percent Removed
0	990	0	0	200	0	0	1,000	0
2	730	26	1	83	59	2	652	35
5	667	33	2	60	70	5	288	71
8	615	38	3	36	82	18	41	96
28	606	39	5	17	92	38	44	96
43	612	38	19	3	98	53	33	97

*No NaOH in aqueous phase.
**Na in final aqueous phase = 43 ppm.

PHOTOGRAPHY

Hydrogen Peroxide Treatment of Spent Fix Solution

R. Fisch and N. Newman; U.S. Patent 3,721,624; March 20, 1973; assigned to Minnesota Mining and Manufacturing Company describe a process for treating spent photographic processing solutions to reduce their biological oxygen demand and toxicity and permit their disposal with minimal pollution of the environment. The disposal of spent photographic processing baths can introduce harmful and polluting chemicals into the lakes and streams. In recent years, color photographic processing has used bleach-fix, hypo or fix baths containing large quantities of thiosulfate (a reducing agent) and color developers containing phenylenediamine compounds as well as hydroquinone, ascorbic acid, hydroxylamine, hydrazine, benzyl alcohol and/or other organic compounds. Stabilizer baths may contain formaldehyde, acetic or citric acid, and other compounds. Wash and other processing baths also tend to include such materials, depending on their location in the photographic processing sequence.

In accordance with this process, the spent fix (including bleach-fix) solution from color photographic processing, such as may be used in processing color print paper, is oxidized by treatment with an oxidizing agent and is then combined with spent phenylenediamine color developer solution to form a precipitate or sludge which is separated from the combined solutions, thereby lowering the biological oxygen demand and the organic content of the combined solutions. By following this sequence, the thiosulfate hypo or fixing agent is reduced to sulfate, and the excess amounts of sulfate then serve to form the insoluble amine salt with the phenylenediamine color developer, which precipitates and can readily be separated from the remaining solution by conventional means, such as filtration. The efficiency of the sequential process is achieved by using the by-product of thiosulfate oxidation to insolubilize the toxic phenylenediamine color developers.

Although any suitable oxidizing agent can be used to oxidize the thiosulfate fix solution, the peroxy compounds are preferred because of the relatively nonpolluting nature of their degradation products. In particular, hydrogen peroxide is not only an effective oxidizing agent but also decomposes to form water. The following examples illustrate the process.

Example 1: To 50 ml of the bleach-fix effluent from a color print paper processing line were added 40 ml of a 15 wt percent solution of hydrogen peroxide. After thorough mixing had occurred, 100 ml of the effluent from a color developer containing a phenylenediamine color developer, specifically 4-amino-3-methyl-N-ethyl-N(beta-methane sulfonamido) ethyl aniline (CD-3) was added with mixing. A precipitate formed and was removed by filtration. The resulting effluent had a considerably lower oxygen demand than the original bleach-fix effluent and a substantial amount of the organic matter was removed in the precipitation step.

Example 2: The process of Example 1 was repeated with the substitution of potassium peroxydisulfate for the hydrogen peroxide. Similar results were achieved, the resulting precipitate appearing to be somewhat more crystalline than the precipitate in Example 1.

Destruction of EDTA by Chlorination

According to a process described by *T.W. Bober, T.J. Dagon and I. Slovonsky; U.S. Patent 3,767,572; October 23, 1973* waste photographic processing solutions which contain EDTA (ethylenediaminetetraacetate), such as exhausted ammonium iron-EDTA bleaching or bleach-fixing solutions, are chlorinated to destroy EDTA and thereby increase the biodegradability of the solution. Chlorination can be effected by introduction of chlorine gas or by the use of hypochlorite solution. Since EDTA and complexes thereof account for a large portion of the oxygen-consuming material present in photographic processing effluent, a significant source of water pollution is substantially eliminated by this method. The following examples illustrate the process.

Example 1: Chlorine gas is supplied from a gas cylinder into 1 liter of a 1.56 M solution of NH_4Fe·EDTA (90 milliliters/liter) at a regulated flow rate of 0.6 liter per minute. The pH of the reacting solution is kept above 11.0 by periodic addition of concentrated (16 N) sodium hydroxide solution. The EDTA, EDTA-complex and iron concentrations are measured at regular intervals. The iron precipitates as the hydroxide and is easily filtered or otherwise removed as a precipitate by filtration after about 30 minutes. Chemical oxygen demand (COD) drops from an initial value of 38,300 milligrams/liter to a final value after 2.0 hours of treatment of 13,000 milligrams/liter and all EDTA was destroyed as proven by analysis.

Example 2: Treatment of a solution of Na_4·EDTA (50 grams/liter) results in complete destruction of EDTA after a period of about 90 minutes, paralleling the results obtained in Example 1.

Addition of large volumes of caustic initially rather than periodically throughout the evaluation does not speed up EDTA degradation, however, it does provide a secondary advantage in that nearly all of the iron hydroxide precipitate can be removed prior to chlorination, thus avoiding plugging of the sintered glass dispersion tubes conventionally used to bubble the chlorine gas into the solution by the precipitate. Further, heating the solution prior to chlorinating, does not appear to speed up degradation of EDTA.

MEAT PROCESSING

Coking Process for Animal and Dairy Wastes

According to a process described by *H.V. Hess and E.L. Cole; U.S. Patent 3,671,403; June 20, 1972; assigned to Texaco Inc.* animal and dairy wastes are converted to deodorized coke suitable for use as fertilizer or soil conditioner, an aqueous odor-free effluent having low chemical oxygen demand and reduced phosphorus content and an odorless gas suitable for discharge to the atmosphere.

In earlier times, and around small dairies and the like, the waste was collected and spread on the soil for fertilizer. However, the concentration of waste from feed lot operations involving up to 10,000 animals each and larger does not allow this method of disposal; for example, an average sample of cattle waste contains about 90% water so that transportation charges for hauling the waste any distance for fertilizer value would become prohibitive.

In a previous process, organic wastes such as potato peelings, orange peelings, whey liquid, newsprint, sawdust, etc. in water slurries have been coked under pressure to produce an odorless coke, a gas which is suitable for venting to the atmosphere, and a liquid effluent which is not malodorous and which is considerably reduced in COD. Attempts to apply this process to the purification of animal wastes, however, have not been successful since the resulting effluent gases and coke are very malodorous. The liquid effluent, while being somewhat reduced in COD by coking, still contains dissolved organic compounds which contribute to COD and make it undesirable for discharge into receiving lakes or streams. In addition, the liquid effluent has a bad odor due to these dissolved organic compounds.

In the process described in detail below, a pumpable animal or dairy waste stream is preheated, coked to form a gas phase, a liquid phase (or aqueous coker effluent) and a wet coke phase; the wet coke phase is withdrawn intermittently, maintained at a slightly superatmospheric pressure and deodorized with pressurized hot air, thereby simultaneously drying the coke. The new moist hot air is compressed and mixed with the aqueous coker effluent and the product gases with the resulting mixture being oxidized to destroy odor-forming constituents therefrom. This oxidized stream is then heat exchanged with the incoming feed stream and the thus cooled effluent composed of gases and liquids can be finally disposed of. The effluent is low in phosphorus and accordingly does not contribute to algae growth in the receiving body of water. The coke which is recovered is relatively high in phosphorus content which enhances its value as a soil conditioner and fertilizer. The process is illustrated in Figure 7.6.

To provide a basis for showing the advantages of this process over the coking process outlined above, a sample of cattle waste obtained from a commercial dairy and having a chemical oxygen demand of 110,884 milligrams O_2/liter and an organic carbon content of 8.5% was coked in a coke drum at 550°F and 1,000 psig. Treatment of this cattle waste according to the above process produced the following results (except for the oxidation step).

	Weight, grams	Yield, weight percent basis charge	COD, mg./O_2 liter	Percent						pH	Gross heat of combustion, Btu/bbl	Coke surface area, M2/g
				C	S	N	P	K	H			
Liquid phase coking of cow manure:												
Charge—cow manure	710		110,884	8.5								
Products from treating at 550° F., 1,150 psig, 2 hours:												
Aqueous effluent	530	74.7	72,076	2.6	0.014	0.23	0.003	0.30		5		
Wet coke	140	19.7										
Gas (95% CO_2)	28.4	4.0										
Material balance	698.4	98.6										
Dry coke	56	7.9		60.0	0.18	3.8	0.66	0.39	5.8	5	12,295	6
Products from treating at 600° F., 1,875 psig, 2 hours:												
Aqueous effluent	531	74.8	64,547	2.2	0.012	0.23	0.002	0.36				
Wet coke	150	21.2										
Gas (95% CO_2)	33	4.6										
Material balance	714	100.6										
Dry coke	64	9.0		55.2	0.23	3.4	1.25	0.54	5.0		11,812	4

In looking at the results shown in the table, it becomes apparent that the bulk of the phosphorus is rejected in the coke product. In the run at 550°F, 95% of the phosphorus and 60% of the nitrogen is removed from the system in the coke product. At 600°F, 98% of the phosphorus and 66% of the nitrogen is removed in the coke. 53% of the sulfur is removed with the coke at 550°F and 66% at 600°F. The potassium content of the waste stream, as might be expected from the solubility of potassium salts in water, is largely retained in the aqueous effluent from the coking operation. At 550°F coking temperature, 89% of the potassium remains in the aqueous effluent; at 600°F coking temperature, 86%.

It will be noted from the table that the carbon distribution from a run at 550°F is: aqueous effluent, 2.24 grams, 26.4%; dry coke, 4.74 grams, 56.0%; gas, 1.5 grams, 17.6% (total 8.48 grams, 100%). From the run at 600°F, the carbon distribution is: aqueous effluent, 1.81 grams, 21.4%; dry coke, 4.97 grams, 58.6%; gas, 1.70 grams, 20.0% (total 8.48 grams, 100%).

It can be seen from the above (the carbon material balance from feed to products is 99.7%) that between 21.4 and 26.4% of the carbon remains dissolved in the aqueous effluent. These carbon-containing water-soluble compounds are low molecular weight substances not subject to coking under the rather mild coking conditions employed. These materials, of course, contribute to the high chemical oxygen demand of the aqueous effluent coming from the coking operation. At 550°F the aqueous effluent from the coker still contains 56% of the COD of the charge, and at 600°F the aqueous effluent from the coker contains 50.5% of the COD of the charge. This high COD of the aqueous effluent from the coking operation also results in the effluent having a very bad odor. This odor as well as the high COD makes it mandatory that additional treating be carried out before the effluent could be dumped into waterways. The wet coke produced in the operation, because it is wet with the aqueous coker effluent, also has a foul odor and thus should also be treated before use as fertilizer or soil conditioner. Gases produced in the coking operation, even though they are roughly 95% CO_2, also possess a foul odor and thus would require treatment before venting to the atmosphere.

FIGURE 7.6: CONVERSION OF ANIMAL WASTE TO DEODORIZED COKE

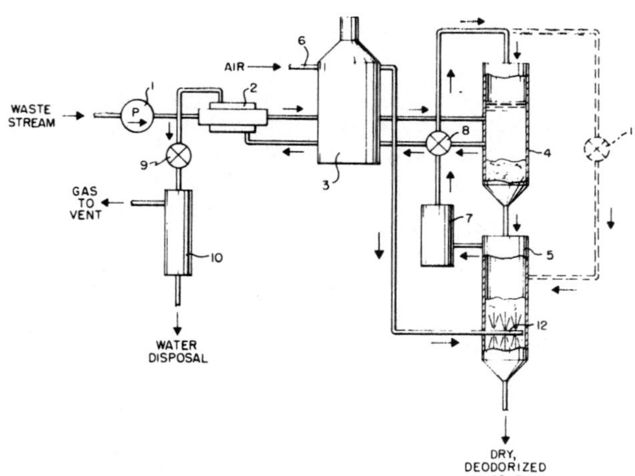

Source: H.V. Hess and E.L. Cole; U.S. Patent 3,671,403; June 20, 1972

Referring to Figure 7.6, the waste stream is pumped up to system pressure in pump **1** through the heat exchanger **2** where it is in heat exchange with the oxidized effluent and through the heater **3** where the cokable materials are coked, and from there to a decanter (held at system pressure) **4** where the system separates into a gas phase, a liquid phase (called aqueous coker effluent) and a wet coke phase. The wet coke phase is drawn down intermittently into vessel **5** which is maintained slightly above atmospheric pressure. Coke is now in the vessel and to dry it and remove malodorous constituents, air is brought in at slightly above atmospheric pressure, preferably 30 to 60 psig, through the top part of the heater **3** where it is heated to 250° to 350°F (a temperature sufficient to dry and deodorize the wet coke) and passes through the sparger **12** shown at the bottom of vessel **5**. The temperature range indicated suffices to dry and to deodorize the wet coke but is not sufficiently hot to burn the coke. The flow rate of hot air through the coke is sufficient to expand the coke bed and maintain it in an expanded condition above the sparger. In any case, sufficient air is used so that there will be an excess of oxygen available for burning the COD from the aqueous effluent and gases.

Next, dry, deodorized coke is removed intermittently from the bottom of the vessel for use as fertilizer or soil conditioner. The warm, moist and now smelly air from the vessel is compressed in compressor **7** to system pressure and mixed with aqueous coker effluent from vessel **4** and gas from this vessel in mix valve **8**, and passes through the lower part of heater **3**. This air-water-gas system is oxidized in the heater whereby the chemical oxygen demand is removed from the gas and the aqueous coker effluent while the odor-forming constituents are destroyed. This oxidized stream is passed through heat exchanger **2** in heat exchange with the raw waste feed and thence through a pressure reducing valve **9** to a gas-liquid separator **10**. The gases and liquid from the separator may be discharged to the atmosphere and ditch respectively. In another example, shown in dotted lines, the gas from vessel **4** can be passed through a pressure reducing valve **11** into the vessel **5** where it is mixed with the gases obtained in drying the coke.

The effluent from the coker (produced at conditions of 550°F and 1,150 psig) has been oxidized with air at 550°F and autogenous pressure to produce an oxidized effluent with a COD of 3,802 milligrams O_2/liter and a pH of 7.6. In considering the coker effluent alone (which had a COD of 72,076 milligrams O_2/liter) this represents a COD reduction of 94.7%, or basing it back on the COD of the charge it represents a reduction of upwards of 96%. The odor is also substantially removed. This oxidation was carried out in a batch operation. Operation on a continuous basis with excess air leads to essentially complete removal of COD.

The process is also suitable for treatment of other animal wastes such as pig wastes. The effluent from poultry feeding units can likewise be handled. A portion of the water produced from the described operation also can be used to wash down dairy barns, feed lots, and the like and then, in a sense, recycled through the whole system.

Reduction of BOD

J. Hollo, J. Toth and I. Zagyvai; U.S. Patent 3,738,933; June 12, 1973; assigned to Tatabanyai Szenbanyak, Hungary describe a process for the purification of sewage and, if required, for the recovery of the inherent protein from sewages. Sewages of the kind to be so treated are mainly found in slaughter houses, meat processing plants, preserve factories, the milk industry, the fermentation industry, and the like.

The sewages of the food industry have a very considerable biochemical oxygen demand (3,000 to 15,000 milligrams/liter BOD_5) as well as a significant content of mechanical impurities, i.e., of suspended organic matter. Owing to their high contamination these sewages are not suitable to be transferred directly to natural or artificial recipients, so that the dissolved and suspended organic matter content is generally first removed by means of various aerobic or anaerobic biochemical decomposition methods. These biochemical methods are lengthy and complicated; they require costly equipment, considerable power input, and significant length of time.

In this process, the biochemical oxygen demand of sewage of the type of meat, dairy and fermentation waste, is reduced by adding to the sewage at least 120 g/m^3 of a water-soluble aluminum salt or bivalent or trivalent iron salt, 0.5 to 1 kg/m^3 of bentonite or kaolin and 5 to 10 g/m^3 of polymers or copolymers of acrylic acid-acrylic amide in the form of an aqueous solution. The suspension is stirred and the pH adjusted to exceed 10 by addition of a basic calcium compound such as lime milk or calcium hydroxide. The precipitate thus obtained is separated by setting and transferred to a conical tank for treatment with carbon dioxide until the pH is no higher than 7, and the resulting precipitate is filtered and sterilized at a temperature above 130°C under pressure to produce a biochemical culture medium or animal feed. The process may be performed within 1 to 1.5 hours with the remarkable result that a sewage having a BOD_5 of 4,000 to 11,000 mg/l and containing a large amount (4,000 to 10,000 mg/l) of suspended solid matter is obtained practically free of contaminations, that is, containining only about 200 mg/l of impurities and with a BOD_5 value of 300 to 400 mg/l. A further advantage of the process is the possibility of recovering the protein contained in the sewage, which may be processed together with the other waste proteins, into a biochemical culture medium or an animal feed. The process is not sensitive to a variation of the sewage composition, it does not require costly and voluminous equipment, and may be performed with high efficiency and intensity using inexpensive chemicals.

Electrolytic De-Emulsification of Meat Packinghouse Wastes

A process described by *H.T. Anderson; U.S. Patent 3,673,065; June 27, 1972; assigned to Swift & Company* is directed to a method of treating wastewater containing appreciable amounts of fats, grease, oily materials and polar materials such as proteins. The process is especially useful in connection with packinghouse and edible oil operations, but it can also be advantageously used for cleaning other types of fat-contaminated water, such as from industrial plants using mineral oils.

It has been the usual practice in the past to run the wastewater from the packinghouse to a settling tank or basin having baffles, where the water would set for about one-half hour or more and the free fat would rise to the top and be skimmed off. The emulsified fat would, of course, remain in the water and would accompany it to the sewers. Various means such as aeration and complex apparatus have been employed in attempts to de-emulsify the wastewaters. Usually, however, unless the emulsified oil was very valuable, no effort was made to separate it from the water that was passed to the sewers and hence to the rivers and seas.

In a few instances, solvent extraction and other unit processes are used to recover the valuable oil. Also, in processes where the water is reused, the oil can be removed from the system by coagulation with aluminum sulfate and alkali, followed by filtration through a nonsiliceous filter medium. The oil is caught in the floc formed, and the floc, with the entrained oil, strained out of the condensate by the filter. Periodical backwashings of the filter with hot caustic soda are required. It should be noted that those processes used to completely remove the oil from the water are clearly uneconomical for use in cleaning up wastewater from packinghouse and edible oil operations. Yet, on the other hand, it is desirable to remove those low grade oils from sewage as they may be used an industrial cutting oils, etc. Further, it has become necessary to clean up the water prior to its discharge into the rivers and seas.

This process is concerned with the treatment of emulsified fat-water systems and comprises impressing direct electrical current in such a manner that a carefully defined, relatively deep anolyte stream (i.e., a path of liquid in the immediate neighborhood of the anodes during electrolysis) is formed near the bottom of a receptacle at the entry end and becomes narrower in depth as it progresses upwardly to near the water surface at the opposite end or exit end of the receptacle. The anolyte stream, having a very low pH due to multiple anodes, breaks the emulsified fat enabling it to rise to the surface and be skimmed off. The system comprises multiple, submerged anodes, usually approximately equal distances apart, parallel to the longitudinal walls of a rectangular tank or other receptacle but inclined upwardly. In a circular tank, the anodes are parallel to the walls but substantially vertically mounted.

In establishing the electrical circuit, anodes are placed in contact with the wastewater. The walls of the receptacle function as the cathode with the current flowing through the liquid from the anode to the cathode. A plurality of rod-shaped anodes starting just below the intake opening (point of entry of the wastewater) of the tank are spaced progressively upward approaching the water surface in the tank at the opposite end. In a circular vessel, anodes are also parallel to the tank wall and usually perpendicular to the bottom. They are low enough to produce a low pH anolyte from the point of wastewater entry to the upper region of the tank. While preventing re-emulsification of fat, they also prevent a potential drop in protein or cellulose fiber which would permit these solids to settle out of reach of the effective skimming means. All anodes are far enough below the water surface so that they do not interfere with the movement of skim bars. The following examples illustrate the process.

Example 1: A steel tank of about 60 ft long by 10 ft wide and 6 ft deep was approximately three-quarters filled with packinghouse wastewater containing principally fat, proteins and cellulose. Ten high silica-iron anodes, each 1½" in diameter and 5 ft long were connected in parallel and positioned below the water surface in two rows parallel the longitudinal walls of the tank. The anodes were about 2½ ft from the longitudinal walls and were inclined progressively upward from inlet end to outlet end with the two anodes nearest the inlet end positioned so as to be slightly below the point of entry of the incoming wastewater. An electrical wire was connected to a high point of the tank wall, which acted as the cathode, and grounding this to the rectifier completed the circuit. Using a rectified DC current to energize the anodes, various voltages were applied. Voltages between 6 and 9 volts and a current of between 20 and 35 amps resulted in an anolyte stream with fat particles and other solids floating to the surface where they were removed by mechanical sweeps.

Example 2: Wastewater from a meat packing plant was clarified in an electrical setup similar to that used in Example 1. A current of 20 amps and a potential of 9 volts was impressed on the system. Solids quickly rose to the surface and were skimmed off after one-half hour and one hour. After one-half hour there was about 25% reduction in total solids in the wastewater effluent and about a 35% reduction in one hour.

Example 3: Two steel catch basins side by side and equipped with mechanical sweeps were filled with the same packinghouse wastewater. One of the tanks was equipped with an impressed current system. In this tank, which is about 60 ft long by 10 ft wide and 6 ft deep, were positioned two rows of five anodes each made from high silica iron and being 1½" in diameter and 5 ft long. The anodes were placed parallel to the longitudinal wall and inclined upwardly toward the exit end of the tank. The anodes were placed approximately 3 ft from and parallel to the longitudinal walls. Using a potential of about 5.8 volts and a current of 22 amps, a definite difference was noted between the two basins. Fat and other solids rose to the top in the basin equipped with anodes. In order to increase the amount of solids, about 300 ppm of sulfuric acid, 50 ppm of ferric sulfate and 2 parts of a cationic flocculant were added. Skim bars supporting rails ¾" below the surface area were not visible in the control unit. However, not only were rails visible in the test cell, but the water was clear to such an extent that one could see the moving parts near the bottom of the catch basin.

OTHER SPECIFIC PROCESSES

Polychloroprene Rubber

R.D. Pruessner; U.S. Patent 3,778,367; December 11, 1973; assigned to Petro-Tex Chemical Corporation describes the separation of emulsifiers which are salts of the condensation product of naphthalene sulfonic acid and formaldehyde from process water by contacting the process water with an amine having at least two amine groups separated by an aliphatic radical having no more than 5 carbon atoms with the separation being made at a pH of less than 7.0. For example, polychloroprene is produced by emulsion polymerization using these salts as surfactants with the salts being washed out into process water, which is contacted with polyethyleneamine, and the pH of mixture lowered to 4.0 to effect separation of the salt. The following examples illustrate the process. All percentages are by weight unless expressed otherwise.

Example 1: A mercaptan-modified neoprene latex is prepared using the following recipe.

	Concentration in Parts by Weight
Polymerization Charge:	
Chloroprene	100
2,6-Ditertiary-butyl para-cresol	0.1
Resin-721S*	4.0
Lomar PW**	0.7
Deionized water	100
Sodium hydroxide (100%)	0.6
n-Dodecyl mercaptan (100%)	0.235
Initial Catalyst:	
Sodium hydrosulfite	0.0294
Deionized water	0.588
Pumped Catalyst:	
0.358% potassium persulfate in aqueous solution used as required to maintain the polymerization rate.	

*A disproportionated wood resin, purified by distillation.
**Lomar PW is the sodium salt of the condensation product of naphthalene sulfonic acid with formaldehyde and is 87% by weight actives.

The latex is acidified with acetic acid and fed to a freeze roll which rotates partly immersed in the latex. This freeze roll is cooled to a temperature of -15°C by circulating brine. In this process the latex is frozen on the drum and is coagulated as the drum revolves. The coagulated film is stripped from the roll by a stationary knife and is placed on a continuous belt where it is thawed and washed. This process water is then treated with polyamines in accordance with this process.

The process wash water is treated with Dow Amine Residue E-100 having the following analysis.

Ethylenediamine (area percent)	0.3
Piperazine	0.8
Diethylenetriamine	0.8
Aminoethylpiperazine	1.0
Nitrilo trisethylamine	0.1
Tetraethylenetriamine	0.7
Diaminoethyl piperazine	0.2
Piperazo ethyl ethylenediamine	0.4
Unknown	0.4
4-Aminoethyltriethylenetetramine branched	0.7
Tetraethylenepentamine	4.4
Aminoethylpiperazinoethylethylenediamine (cyclic)	1.6
Piperazinoethyldiethylenetriamine (cyclic)	1.8
Pentaethylenehexamine	45.2
Heavies (higher ethyleneamine polymers)	41.8
Amine number ring KOH/gram	1,232

The process wash water contains 800 ppm of the condensation product salts (Lomar PW) based on the process wash water. The amine is intimately mixed with the process wash water in an amount of 0.5 part of amine composition of the analysis shown per 1.0 part of condensation product salts in the process wash water. While the wash water is being agitated, the amine composition is gradually added to the wash water over a period of 30 minutes, and thereafter the mixture is vigorously agitated with a propeller mixer for one hour, during which time the temperature is maintained at 40°C. The pH of the mixture is then lowered to 3.5 by the addition of concentrated sulfuric acid while continuing agitation. The agitation is stopped and the reaction product is allowed to settle out in a settling tank, and a dilute solution of reaction product is removed from the bottom of the tank. The clear water overhead is pumped to a biooxidative degradation unit pond for further treatment to remove biodegradable components from the water. The slurry from the bottom of the settling tank is evaporated to remove a portion of the water and incinerated.

Examples 2 and 3: The following examples utilize the process of Example 1 with the noted exceptions. The same process water is treated in each instance.

Polyamine	Acid for pH	Temp. of Reaction
$H_2N(CH_2)_2NH_2$	37% HCl	25°C
$H_2N-CH_2-CH(CH_3)-NH-CH_2-CH_2NH_2$	15.6 N HNO_3	20°C

Example 4: The process of Example 1 is repeated with the exception that the condensation product salt is an ammonium salt instead of the sodium salt.

Examples 5 through 14: A series of examples are run using a 1,000 ml beaker equipped with a Bird Uniform mixer. The process is conducted at approximately 25°C. 500 ml of water solution containing 1,000 ppm of the sodium salt of the condensation product of naphthalene sulfonic acid with formaldehyde (Lomar PW) are used and the polyamine is added in the stated amount on the total weight of composition. The amine is added while the solution is being stirred rapidly at 40 rpm. Hydrochloric acid is then added to lower the pH to 3.5 and the mixing speed is reduced to 5 rpm to allow the precipitate to agglomerate. The water is poured off of the agglomerate and the percent removal of the condensation product salt from the water is determined by total carbon analysis.

Polyamine	p.p.m. amine	Percent removal
Tetraethylenepentamine	400	69
Polyamine spec. H (Union Carbide)	100	68
Dow E-100 (analysis in Example 1)	100	69
Dow BDE-25 1065	100	63
Dow BC 3565	100	63
Dow CE 46	100	67
Dow BE 2575	100	65
Jefferson AL-7	700	49
Analine		(1)
Nalco	700	57

1 No precipitate detected.

Although this process has been particularly described for the separation of condensation product salts from neoprene process water, it can be used also for the separation of condensation product salts from other aqueous compositions where the water contains less than 5% total solids including the condensation product salts. The process is useful for the separation of the condensation product salts from other polymerization processes. The examples of other processes are emulsion or suspension polymerization processes such as for butadiene-styrene rubber (GRS) and polyvinylchloride or polyvinylacetate.

Chlorinated Organic Pesticides

K.H. Sweeny and J.R. Fischer; U.S. Patent 3,737,384; June 5, 1973; assigned to the U.S. Secretary of the Interior have found that certain metallic couples act as efficient reducing and dehalogenating agents when reacted with halogenated organic compounds under mildly acidic conditions. Metallic couples found to be effective in the reaction comprise a combination of a minor amount of catalytic metal, such as copper or silver, with a major amount of a reductant metal such as zinc, aluminum, iron, cadmium and magnesium. Couples comprising one or more catalytic metals with two or more reductant metals are also effective. Examples of metallic couples found to be effective in the reaction include zinc-copper, zinc-silver, iron-copper, aluminum-copper, magnesium-copper, cadmium-copper and zinc-iron-copper. The reaction is relatively rapid at room temperature but higher temperatures may be used to advantage. Mildly acidic conditions are necessary with a preferred pH range of 1.5 to 4.

Previous work with the zinc degradation of DDT and similar pesticides, described in U.S. Patent 3,640,821, tended to indicate that reduction of DDT may proceed by way of a complex, two path reaction. A portion of the p,p'-DDT appears to be rapidly converted to the ethane derivative, 1,1-bis(p-chlorophenyl)-ethane referred to as DDEt, losing all three aliphatic chlorine atoms. Another portion appears to be reduced in a stepwise and much slower fashion. This reaction appears to progress from p,p'-DDT to p,p'-DDD [2,2-bis(p-chlorophenyl)-1,1-dichloroethane] to DDMS [2,2-bis(p-chlorophenyl)-1-chloroethane] and finally to DDEt. The latter product, DDEt, is a preferred reaction product since it appears to have far less physiological activity than DDT or the other intermediate products.

It has been found that the use of particular metallic couples as the reducing agent for halogenated organic compounds, particularly such pesticides as DDT, results in substantial advantages compared with the use of zinc alone. A marked increase in reaction rate is observed but, more importantly, the proportion of relatively innocuous decomposition products increases. It is expected that the process will prove most useful in the decomposition of halogenated organic residues to produce less highly halogenated products. Specific uses contemplated include the decomposition of chlorinated organic pesticides such as DDT found in process wastewaters and even in soils. The following examples illustrate the process.

Example 1: A zinc-copper couple was formed by adding an aqueous solution of cupric chloride to an acetone suspension of technical grade zinc dust. Copper addition was in the amount of 1 meq per gram of zinc. The usual grey color of zinc immediately darkened to a brown to black appearance indicating copper precipitation on the zinc particles. A suspension of the zinc-copper couple was then acidified to 0.5 normal using acetic acid and was reacted at 25°C on an equal weight (metal to pesticide) basis with DDT. Comparative tests using zinc alone under identical conditions were also performed. Analyses for residual DDT and for several of the major decomposition products of the reaction were performed periodically after initiation of the reaction. Results of these analyses are presented as Figures 7.7a through 7.7c.

Figure 7.7a shows the rate of formation of bis(p-chlorophenyl)ethane (DDEt) using a zinc-copper couple as the reductant as compared to the rate of formation using zinc alone. As may be seen, the amount of DDEt product formed in an equivalent time was 1.9 to 2.7 times greater in the samples in which the zinc-copper couple was used compared to those in which zinc was the reductant. Since DDEt is considered to be the least hazardous and most desirable degradation product of DDT, the advantages of the copper-zinc couple are evident.

Figure 7.7b depicts the rate of formation of bis(p-chlorophenyl)-dichloroethane (DDD) as a degradation product of DDT. DDD is considered to be a highly undesirable product since it, like DDT, is believed to affect the reproductive mechanisms of birds. Those samples in which a zinc-copper couple was used as the reductant produced substantially less DDD than did the samples using zinc alone.

In Figure 7.7c there is shown the rate of formation of bis(p-chlorophenyl)chloroethane (DDMS) as a degradation product of DDT. Concentration of this compound was reduced by a factor ranging from 1.8 to 2.8 when using the zinc-copper couple as compared to the use of zinc alone as the reductant.

Example 2: A series of tests were run varying the ratio of copper to zinc used in forming the zinc-copper couple. In these tests, the efficacy of the reduction of DDT after 15 minutes reaction at 25°C was compared. In all cases, DDT was reacted with an equal weight of zinc metal and the acidity of the reaction medium was adjusted to 0.5 normal at the reaction start. Proportion of copper was varied from 0.1 to 10.0 meq per gram of zinc. Over the entire range of copper concentration tested, the zinc-copper couple proved superior to the use of zinc alone. Best results were obtained, however, at a copper to zinc ratio of about 1 meq of copper per gram of zinc.

Example 3: An aluminum-copper couple was prepared by adding 1 meq of cupric chloride in aqueous solution to 1 gram of aluminum powder. The couple was then reacted with an equal weight of DDT at 25°C in a reaction medium acidified to 0.5 normal using either acetic or sulfuric acid. Essentially all of the DDT was converted, but the major reaction product was different than that obtained using either zinc or a zinc-copper couple as the reductant. The acetone-soluble portion of the reaction product accounted for only about 5.3% of the DDT converted. Remaining product was in the form of a voluminous precipi-

FIGURE 7.7: DECOMPOSITION OF ORGANIC HALOGEN COMPOUNDS USING METALLIC COUPLES

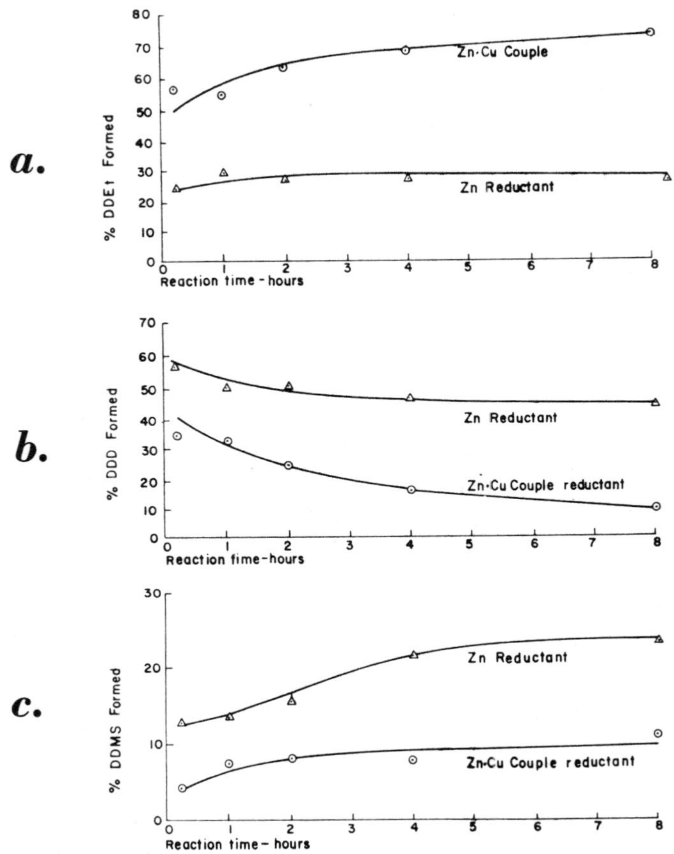

Source: K.W. Sweeny and J.R. Fischer; U.S. Patent 3,737,384; June 5, 1973

tate insoluble in ethanol and hexane as well as acetone, but soluble in warm benzene. When extracted and recrystallized, a product with a melting point of 267°C was obtained in an amount equal to 86.2% of the initial weight of the DDT. Infrared analyses showed the presence of C—Cl bonds but no carbonyl or phenolic hydroxyl. It was hypothesized that the compound might be tetra(p-chlorophenyl)tetrachlorobutane formed by reductive coupling of two DDT molecules according to the following reaction.

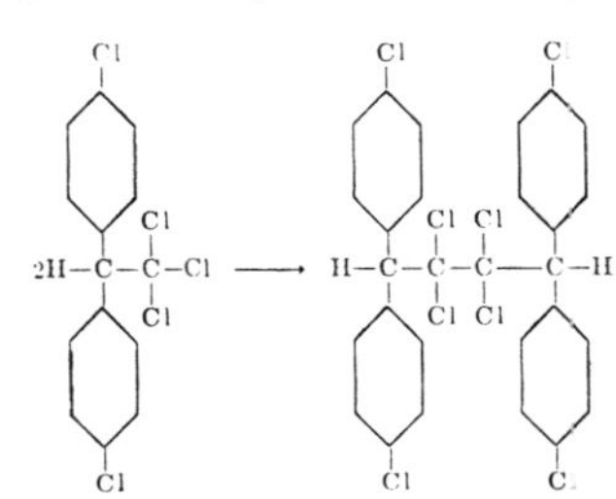

Elemental analysis of the compound having the elemental formula $C_{28}H_{18}Cl_8$ is consistent with this hypothesis. Theoretical: C, 52.6%; H, 2.82%; Cl, 44.5%. Found: C, 51.8%, 51.2%; H, 3.0%, 2.6%; Cl 42.7%, 43.0%. Nuclear magnetic resonance spectroscopy gave results which are also consistent with the hypothesized structure. The postulated compound is known in the literature and has been prepared by reducing DDT with hydrogen using a palladium catalyst. Reported melting point of 270°C compared closely with the 267°C value obtained.

Similar results were obtained when an aluminum-copper alloy was used as the reductant. In contrast, when aluminum powder alone was contacted with DDT under identical conditions, essentially no reaction occurred.

Ozone Treatment of Waste Stream from Silicone Plant

H. Lapidot; U.S. Patent 3,732,163; May 8, 1973; assigned to General Electric Company describes a process for the ozone treatment of a liquid material which includes introducing a major portion of the liquid into a first ozonation zone; introducing the remainder of the liquid into a second ozonation zone; introducing an ozone-containing gas into the first ozonation zone; removing a gaseous stream from the first ozonation zone and conducting it to the second ozonation zone; removing a gaseous stream from the second ozonation zone and conducting it to an ozone-generating zone where the ozone-containing gas is produced. The second ozonation zone is smaller in size than the first ozonation zone but provides more efficient liquid-gas contact than does the first ozonation zone.

The liquid material which can be subjected to the ozonation process can be any liquid which is or contains constituents susceptible to treatment with ozone. Of particular importance are industrial wastewater streams and particularly those streams which contain organic contamination. Specifically the liquid can be the aqueous waste stream from an industrial silicone polymer producing plant. Such an aqueous waste stream from a silicone polymer plant typically contains water-soluble organic contaminants such as methanol, acetone, butanol and isopropanol; toluene; benzene; xylene and the like and a COD of 30 to 1,500 ppm. When the liquid material treated according to this process is a wastewater stream from a silicone polymer plant, it is preferred that such stream has a pH of at least 12, and it is most preferred that it has a pH of between 12 and 12.3. The following examples illustrate the process.

Example 1: 1,200 gallons per minute of an industrial wastewater stream from a silicone polymer plant having a COD of 90 ppm, a temperature of 30°C and a pH of 12 are split up into a first stream and a second stream. The size of the first stream is 1,080 gallons per minute and the size of the second stream is 120 gallons per minute. The first stream is fed to a packed column having an 8 foot diameter and being 40 feet high and being packed with 2 inch ceramic Intalox saddles. 3,300 pounds per hour of a gas containing 5% by weight ozone, 80% by weight oxygen and 15% by weight nitrogen are added to the first packed tower at the lower vapor inlet of the tower. About 1,080 gallons per minute of a liquid product having a COD of 10 ppm are removed from the bottom of the first packed tower. A gaseous stream containing 0.5% ozone is removed from the overhead of the first packed tower and is conducted to the bottom portion of a second packed tower having a diameter of 2 feet and a height of 15 feet and being packed with one-half inch ceramic Intalox saddles.

Meanwhile the second portion of the wastewater stream is sent to a heat exchanger whereby it is cooled to a temperature of 2°C. The cooled wastewater liquid is then conducted to the upper portion of the second packed tower. About 120 gallons per minute of a liquid product having a COD of 15 ppm are removed from the second packed tower. This stream is mixed with the product stream from the first packed tower. A gaseous stream is removed as the overhead from the second packed tower and is conducted to a deentrainer. A gaseous stream is removed as the overhead from the deentrainer. This stream contains about 16.65% by weight nitrogen, 1% by weight water vapor, and 0.01% by weight ozone with the remainder being oxygen. 56.3 pounds per hour of this stream are purged. The remainder of the stream is then conducted to a silica gel dryer and then mixed with 186.6 lb/hr of oxygen. The gaseous stream removed from the silica gel dryer is essentially anhydrous and after adding the mentioned amount of oxygen contains about 85% of oxygen and about 15% nitrogen. The gaseous stream is then conducted to an ozone generator. 3,300 lb/hr of a gas containing 5% by weight ozone, 80% by weight oxygen, and 15% by weight nitrogen are removed from the corona discharge generator and are conducted to the first packed tower.

Example 2: The procedure of Example 1 is repeated except that the wastewater stream initially has a COD of 180 ppm, and contains about 59 ppm of suspended solids, 80 ppm of methanol, 55 ppm of ethanol, 10 ppm of acetone, 10 ppm of isopropanol, and 40 ppm of other unidentified impurities. The liquid product obtained from the process has a COD of 42 ppm and contains 4 ppm of suspended solids, 0 ppm of methanol, 0 ppm of ethanol, 10 ppm of acetone, 0 ppm of isopropanol, and 0 ppm of the other unidentified impurities.

Removal of Iron Carbonyl

According to a process described by *R.H. Callighan and J.O. Hawthorne; U.S. Patent 3,780,163; December 18, 1973; assigned to U.S. Steel Corporation* liquids and gases containing iron carbonyl contaminants may have these carbonyls reduced by reaction with up to stoichiometric amounts of ozone to convert the carbonyls into ferrous carbonate.

The extent to which iron carbonyls are removed from the streams containing the carbonyl depends upon the amount of ozone added to the stream. When a stoichiometric amount of ozone is added to the stream, substantially all of the iron carbonyl is converted into ferrous carbonate. In some cases, a minimum amount of carbonyl can be tolerated so it is only necessary to convert a portion of the carbonyl present. This may be done by using less than a stoichiometric amount of ozone for conversion of the carbonyl into the carbonate. By controlling the amount of ozone, the degree of reduction of iron carbonyl concentration in the stream is also controlled.

The reaction temperature may vary over a wide range. Temperatures as low as -70°C or as high as 200°C are particularly useful. At temperatures above 250°C, the ferrous carbonate will decompose exothermically to produce iron oxide. Generally, the ozone is introduced into the gas or liquid stream at the ambient temperature of the stream.

The reaction may take place under pressures from atmospheric up to 10 atmospheres or more. Pressures of one atmosphere are convenient because large numbers of industrial processes yielding gaseous or liquid streams having the iron carbonyl contaminants are performed at this pressure. The ozone should be thoroughly mixed with the gas or liquid stream into which it is introduced.

The ferrous carbonate which is the product of the reaction between ozone and the iron carbonyl is an amorphous solid that can be removed from the stream in which it is formed. In the case of liquid streams, filtration or centrifugation easily remove the ferrous carbonate from the liquid body. At temperatures above 250°C, ferrous carbonate will decompose to produce iron oxide. The iron oxide may also be removed by filtration or centrifugation or other conventional methods such as electrostatic precipitation. The following examples illustrate the process.

Example 1: Iron pentacarbonyl was dissolved in ethyl acetate to give a solution containing 1.08% by weight of iron pentacarbonyl or 0.485 mol percent of ethyl acetate. This solution was cooled to -35°C and then reacted with an ozone-oxygen stream, this stream had about 5% ozone. The reaction was continued until an amount of ozone calculated to give a 3 to 1 mol ratio for iron carbonyl was introduced. A brown precipitate was filtered and dried from the solution. This precipitate was amorphous ferrous carbonate.

Example 2: A gas mixture containing 30% methane, 6% carbon monoxide, 0.46% nitrogen and the remainder essentially hydrogen with three parts per million by volume of iron pentacarbonyl was reacted with an amount equivalent to 9 ppm by volume of ozone. After mixing, no iron carbonyl could be detected in the exit gas and a brown deposit was removed by gravity separation from the gas, which deposit was identified as ferrous carbonate.

Example 3: The procedure of Example 2 was repeated except that the gas contained 10 parts per million by volume of 1,3-butadiene. When using 9 ppm by volume of ozone, the iron pentacarbonyl was still completely removed in the form of ferrous carbonate while the butadiene was unaffected.

Example 4: The procedure of Example 2 was repeated except that the gas contained 40 parts per million by volume of dimethyl sulfide. Again, when using 9 ppm by volume of ozone, the iron pentacarbonyl was still completely removed without affecting the dimethyl sulfide.

Continuous Separation of Polyolefin Slurries

A process described by *M. Koga and K. Ando; U.S. Patent 3,661,260; May 9, 1972; assigned to Mitsubishi Petrochemical Company Limited, Japan* relates to a method and apparatus for continuously separating a process mixture into an oil slurry phase and an aqueous phase. In the chemical industry, there are instances where it is necessary to process mixtures each composed of an oil slurry phase disposed as an upper layer and an aqueous phase disposed as a lower layer when left standing and caused to settle. For example, in some cases in the process of producing a polyolefin, it is necessary to separate a process mixture into an oil slurry phase composed of a hydrocarbon slurry in which a polyolefin is suspended and an aqueous phase.

Such a separation in a continuous manner is not easy in many cases. The reason for this is that, while it should be possible to separate the light oil phase and the heavy aqueous phase by leaving the mixture thereof at rest to settle for a certain time, a problem arises on this point in the case where a slurry is being processed. More specifically, the specific densities of the dispersed phase and the dispersing medium within an oil slurry phase are different in many instances. For example, in a polyolefin slurry, the specific gravity of the dispersed phase, namely the polyolefin, is of the order of from 0.8 to 1.0 while that of the hydrocarbon dispersing medium is of the order of from 0.6 to 0.8. Consequently, when the mixture is left standing for an amply long settling time, sedimentation of the polyolefin occurs. On the other hand, when the settling time is shortened, or the oil slurry phase is agitated, the separation of the two phases becomes incomplete.

According to the process there is provided, in the separation of a slurry from a mixture of an oil slurry phase and an aqueous phase which stratify into respective and contiguous upper and lower layers when left standing at rest, a method characterized by the steps of introducing the mixture continuously into a lower-type separation chamber, causing the mixture to be retained for a certain time to stratify into the upper and lower layers, and continuously drawing off slurry from the upper layer and aqueous phase from the lower layer in a manner to maintain a state of dynamic equilibrium within the chamber as a portion of the slurry thus drawn off is continuously discharged as a slurry product and the remainder of the drawn off slurry is recirculated into the upper-layer phase within the chamber.

In the process, the dynamic equilibrium state within a separation chamber is so maintained that there is substantially no fluctuation of an interface between an oil slurry phase and an aqueous phase, and the slurry phase is recirculated to aid the prevention of sedimentation of the dispersed phase within the oil slurry phase (referred to as the slurry phase) thereby to make possible continuous separation.

The principle of the process is considered with respect to one example of apparatus diagrammatically indicated in Figure 7.8a in which a separation chamber **16** is used and contains an aqueous phase **2** constituting a lower layer and a slurry phase **1** constituting an upper layer above the aqueous phase with an interface **14**.

FIGURE 7.8: CONTINUOUS SLURRY SEPARATION

a.

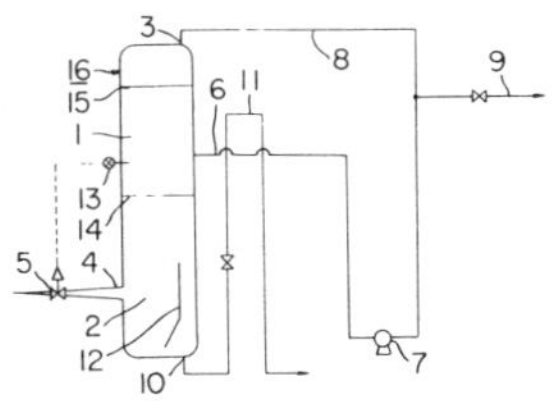

b.

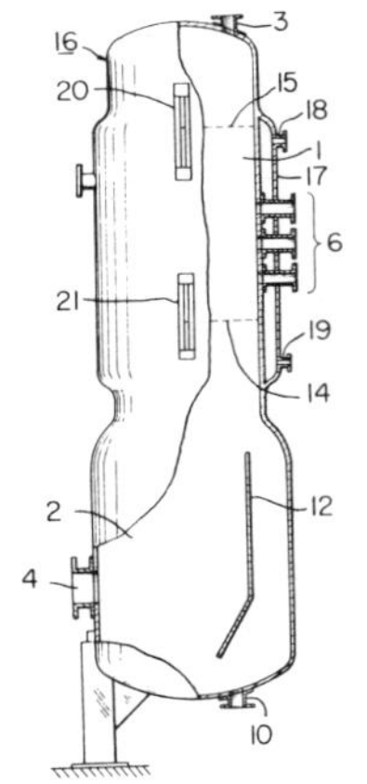

Diagrammatic Flow Chart

Separation Chamber

Source: M. Koga and K. Ando; U.S. Patent 3,661,260; May 9, 1972

The separation chamber is provided with a slurry phase drawoff outlet **6** connected to the chamber in the region thereof containing the upper-layer slurry phase **1**. An aqueous phase draw-off outlet **10** is connected to the bottom of the chamber. A supply inlet **4** is provided for supplying into the chamber a mixture of slurry and aqueous phases, and a slurry phase recirculation line **8** is connected to the top of the chamber.

While the recirculation of the upper-layer slurry phase is a unique and prominent feature in the practice of this process, it is not absolutely necessary to carry out independently and separately the drawing off of the upper-layer slurry phase for this recirculation and the discharging of the upper-layer slurry phase as a slurry product. That is, it is not necessary to provide a slurry phase draw-off outlet in the separation chamber for each of these purposes.

Accordingly, the discharging of the slurry product may be carried out from the slurry recirculation system or circuit as indicated in Figure 7.8a, or, alternatively, a portion of the slurry may be diverted from a pipe line for discharging the slurry product and returned to the separation chamber thereby to form a slurry-recirculation system (the slurry phase draw-off outlet confronting the upper-layer slurry phase being of one kind in each of these two cases).

It has been found that, for preventing clogging due to sedimentation and accumulation of the dispersed phase within the slurry phase **1**, an appropriate control step is to hold the slurry rising linear velocity from the interface **14** formed between the two phases after stratification to the slurry phase draw-off outlet **6** in a range of from 0.05 to 3 cm/sec. For this purpose of controlling the slurry rising linear velocity, the use of a separation chamber **16**, of the vertically-long tower type, is suitable and convenient.

The term tower-type is used here to designate a separation chamber of a shape such that its height is substantially greater than its diameter or largest cross-sectional area. Provided that these conditions are fulfilled, the cross section of the tower-type separation chamber may be of any convenient shape such as an ellipse or polygon in addition to a circle.

Furthermore, the tower need not be of the same diameter from its upper part to its lower part. For example, for prolonging the residence time (or retention time) of the aqueous phase, the chamber diameter of at least the lower part of the region in which the aqueous phase **2** resides can be made greater than the chamber diameter of the region in which the slurry phase **1** resides, as illustrated by one specific example in Figure 7.8b.

It has been found that the mixture supply inlet **4** is preferably provided at a level at least 0.5 of a meter below the interface **14** so as not to cause undue disturbance of the interface and thereby to prevent admixing of the aqueous phase into the slurry phase. While a suitable position for the supply inlet **4** depends on the rate of supply of the mixture, it is desirable in many cases that this inlet be positioned at least 0.8 of a meter below the interface, a suitable range being of the order of from 1.0 to 3.5 meters. A distance greater than 5 meters is probably economically disadvantageous because the total tower height would become unnecessarily great.

It is desirable that the diameter of the supply inlet **4** be so selected that the linear velocity of supply of the mixture of the two phases will be of the order of from 0.1 to 20 cm/sec. As long as the supply speed is within this range, the separation of the slurry and aqueous phases can be accomplished with a compact apparatus. More specifically, when this supply speed is excessively high, it is necessary to use a separation chamber having a part for the aqueous phase **2** of an unnecessarily large volume in order to avoid disturbance of the interface and thereby to prevent admixture of the slurry phase into the discharged aqueous phase.

Accordingly, a supply inlet **4** formed in the shape of a hollow cone, diverging away from the tower as indicated in Figure 7.8a to lower the supply linear speed is desirable. For the same reason, it is also advantageous to make the chamber diameter at its part containing the aqueous phase **2**, particularly in the vicinity of the supply inlet **4**, greater than that of the part of the slurry phase **1**.

It has been found that the height of the slurry-phase draw-off outlet **6** is desirably at least 0.3 of a meter above the interface **14**. In general, a value of this height of at least 0.5 of a meter is preferable in many cases. If this height is too short, there will be the possibility of the aqueous phase becoming admixed in large quantity into the slurry being drawn off. On the other hand, it is desirable that this height be limited to 1.5 meters in order to limit fully the sedimentation of the dispersed phase within the slurry phase.

Generally stated, the slurry-phase draw-off outlet **6** is preferably disposed from 0.5 to 1.5 meters above the interface **14**. When a plurality of slurry-phase draw-off outlets are used in the separation chamber **16** to draw-off the upper-layer slurry phase, as shown in Figure 7.8b, this height condition is preferably applicable to each of the outlets.

Recirculation of the slurry phase is necessary for suppressing sedimentation of the dispersed phase within the slurry phase **1** and, in the specific example illustrated in Figure 7.8a, for drawing off slurry phase to become the process product. The slurry-phase recirculation is accomplished by means of a pump **7** through the slurry-phase recirculation line **8** preferably at a flowrate such that a quantity of slurry which is from 5 to 100 times the volume of the upper-layer slurry phase **1** present in the separation chamber **16** under a state of dynamic equilibrium passes through the outlet **6** every hour.

With a flowrate corresponding to less than 5 times the volume of the slurry phase **1**, the suppression of the sedimentation of the dispersed phase within the slurry phase is insufficient. On the other hand, a flowrate corresponding to more than 100 times the volume of the slurry phase **1** gives rise to a violent flow of the slurry phase **1** and a great disturbance of the interface **14** and slurry-phase surface **15**, whereby the process operation becomes unstable. While the inlet for introducing the recirculated slurry phase may be above or below the slurry phase surface **15**, it should be positioned well above the slurry draw-off outlet **6**. An empty space **3** of a height of the order of 1 meter above the slurry-phase surface **15** has been found to be suitable.

The draw-off outlet **10** for draining the aqueous phase **2** is provided below the interface **14** in the side wall of the separation chamber **16** or in the bottom thereof. In order to prevent admixture of the slurry phase supplied through the supply inlet **4**, the aqueous-phase draw-off outlet **10** is preferably positioned below the supply inlet **4**, for example, in the bottom of the separation chamber **16** as indicated in Figure 7.8a. In addition, the provision of a baffle plate **12** is also desirable.

The aqueous phase drained out of the separation chamber **16** through the outlet **10** can be conveniently discharged in an automatic manner by means of an overflow pipe **11** as shown in Figure 7.8a, utilizing a hydrostatic head corresponding to the difference between the heights of the top of the overflow pipe and the slurry-phase surface **15**. Discharging by means of a pump is, of course, also possible.

The process operation is carried out in a manner such that a state of dynamic equilibrium is sustained within the separation chamber. Such an operation can be accomplished by effecting process control so that the rates of feed of the slurry phase and the aqueous phase supplied as a mixture are substantially equal to the rate of discharging of the slurry product and the rate of aqueous-phase draw-off, respectively. In this operation, it is preferable that fluctuations of the level of the interface **14** between the two phases **1** and **2** be held within a range of ± 10 cm.

Example: A mixture composed of 100 parts of polypropylene in powder form, 200 parts of heptane for industrial use, and three parts of an alcohol (all parts being by weight) was separated into a slurry phase and an aqueous phase (an aqueous solution of the alcohol) by means of an apparatus which was essentially the same as that illustrated in Figure 7.8a. The separation chamber **16** of this apparatus was of an overall height of 4,500 mm, a diameter in the vicinity of the two-phase mixture supply inlet **4** of 500 mm, and a diameter from the top to a level 3,700 mm therebelow of 210 mm and the height of the overflow pipe (corresponding to pipe **11**) was 3,200 mm. A level gauge of differential pressure type was installed in the chamber at a part thereof containing the slurry phase.

Slurry, as a product, was drawn off through a slurry discharge pipe line (corresponding to pipe line **9** in Figure 7.8a) at a constant flow rate of 215 l/hr, while the aqueous phase was discharged in a natural manner through the overflow pipe. The height level of the slurry-phase surface **15** was maintained constant at 3,700 mm by a differential pressure level gauge (corresponding to gauge **13**) and a supply control valve **5**. After dynamic equilibrium was attained, the interface **14** between the two phases was maintained at a level of 1,700 mm, the slurry draw-off outlet **6** then being 1,000 mm above and the two-phase mixture supply inlet then being 1,300 mm below the interface level.

The slurry recirculation flow rate was 1,000 l/hr, which corresponds to approximately 14.5 times the quantity of the residence or retained slurry. The mixture supply pipe line downstream from the supply control valve **5** was formed with a divergent flare, whereby the diameter at the supply inlet was 100 mm (4 inches).

During the operation of the apparatus under the conditions as described above, the average rate of supply of the mixture of the two phases was 315 l/hr. It was found that during this operation no slurry (heptane and polypropylene) was admixed with the discharged aqueous phase, and moreover, only 0.05% of water was detected in the slurry product, whereby it could be concluded that the separation of the slurry and aqueous phases was thorough. Furthermore, there was no instance of malfunctioning of the apparatus or misoperation of the process during one week of continuous operation.

Organic Cyanides

J.-P. Zumbrunn; U.S. Patent 3,715,309; February 6, 1973; assigned to L'Air Liquide, SA pour l'Etude et l'Exploitation des Procedes Georges Claude, France describes a process for detoxicating wastewaters which contain organic cyanide impurities, particularly nitriles. The process involves rapid degradation of the nitriles into amides and then into salts of organic acids by treatment with hydrogen peroxide in alkali medium, and the per compounds.

This process, which has the advantage of saturating the effluents with oxygen, permits the total purification to a residual content lower than 0.5 ppm of any effluent coming from chemical synthesis plants, such as the factories for the production of plastic materials or synthetic fibers, and containing organic nitriles. The following examples illustrate the process.

Example 1: To 1 liter of aqueous solution of an effluent containing 50 ppm of acrylonitrile are added 5 grams of sodium hydroxide and 0.25 ml of 35% hydrogen peroxide, i.e., a molar ratio $H_2O_2/CH_2{=}CH{-}CN = 3$. For reaction temperatures which are respectively 25°, 40°, 60°, 80° and 90°C the disappearance of the acrylonitrile final content below 0.5 ppm, checked by vapor phase chromatography is obtained in: 3½ hr, 2 hr, 30 min, 20 min and 10 min.

Example 2: The same solution as that described in Example 1 is treated under the same conditions, but with a quantity of sodium hydroxide reduced from 5 to 0.5 grams. The complete detoxication is obtained for the same temperature range in 7 hr, 2 hr, 1 hr, 30 minutes and 15 minutes.

Example 3: Using the conditions of Examples 1 and 2, industrial wastewaters charged with 50 ppm of acrylonitrile are treated, but the H_2O_2/RCN ratio is brought from 3 to 5. The reaction is twice as fast.

Example 4: 1 liter of an aqueous solution containing 1 gram of acrylonitrile and 5 grams of sodium hydroxide is treated with hydrogen peroxide in accordance with a molar ratio of $H_2O_2/RCN = 3$, i.e., an excess of 50% with respect to the theoretical. At 60°C, the acrylonitrile has disappeared in 30 minutes. At this moment, already more than 50% of the amide corresponding to the introduced nitrile still remain and the amide has completely disappeared after 6 hr. A comparison test, under the same conditions, but without hydrogen peroxide, shows that the amide content after 6 hr is 60% of the value corresponding to the introduced nitrile.

Example 5: To 1 liter of aqueous solution containing 2 g/l of oxodipropionitrile are added 0.5 gram of sodium hydroxide and 7 ml of 35% hydrogen peroxide, i.e., an H_2O_2/RCN ratio of 2.6. The detoxication is complete in less than 30 min at 60°C.

Example 6: 1 liter of an industrial wastewater coming from a plant for the preparation of ABS elastomer (acrylonitrile-butadiene-styrene) and containing 0.5 gram of acrylonitrile and 5 grams of oxodipropionitrile is treated with 22 ml of 35% hydrogen peroxide at the temperature 60°C, i.e., an H_2O_2/RCN ratio = 2.8. Complete detoxication is obtained in less than 30 min.

Organically Polluted Water

E.O. Box, Jr. and F. Farha, Jr.; U.S. Patent 3,823,088; July 9, 1974; assigned to Phillips Petroleum Company describe a process for continuously converting organic contaminated aqueous streams to fresh potable water by contacting either in the liquid phase or gaseous phase an aqueous stream so contaminated with a zinc aluminate catalyst promoted with at least one metal active for initiating oxidative reactions under conditions so as to produce water essentially free of pollutants.

It has also been found that the activity and life of metal promoted zinc aluminate oxidation catalysts for the removal of organics from aqueous streams can be increased by incorporating a rare earth metal in the catalyst system. Particularly effective catalyst compositions that have been prepared include Cu-zinc aluminate, Ni-zinc aluminate, Cu-Ni-zinc aluminate, Cu-Mn-zinc aluminate, Bi-zinc aluminate and Ce-zinc aluminate.

The process is particularly applicable for feeds in which the organic materials are hydrocarbons and/or oxygenated hydrocarbons. The organic materials can be present in the predominantly aqueous stream in a broad range of concentrations, but will ordinarily be present in amounts less than about 10 weight percent. It is generally more convenient to characterize the aqueous feeds by the total carbon content. Consequently, such feeds can contain from about 10 to 100,000 ppm carbon, more frequently 25 to 10,000 ppm carbon, still more frequently 100 to 5,000 ppm carbon. Such aqueous waste streams can be derived from any source such as chemical or biological sources. For best results, the aqueous waste streams will contain relatively little nonvolatile and/or nonoxidizable inorganic materials. The following examples illustrate the process.

Example 1: Preparation of Zinc Aluminate – A 1,700 gram quantity of finely divided commercial flame-hydrolyzed alumina was slurried with 7 liters of distilled water. A 1,450 gram quantity of finely divided zinc oxide was then added and the resulting slurry was ball-milled for 1 hour. The slurry was poured into shallow pans and dried overnight at 275°F in a forced draft oven. The dried cake was broken up and screened into 40-mesh particles (3,025 grams) and mixed with 182 grams of polyethylene powder (tableting lubricant) in a mill, and formed into ⅛ x ⅛ inch tablets in a tableting machine. The tablets were then calcined by a heating sequence which included 1 hr at 800°F, 1 hr at 1000°F, 1 hr at 1100°F, and 3 hr at 1800°F. The tablets were broken up and screened to 14/20 mesh size granules prior to impregnation with the promoter metals.

Several variations of the above described procedure were carried out in which minor amounts of tin were incorporated into the zinc aluminate. This was done using the same procedure except that minor amounts of tin oxide powder were added to the initial aqueous slurry of alumina and zinc oxide. Sufficient tin oxide was added to provide zinc aluminate compositions which contained, after calcination, 0.4, 1 and 4 weight percent tin calculated as the metal and based on the zinc aluminate.

Example 2: Preparation of Metal-Promoted Zinc Aluminate Catalysts – A 38 gram quantity of zinc aluminate prepared as in Example 1 was impregnated with an aqueous solution of palladium dichloride. A 14.2 cc quantity of a $PdCl_2$ solution containing 0.0107 gram Pd/cc was diluted to 25 cc with distilled water. The zinc aluminate granules were soaked with this solution until essentially all the liquid was absorbed. The impregnated granules were then dried at 248°F, calcined in air at 1100°F. Other zinc aluminate catalysts were impregnated with water-soluble compounds of the promoting metals in a similar fashion to prepare active zinc aluminate catalysts promoted with copper, nickel, manganese, bismuth, cerium, platinum, iridium, rhenium, and combinations of these.

Example 3: Air-Oxidation of Aqueous Wastes – Catalysts prepared in Example 2 were used to oxidize aqueous solutions of organic wastes. These aqueous solutions were obtained from process water associated with the catalytic oxidative dehydrogenation of butenes to butadiene. The water contained various amounts of dissolved and suspended by-products such as paraffins, olefins, naphthenes, aromatics, as well as various oxygenated materials such as acids, aldehydes, ketones, alcohols, ethers, and the like.

The catalysts were charged into tubular fixed reactors which were vertically positioned. A mixture of air and wastewater was continuously and downwardly passed through the reactor. Depending upon the conditions of temperature and pressure, some reactions were vapor phase while others maintained a liquid phase in the reaction zone. After a period of time on stream, the effluent of the reaction zone was sampled and analyzed for residual organic content in terms of parts per million carbon. The essential conditions as well as the results of these tests are shown in Tables 1 and 2 which follow.

Runs 1 through 19 in Tables 1 and 2 show that the catalysts of the process are highly effective for the removal of organic matter from aqueous streams by air oxidation. By varying conditions such as feed rate, temperature, pressure, and water:air ratios, the desired level of residual organic level can be achieved up to and including essentially complete organic removal. In every run, the catalysts were found to be physically durable and long lasting. For example, the catalyst of Run 19 showed good results after 407 hours on stream. Runs 8 and 17 show that zinc aluminate containing small amounts of tin, but otherwise unpromoted, shows substantial activity and durability for the reaction.

Comparison Runs 20 and 21 show that a similarly metal-promoted alumina catalyst has neither the activity nor, more importantly, the durability for this severe service. In Run 20, which was the first attempt to use the alumina-supported catalyst, the catalyst failed presumably because it was unable to survive the initial shock of startup. The run was started by pressuring up the heated reactor with air, then introducing the water at about 300 psig. In Run 21, a milder startup was used in which the heated reactor was pressured with both air and water simultaneously. The procedure allowed the catalyst to operate but its activity was less than that of the similar zinc aluminate catalysts of Runs 18 and 19. The durable zinc aluminate catalysts do not exhibit the sensitivity to startup procedures as did the alumina catalyst.

TABLE 1: VAPOR PHASE CATALYTIC AIR-OXIDATION OF AQUEOUS WASTES OVER METAL-PROMOTED ZINC ALUMINATE CATALYSTS

Run no.	Weight percent promoter metal	Temperature, ° F	Pressure, p.s.i.g.	Water: air, mole ratio	Water rate, LHSV	Hours on stream	Carbon in feed, p.p.m.	Carbon removal, percent
1	2 Cu	1,005	35	18:1	17.5	4.5	810	99.4
2	2 Cu	992	45	58:1	61.5	9.9	810	97.5
3	2 Ni	1,002	35	42:1	42.5	2.1	810	97.2
4	1 Cu-1 Ni	1,016	35	31:1	42.5	20.6	760	99.2
5	1 Mn-1 Cu	1,000	35	137:1	53.4	5.5	760	95.4
6	2 Bi	1,006	40	21:1	35.8	4.1	760	98.7
7	2 Ce	1,004	35	7:1	14.2	2.6	760	89.8
8 [a]	None	1,006	35	8:1	7.7	29	800	97.6
9 [b]	0.4 Pt	1,002	36	10:1	28.3	26.5	800	99.0
10 [b]	0.4 Pt	1,000	160	137:1	30.8	31.1	800	99.3

[a] 1 wt. percent Sn had been incorporated into the zinc aluminate during its preparation.
[b] 0.4 wt. percent Sn had been incorporated into the zinc aluminate prior to impregnation with Pt.

TABLE 2: LIQUID PHASE CATALYTIC AIR-OXIDATION OF AQUEOUS WASTES OVER METAL-PROMOTED ZINC ALUMINATE CATALYSTS

Run no.	Weight percent promoter metal	Temperature, ° F	Pressure, p.s.i.g.	Water: air, mole ratio	Water rate, LHSV	Hours on stream	Carbon in feed, p.p.m.	Carbon removal, percent
11	1 Cu-1 Ni	398	275	14:1	5.1	23.2	760	45.7
12	2 Bi	407	310	9:1	7.5	24.5	760	52.6
13	4 Cu-0.1 Pt	484	690	12:1	1.6	53.7	1,053	93.7
14	0.4 Pd	477	665	19:1	1.9	4.5	1,053	89.5
15	0.4 Ir	489	690	32:1	1.2	22.5	1,053	96.8
16	0.4 Re	487	675	29:1	1.2	23.0	1,053	69.5
17 [c]	None	484	670	38:1	2.4	21	1,053	71.9
18 [d]	4 Cu-2 Mn	490	660	11:1	1.2	16.6	2,000	97.7
19	4 Cu-2 Mn	488	670	32:1	1.9	407	890	85.7
20 [e]	4 Cu-2 Mn	490	670	~25:1	~2	~3	890	(f)
21 [e, g]	4 Cu-2 Mn	490	690	26:1	2.3	11.7	890	59.1

[c] 4 weight percent Sn had been incorporated during its preparation.
[d] The zinc aluminate had surface area of 35 $m.^2/g$. All others had about 6 $m.^2/g$.
[e] A commercial alumina (Kaiser) catalyst material used in this run instead of zinc aluminate.
[f] Run was shut down after 3 hours due to disintegration of catalyst into a paste.
[g] Repeat of Run 20 except that, on startup, the reactor was pressured up with both air and water simultaneously.

Example 4: Catalysts were evaluated in a continuous automated test unit at 700 psig, 4 LHSV water containing 1,000 ppm dissolved organic carbon, and 20/1 water to air mol ratio. Both catalysts comprised 4% copper and 2% manganese on zinc aluminate. To one of these was added 1% lanthanum by impregnation with lanthanum nitrate solution. The percent of organics removal is tabulated after several hours on stream.

TABLE 3

	450° F., percent at—		490° F., percent at—	
Time on stream	1% La	No La	1% La	No La
Hours:				
46	80	74	----------	----------
73			91	86
121	75	62	----------	----------
216			93	78
269	76	51	----------	----------
294			90	78
345	59	49	----------	----------

Acid Wastes from Newsprint

J.P. Field; U.S. Patent 3,743,598; July 3, 1973 describes an apparatus for the continuous mixing of chemicals to eliminate acid wastes. Many industries in the normal course of carrying out their business produce an acid waste. One such industry is the newspaper publishing industry. Ecology laws require that acid waste be neutralized before it can be disposed of. Systems for neutralizing acid employ large holding tanks. The acid waste and a suitable base are introduced into the tank, and the solution is held in the tank for a long period of time until the mixture reaches a pH of about 7 at which time the mixture can be disposed of. These holding tanks are expensive and require substantial time to accomplish the neutralizing function.

This process employs a mixing apparatus and method which significantly reduces the time required for neutralization. With the process, the acid waste is not held in a holding tank, but rather is continuously fed through a tank at a rate commensurate with a reasonable drain rate of the acid waste from the chamber in which it is held. Although the mixing apparatus is particularly adapted for acid waste neutralization, it can be readily used for other controlled mixing functions including acid injection.

A mixing apparatus constructed in accordance with the process includes a tank having an acid waste inlet, a base inlet and an outlet. Acid waste is typically in the form of a liquid although it may have undissolved solids, and it may be heavy or sludge-like. The acid waste enters the tank and fills the tank to a predetermined level. To obtain some turbulence and mixing of the acid waste and base, the base is preferably introduced into the tank below liquid level. The base can advantageously take the form of a gas such as ammonia (NH_3) which is soluble in the acid waste and which bubbles up through the acid waste.

A feature of this process is that the injection of the base into the tank is not modulated, but is controlled solely by an on/off valve in response to the pH of the mixture adjacent the outlet. One advantage of this system is that digital operation simplifies valve and controller construction, and this produces a corresponding cost savings.

Figure 7.9a shows an apparatus **11** constructed in accordance with the process. The apparatus is particularly adapted for mixing acid waste and a base to thereby neutralize the acid waste. However, the apparatus may be utilized for other mixing operations and is not restricted to an acid neutralization function. Moreover, depending upon the setting of the instruments employed, the acid and base could be mixed to provide an average pH other than 7, if desired.

The apparatus includes a tank **13** which is constructed of a polyvinyl chloride pipe **15** of cylindrical configuration and a pair of polyvinyl chloride end caps **17**, each of which is in the form of a plate. The end caps are suitably retained on the pipe as by a plurality of elongated bolt assemblies **19** which clamp the end caps against the opposite ends of the pipe. The tank is divided into an inlet zone **21**, intermediate or mixing zone **23** and an outlet or sensing zone **25**. The inlet zone is supplied with acid waste in substantially liquid form through an acid waste inlet **27** formed in the end cap. Acid waste is supplied to the inlet **27** from a source through a conduit **29**, a manual valve **31** and a check valve **33**.

A base is supplied to the inlet zone from a source **35** through a conduit **37**, an on/off automatic control valve such as a solenoid valve **39** and an injector **41** which is a sparge. The injector is an elongated tube having two rows of circumferentially offset apertures or base inlets **43** (Figure 7.9f) with the apertures of the two rows being axially offset. To improve mixing, the apertures open generally downwardly, e.g., at a 45° angle relative to the vertical. The apparatus is particularly adapted, however, to neutralize acid to the type typically produced in publishing newspapers. This acid is typically nitric acid. However, virtually any acid including sulfuric acid and hydrochloric acid can be neutralized. Although various bases can be utilized, a gas is preferred because no pump is required to supply it and because of the agitation effect which it has when it is discharged through the apertures **43**. Ammonia gas can be used to advantage as it is readily soluble in the acid waste and in water.

FIGURE 7.9: APPARATUS FOR CONTINUOUSLY NEUTRALIZING ACID WASTE STREAMS

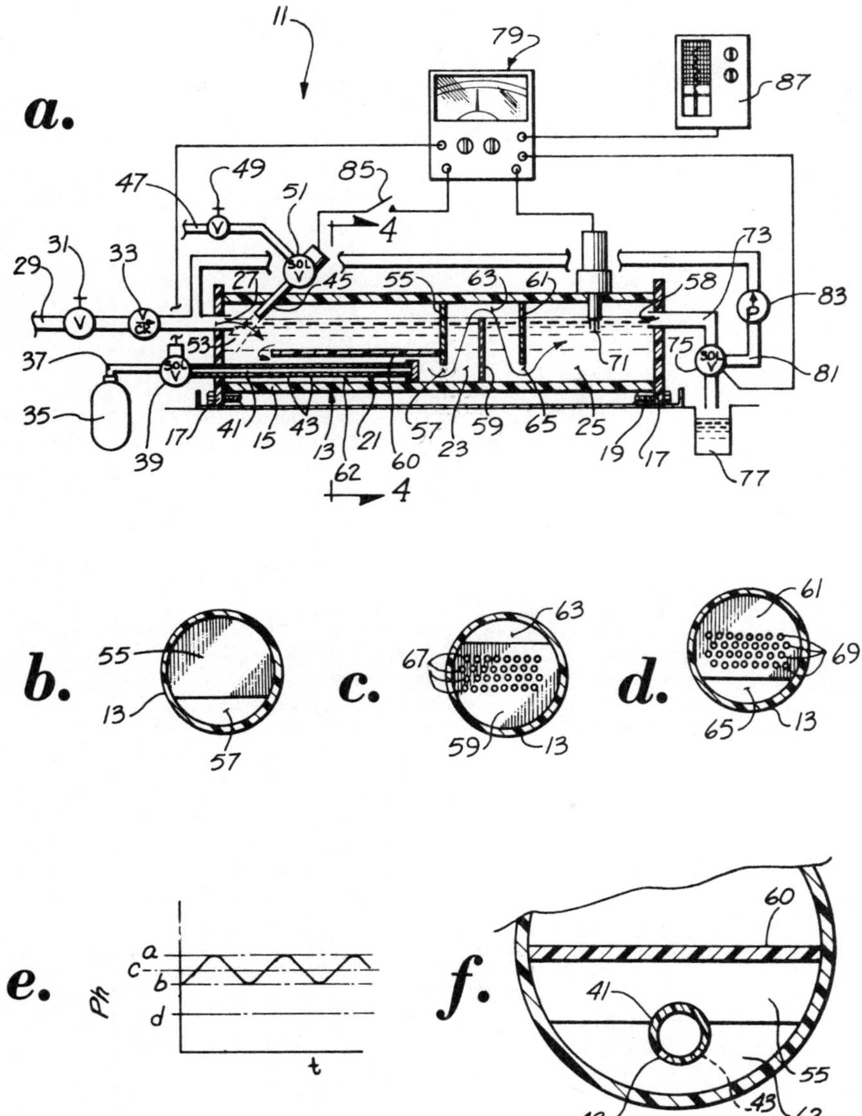

(a) Sectional View of Apparatus
(b)(c)(d) Sectional Views Through Tank Illustrating Three Baffles
(e) pH Values Versus Time at Outlet of Tank
(f) Sectional View Along Line 4–4 of Figure 7.9a illustrating Injector.

Source: J.P. Field; U.S. Patent 3,743,598; July 3, 1973

A nozzle **45** is suitably mounted on the tank **13**. The nozzle is supplied with water from a source (not shown) through a conduit **47**, a manual valve **49** and an on/off automatic control valve such as a solenoid valve **51**. The nozzle is arranged at an angle to direct a stream of water in a pattern **53** which passes just beneath the inlet **27**. This assures that the water from the nozzle will not be directed into the inlet, but will be directed primarily at the acid waste exiting from the inlet. This water stream agitates the acid base mixture, dilutes the acid, provides a cooling effect on the exothermic reaction in the inlet zone **21** and assures that the ammonia gas will be fully absorbed in the liquid phase within the inlet zone and not allowed to escape as a gas.

The inlet zone extends from the end cap **17** on the left as viewed in Figure 7.9a to an imperforate baffle **55** which is in the form of a part circular plate suitably affixed to the interior of the tank **13** and generally perpendicular to the axis of the tank. The baffle closes the tank except for a restricted orifice **57** through which the liquid in the inlet zone **21** must pass en route to an outlet **58** of the tank. Another feature of the process is that the injector **41** extends for substantially the full length of the inlet zone **21** to thereby distribute the ammonia gas throughout the full length of the inlet zone. The injector terminates closely adjacent the restricted orifice **57**.

A plate **60** is mounted on the baffle and extends toward the inlet end of the tank to thereby define along with the baffle a restricted passage **62** in which the sparge **41** is positioned. The baffle and the plate cooperate to completely close off the tank except for the passage **62**. Accordingly, all of the acid waste is routed over the sparge in close proximity thereto to thereby achieve good initial mixing of the acid and the base. The plate also elongates the flow path through the tank.

The baffle and two additional baffles **59** and **61** are in the intermediate or mixing zone **23** with the baffles **55** and **61** defining the extremities of the intermediate zone. The baffles **59** and **61** are suitably mounted on the tank in generally perpendicular relationship to the axis of the tank and define with the tank restricted orifices **63** and **65**, respectively. The baffles **55** and **61** are mounted so as to provide the restricted orifices **57** and **65** at the lower region of the tank and the baffle **59** is positioned so as to provide the restricted orifice **63** at the upper end of the tank. The flow path through the orifices **57, 63** and **65** is as shown by the arrows in Figure 7.9a. The baffles are preferably constructed of sheets of plastic and any suitable number of the baffles may be provided.

The baffles **59** and **61** have small openings **67** and **69**, respectively, therein. Although various numbers and patterns of the openings **67** and **69** may be provided, manufacture of the baffles **59** and **61** is facilitated if the openings in both baffles are identical as to size and position. As shown in Figures 7.9c and 7.9d, the openings **67** and **69** lie entirely within the regions of their respective baffles **59** and **61** which directly confront each other. Stated differently, all of the openings **67** and **69** in both of the baffles **59** and **61** lie above a horizontal plane which includes the lowermost edge of the baffle **61** and below a horizontal plane which includes the uppermost edge of the baffle **59**.

The baffles **55, 59** and **61** serve several important functions. First, the baffle **55** assists in channeling all of the liquid in the inlet zone **21** across the injector **41** in passing from the inlet zone to the intermediate zone **23**. Second, the baffles provide disaligned orifices **57, 63** and **65** to thereby provide an elongated flow path through the tank so that the flow path through the tank is longer than the distance between the inlet **27** and the outlet **58**. Third, the elongated flow path provided by the baffles is tortuous, i.e., requires the fluid to make several changes of direction with consequent production of turbulence and agitation which improves mixing.

Fourth, the openings **67** and **69** allow some of the liquid in the intermediate zone **23** to pass directly through one or both of the baffles **59** and **61**. This means that fluid which flows through the restricted orifices **63** and **65** flows along a different path, i.e., a generally axial path than fluid which flows directly through the openings **67** and **69**. The impingement of these two streams flowing in different directions further agitates the mixture with consequent thorough mixing. Thus, thorough mixing of the acid waste and the base is accomplished without any mechanical agitation or power driven agitators.

A pH sensor **71** is located in the outlet zone **25** closely adjacent the outlet **58** and beneath the liquid level in the outlet zone. With the pH sensor **71** so located, the pH of the liquid at the sensor is substantially identical to the pH of the liquid at the outlet **58** and downstream thereof. The mixture in the outlet zone **25** is continuously removed and allowed to flow through a conduit **73** and a three-way control valve **75** to an open drain **77**.

The apparatus **11** is controlled by a pH controller **79** which may be of conventional design. An electrical input signal is provided to the pH controller by the sensor **71** with a characteristic of such signal varying with the pH of the mixture of liquid adjacent the outlet **58**. The pH controller **79** opens and closes the control valve **39** to thereby control the flow of ammonia from the source **35** into the tank in accordance with the pH of the liquid at the sensor **71**. For example, the control valve **39** may be opened by the pH controller **79** as long as the sensor **71** senses a pH of less than a preselected value and may be closed when the pH sensed is greater than such preselected value.

The preselected value of pH may be adjusted as necessary to produce the desired results, and the valve **39** can be opened at one pH and closed at another, if desired, With this type of control and with normal flow rates through the tank, the valve **39** will be opened and closed periodically to supply bursts of pressurized ammonia gas through the apertures **43** of the injector **41**. Regardless of the duration of the burst, the beginning of each burst further contributes to the agitation of the acid waste and base in the inlet zone **21**.

With this type of control, the pH of the liquid fed to the open drain **77** may vary generally in accordance with the curve shown in Figure 7.9e. As shown in Figure 7.9e, the pH of the liquid at the outlet **58** varies generally in sawtooth fashion between an upper level **a** and a lower level **b** with the average pH being at an intermediate level **c**. For example, the pH of the levels **a, b** and **c** may be 9, 5 and 7, respectively. Over a period of time, the pH of the liquid at the outlet averages out at the average level **c** which in this case is 7. However, the pH controller **79** can be adjusted to maintain the average pH at other values.

The pH controller **79** has a second set point or alarm point so that if the pH of the liquid at the outlet **58** drops below a preselected value **d** which is the minimum acceptable value, an electrical control signal is fed to the control valve **75** so that further flow from the outlet **58** is through a conduit **81** and a pump **83** back into the conduit **29** just upstream of the inlet **27**. With the control valve **75** in this position, there is no flow into the open drain **77**. This recycling or bypassing of the liquid from the outlet **58** will continue until the pH of the liquid at the outlet **58** rises so as to be above the acceptable minimum **d**. The pH level represented by **d** in Figure 7.9e can be adjusted to any suitable value by the pH controller **79**.

In operation of the apparatus **11**, the three-way control valve **75** is normally open to the drain **77**. It is contemplated that the control valve **75** will recycle liquid back through the conduit **81** only in situations where there is a malfunction or in other extraordinary circumstances such as when the source **35** of ammonia is depleted, or when an unusually highly concentrated acid waste is supplied to the tank **13**, etc. Thus, the control valve **75** serves as a safety device to positively prevent dumping of a liquid into the drain **77** which has a pH below the acceptable minimum value **d**.

When the liquid is recycled, the pump **83** is operative to supply the necessary pressure to accomplish the recycling. In addition, the valve **51** is closed by the pH controller **79**. So long as the pH of the liquid adjacent the outlet **58** is within acceptable tolerances, the pH controller **79** has no effect on the solenoid valve **51**, in which event the valve can be opened to supply water to the nozzle **45** by the mere closing of a manual switch **85**. If desired, a recorder **87** can be coupled to the pH controller **79** to make a permanent record of the variation in pH at the outlet **58** with time.

The tank **13** should be installed in a position so that residual of liquid will remain in the tank when the apparatus **11** is not in use. For example, with the outlet **58** located in the end cap **17** as shown in Figure 7.9a, the tank can advantageously be installed with the longitudinal axis horizontal or inclined slightly toward the outlet. The residual level should be sufficient so that the sensor **71** remains submerged.

In operation of the system, the residue in the tank is sufficient so that the sensor **71** is always submerged. The switch **85** is first closed to cause the nozzle **45** to supply water to the tank, and the manual valve **31** is opened to allow acid waste to be fed into the inlet zone **21**. This addition of acid waste to any liquid residue left in the tank will ultimately bring the pH at the sensor **71** below the average level **c** to thereby cause the pH controller **79** to open the control valve **39**. A burst of ammonia gas is then supplied through the valve **39** and the apertures **43** to mix with the incoming acid waste in the manner described hereinabove. The liquid from the inlet flows through the intermediate zone **23** as described hereinabove to the outlet zone **25**.

After a period of time the pH at the sensor **71** rises above the level **c**, at which time the control valve **39** is closed by the controller **79**. The flow through the tank **13** is continuous and under the influence of gravity. The flow of liquid through the outlet **58** is also continuous. Accordingly, there is no holding time absent an alarm condition which would actuate the control valve **75** to cause recycling.

In the above described operation of the apparatus, it is assumed that operating conditions will cause the solenoid valve **39** to be cycled open and closed repeatedly so that ammonia gas is periodically supplied through the apertures **43**. While the apparatus is capable of functioning in this manner, it should be understood that the opening and closing of the solenoid valve **39** is controlled by the sensor **71** and the pH controller **79** in accordance with the pH of the mixture adjacent the outlet **58**. Accordingly, the solenoid valve **39** may cycle on and off infrequently, or may remain open continuously, or be otherwise controlled by the pH controller **79** to achieve preselected pH conditions adjacent the outlet.

Nitric Acid from Radioactive Waste

S. Drobnik; U.S. Patent 3,673,086; June 27, 1972; assigned to Gesellschaft fur Kernforschung mbH, Germany describes a method of removing nitric acid, nitrate ions, and nitrite ions from aqueous waste solutions, especially from intermediate and high level radioactive solutions in which the waste solutions are treated with reducing agents.

The process involves feeding the aqueous waste solution into a heated, preferably boiling aqueous solution of a reducing agent. In another case, the offgases generated in a reaction between a waste solution containing radioactive iodine and the reducing agent, which contain no higher nitric oxides, are led through a filter before they are discharged into the ambient air. The reducing agent employed is an organic substance, i.e., 98% formic acid, in at least twice the molar ratio or more relative to the hydrogen ion concentration of the nitric acid-bearing waste solution.

The advantages gained with the process in particular are the fact that the treatment of inactive industrial wastewaters containing nitric acid, nitrates or nitrites avoids the expensive neutralization step and thus the great increase in salt content of the wastewater and renders unnecessary any dilution to harmless nitrate concentrations with respect to the required prevention of pollution of surface waters used for drinking water supply. Moreover, the removal of nitric acid, nitrate and nitrite ions will produce no higher nitric oxides and produces carbon dioxide, laughing gas (nitrous oxide) and only small quantities of NO and N_2, e.g., 78 vol percent of CO_2, 16 vol percent of N_2O, 5 vol percent of NO and 1 vol percent of N_2.

Most of the NO in this case is generated only towards the end of the process. The gas composition is independent of the acid concentration in the starting solution. This gaseous mixture may be discharged into the ambient air without any additional treatment, e.g., cleaning.

The reaction is performed in a reducing medium, i.e., the waste solution is introduced into a given solution of the reducing agent, preferably into the given 98% formic acid, in a batch process or, if the process is carried out continuously, with a small excess of formic acid. The reaction is exothermic and very fast, occurring chiefly according to the equation:

$$2HNO_3 + 4HCOOH \longrightarrow N_2O + 4CO_2 + 5H_2O$$

Heavy metal ions present in the water accelerate the reaction so that the induction period, that is the time between the startup of introduction of the wastewater into the formic acid heated to 100°C and the beginning of the reaction, is only a few seconds. This is an advantage, especially if nitric acid, nitrate ions, and nitrite ions are removed from radioactive wastewaters, where any risk must be avoided and the reaction must be easily controlled. The following examples illustrate the process.

Example 1: 1,000 ml of a synthetic model solution were introduced into 375 ml of 98% formic acid in a metering time of 4 hours. The model solution had this composition:

	Mol
HNO_3	4
$Fe(NO_3)_3$	0.2
$Cr(NO_3)_3$	0.05
$Ni(NO_3)_2$	0.05
$Al(NO_3)_3$	0.05
$Ca(NO_3)_2$	0.01

Since salts were used which had been prepared for technical purposes, the hydrogen ion concentration and the nitrogen content were analyzed before the treatment:

H^+-ions	5 mols	N (69 mg N/ml)	4.92 mols

After the reaction time, 1,080 ml of solution with 4.5 mg N/ml were obtained, i.e., 93% of the overall nitrate ion concentration had been destroyed. The product solution had a pH of 3.5 (hence, all the nitric acid had been destroyed) and contained a fine-grained brown precipitate. The volume reduction relative to the sum total of the reaction solutions was 21.5 vol percent. Some 300 liters of offgas were developed during the reaction.

Example 2: 1,000 ml of a radioactive waste solution produced in the decontamination of equipment of the composition:

1.4 mol H^+ (given as HNO_3)
1.86 mol N (26 mg N/ml)
1 mg/ml Arkopal N-130 (non-ionogenic tenside)
27.7 mg/ml dry residue (dried at 105°C)
3 x 10^{-3} μCr/ml α-activity
1.7 x 10^{-3} μCr/ml β-activity

were introduced into 107 ml of 98% formic acid within 4 hours. The volume of the product solution was 1,060 ml (volume reduction 4.2%, contained 0.63 mg N/ml and had a pH of 2.5. Hence, 97.5% NO_3^- ions had been destroyed, which produced 85 liters of offgas. Foaming, which had been anticipated because of the presence of Arkopal, did not occur. No volatile radionuclides were detected in the offgases.

Wool Scouring and Dyeing Wastewater

H. Fujimoto and M. Fukuda; U.S. Patent 3,677,940; July 18, 1972; assigned to Kurita Water Industries Ltd., and Unitika Ltd., Japan describe a method of treating the wastewater arising from the process of scouring raw wool, and the wastewater arising from the process of dyeing, in general, simultaneously and efficiently.

According to the process, the simultaneous treatment of a mixture of the scouring wastewater and the dyeing wastewater has such a merit that such ingredients as earth and sand, excrement of sheep, oil and fat contained in the scouring wastewater function as an assistant coagulant and settle while absorbing dyes and other ingredients of the dyeing wastewater at the time of forming the aggregated floc, so that not only is it possible to remove the colored substances, but also the sludge which settles while absorbing the colored substances as above is compact, small in sludge volume, and possessed of superb separability. Accordingly, the sludge thus formed can be easily separated through sedimentation, flotation, filtration and the like, and the quality of the treated water becomes quite satisfactory.

The scouring wastewater generally contains a variety of foreign elements mentioned above in so high a concentration that it is usually necessary to dilute it to an appropriate concentration in order to perform an aggregation treatment. According to the process, an appropriate dilution of the scouring wastewater required as above is achieved by using the dyeing wastewater. In this connection, it is advisable that the scouring wastewater be subjected to a preliminary treatment for the purpose of separating solid matters therein in a sedimentation tank prior to mixing with the other wastewater.

The mixing ratio of the two wastewaters can be determined at will depending upon the concentration as well as the composition thereof; although it is generally in the range of 4:1 to 1:4, there is no necessity for strictly adhering to the foregoing mixing ratio as long as there is no excessive deviation. As the coagulant consisting of inorganic metallic salt for use in this process are applicable such coagulants as an aluminum salt such as aluminum sulfate, aluminum chloride and polyaluminum chloride, an iron salt such as ferric sulfate, ferric chloride and ferrous chloride, a zinc salt such as zinc chloride and zinc sulfate and, titanium slag, and capable of releasing polyvalent metallic cations.

Any of these coagulants is made into an aqueous solution of 10 to 40% or thereabout, and the aqueous solutions added to the mixture of wastewaters to obtain the concentration thereof in the range of 100 to 1,200 ppm and subjected to stirring. On this occasion, it is necessary to adjust the pH in the range of 5 to 11 as required for generating the floc of hydroxide through hydrolysis of the added metallic salt within the mixed liquid, and, for this purpose, an alkaline agent such as sodium hydroxide, calcium hydroxide and the like is added in the liquid. The preferable pH range is 5 to 8 for the aluminum salt, 5 to 11 for the ferric salt, 6 to 9 for the ferrous chloride and 8 to 11 for the zinc salt.

In this state of the mixed liquid, the floc of hydroxide produced still exists in the form of fine particles, and this floc of hydroxide can be made to agglomerate further by adding an anionic organic coagulant to the mixed liquid, resulting in a large sized, compact sludge with superb separability. As for the organic coagulant to add, it being for the purpose of effecting aggregation of the cationic metallic hydroxide, any anionic organic coagulant will do: for instance, such anionic organic coagulants as partially hydrolyzed polyacrylamide, polyacrylic acid or salt, anion-modified polyacrylamide, sodium alginate, sodium salt of carboxymethylcellulose, etc. are all applicable. The amount of this coagulant to be added is in the range of 1 to 5 ppm, and after adding, it is mixed by stirring.

The sludge thus obtained is left standing so as to let it settle and separate and it can be subsequently subjected to filtration for dehydration. It is also possible to have the sludge floated and separated in a concentrated state by making it float by means of gas-bubbles blown into the mixed liquid by taking advantage of a lot of surface active agent contained in the mixed liquid adhering to gas-bubbles.

In the method of separating the sludge by flotation, a mixed liquid, after the aggregation process, is introduced into a flotation tank in which the sludge is made to float by virtue of the air blown into the tank from its lower parts. On this occasion, a method of pressure flotation, which comprises the steps of forming a pressurized water by dissolving air in some water, e.g., a separate portion of the mixed liquid or treated wastewater under elevated pressure, mixing the pressurized water with the mixed liquid to be treated, and obtaining fine gas-bubbles by applying reduced pressure, is particularly suitable. According to this method of pressure flotation not only the fine gas-bubbles ascend accompanied by the sludge adhering thereto but also the surface active agents as well as organic matters contained in the mixed liquid ascend in the state of being concentrated on the surface of such gas-bubble, so that the sludge thus separated comes to be of high concentration while the concentration of the impurities in the treated water becomes extremely low.

Example 1: The scouring wastewater was left to settle naturally for 10 hours and mixed thereafter with the dyeing wastewater in an amount of four times as much as the former. The dyeing wastewater arising from dyeing contained acid mordant dye for wool, potassium bichromate mordant dye, reactive dye and azo-type dispersed dye for synthetic fiber and presented a mixed color of red, green, blue and black. Aluminum sulfate in the amount of 400 to 600 ppm, was first added to the mixed wastewater and next calcium hydroxide in the amount of 350 to 400 ppm was added to the same to thereby adjust the pH to be 6.5. After about 5 minutes at the concentration with rapid stirring, 30% hydrolyzed polyacrylamide was added at a rate of 2 ppm and mixed by about 1 minute stirring.

Upon introducing the thus treated mixed wastewater into a flotation tank, pressure water which was prepared by dissolving the air in a separate portion of the treated wastewater by applying a pressure of 3 kg/cm^2 was blown into the stock in the tank, whereby the sludge was separated by the aforementioned method of pressure flotation. The flotation rate of the sludge in this case was 10 to 15 m/hr, the concentration of solid matter in terms of separated sludge having a sludge-volume of 6 to 7 vol percent after 10 minutes was 3.8 weight percent, and the comparative quality of the mixed wastewater and the treated water was as shown in the following table.

		------ Treated Water -----	
	Mixed Wastewater	**Example 1**	**Example 2**
Appearance	*	**	**
pH	4.6	6.5	7
Transparency (degree)	<0.5	10-12	13-14
COD (ppm)	800-900	100-130	130
SS (ppm)	150-300	10-30	20-30
Oil and fat (ppm)	870	10-15	10-15

*Turbid with reddish brown color.
**Slightly turbid with light yellow color.

Comparative Experiment 1: In case the treatment was conducted of the same scouring wastewater in the same way as Example 1, except for diluting the wastewater with tap-water in the amount of four times as much as the former, the resultant sludge-volume was about 50% (in case the scouring wastewater was undiluted, it was about 80%) and the sludge was inseparable.

Comparative Experiment 2: In case of the individual treatment of the dyeing wastewater, although there was formed the floc of aluminum hydroxide, there was little change in the color of the treated wastewater compared with the wastewater before treatment, that is, it was impossible to effect decoloring.

Comparative Experiment 3: The treatment was conducted in the same way as in Example 1 except for omitting application of the partially hydrolyzed polyacrylamide. In this case, the floc of aluminum hydroxide was kept in an overall suspension and the flotation rate was low.

Example 2: By adding first 400 ppm of ferric chloride and next 450 ppm of calcium hydroxide to the same mixed liquid as in Example 1, the pH was adjusted to be 7. Subsequent to about 5 minute rapid-stirring of the mixed liquid processed as above, 30% hydrolyzed polyacrylamide was added to the mixed liquid and mixed therein by stirring for about 1 minute. Thereafter, the mixed liquid was introduced into a flotation tank in which the sludge was separated through the pressure flotation process utilizing pressurized water in the same way as in Example 1. The flotation rate of the sludge in this case was

5 m/hr, the sludge-volume after 10 minutes was 7 to 8 vol percent, and the quality of the treated water was as shown in the table on the previous page.

Streams from Railroad Equipment Servicing Facilities

W.F. Burns and R.B. Martin; U.S. Patent 3,707,464; December 26, 1972; assigned to Petrolite Corporation describe a process for clarifying an oil-solids contaminated aqueous stream such as may arise from the operation of a railroad equipment servicing facility. A primary clarifier gravitationally separates the stream into settled solids, a merchantable oil product and a clarified water stream. The clarified water stream is treated by creating a flocculation product for removing solids and oil to produce a clear water stream delivered to a subsequent utilization. Settled solids may be returned to the primary clarifier. The flocculation product is accumulated in a sludge cleaner to be periodically converted into water and oil phases. The oil phase is returned into the primary clarifier.

The water phase is reformed into an oil-free flocculation product and separated from a water filtrate returned into the primary clarifier. Solids from the primary clarifier are accumulated in a sand cleaner to be periodically cleaned to separate oil-free solids from a solids-free water-continuous filtrate returned to the primary clarifier. The reformed flocculation product and oil-free solids are separated in a solids dewatering vessel from a water filtrate returned into the primary clarifier. The relatively water-free solids are then delivered to a subsequent utilization such as landfill.

One wastewater stream especially suited for processing by the steps of this process is produced by the operation of railroad equipment servicing facilities. Such a waste stream includes the run-off from engine washing, fueling and sanding aprons, diesel shops, round house pit drainage and miscellaneous other water streams released in servicing and maintenance of railroad locomotives and related equipment. These waste streams contain free oil, emulsified oil, sand, silt, grease, carbon and various other solids.

Dissolved substances such as lye and other water-soluble alkaline, acidic and salt materials may be present as a result of cleaning operation for servicing railroad rolling stock. The resultant wastewater stream has a pH of above 10 and cannot be merely settled in a hope of providing a sufficiently clear water stream acceptable for disposal in public water systems such as streams and rivers. The process will be described specifically for processing such a waste stream from a railroad servicing facility but it will be apparent it is equally applicable to other types of wastewater streams.

Activated Carbon as Catalyst to Remove Hydrogen Sulfide

B. Hoke; U.S. Patent 3,817,862; June 18, 1974; assigned to Fried, Krupp Gesellschaft mit beschrankter Haftung, Germany describes a process for treating the wastewaters from the paper, leather, textile, cleaning, food, and plastics industries, containing emulsified and dissolved, oxidizable, organic and possibly inorganic (such as H_2S) substances and being substantially free of cyanides, nitrite, hydrazine, dimethyl hydrazine, and nitrogen tetroxide (Aerozin).

Such waters are contaminated by a multitude of chemical substances, such as organic acids, esters, alcohols, aldehydes, ketones, phenols, phenol derivatives, solvents, disperse dyes, acid dyes, metal-complex dyes, printing dyes, wetting agents, sulfites, and sulfides, as well as nitrogen compounds and the like. Such waters, because of the ability of their contaminants to overload the accommodating capacity of the natural waters must necessarily first be subjected to thorough cleaning and detoxification before being released.

The process involves adding wastewater to catalyst selected from the group consisting of carbon and ion exchanges, and forcing air through the resulting mixture of wastewater and catalyst. The forcing of air through the mixture has the function of increasing the interfacial area between the air and the resulting mixture of wastewater and catalyst over the interfacial area resulting when the mixture is at rest with simply its upper liquid surface lying in and forming the air-mixture interface. A preferred catalyst is activated carbon. However, porous to microporous ion exchangers can be used, it being preferred to use microporous ion exchangers which sorb organic substances. When using activated carbon as catalyst, it is preferred to limit its quantity to a maximum of 20 weight percent as based on the weight of wastewater in the mixture, and to filter the catalytically oxidized product through an activated carbon filter.

It has been discovered that these catalysts cause the catalytic oxidation of the contaminants in the wastewaters from the paper, leather, textile, cleaning, food and plastics industries. Catalytic oxidation has already been proposed for detoxifying waters loaded with hydrazine, dimethyl hydrazine, nitrogen tetroxide, and cyanide. It was, however, surprising that such a method also leads to the complete detoxification and cleaning of waters contaminated with emulsified and dissolved organic substances of the above named types.

The following examples are illustrative of the process. Used for measuring the efficiency of the water cleaning were the chemical oxygen demand, which is referred to by the letters COD, and the permanganate consumption, which is referred to by the letters PV. The chemical oxygen demand (COD) is equal to the biochemical oxygen demand 20, this being referred to by the symbol BOD_{20}; PV is equal to the $KMnO_4$ consumption.

Used as a reaction chamber was an approximately 60 liter plastic tank equipped with six aerating nozzles for increasing the interfacial area between the air and the mixture of wastewater and catalyst in the tank. These nozzles were mounted on the floor of the tank and assured a uniform distribution of the air being blown into the water. The feeding of the chamber with wastewater during the flowthrough tests, such as in Example 3, was done using a centrifugal pump, which brought about 15 liters per hour continuously from a storage vessel into the reaction chamber. The inlet pipe opening into the reaction chamber lay below the water level.

Using a second pump, a part of the water could be taken from the floor of the reaction chamber and fed back to the water in the chamber by means of a spray pipe directed at the upper surface of the water. The purpose of this facility was to cope with any strong foaming caused by the aeration. It would additionally be expected to increase the amount of air absorbed in the water by serving the function of increasing the interfacial area between the air and the mixture of wastewater and catalyst.

The outlet from the reaction chamber was through a slitted plastic nozzle. In order to remove carbon particles resulting from catalyst wear, and dyes, the water flowing out of the chamber was allowed to free-fall through an activated carbon filter.

Aeration of the water was done using a controllable laboratory blower, and the air reached the six nozzles from an annular pipe around the tank. In order to test whether the addition of chlorine would influence the COD of the water and destroy the dyes in the water, a sodium hypochlorite solution was added to the wastewater.

Example 1: The reaction chamber was filled with 30 liters of wastewater. This wastewater was collected at the inlet to a biological clarification plant over a period of 24 hours using a device for giving an average sample of the water running through the inlet during an entire course of production for 1 day. Activated carbon was used as catalyst. This same activated carbon was kept in the reaction chamber during all subsequent tests in order to exclude adsorptive phenomena. The weight of the carbon amounted to 1,250 grams, which give a specific loading of 42 grams of carbon per liter of wastewater.

To increase the effect of the air bubbled through the water and to provide better mixing, a stirrer was installed. Thus, it became quickly apparent that the amount of air required to produce the necessary turbulence in the water was larger than the maximum amount which could be supplied by the laboratory blower. With the stirrer in operation, a better movement of unreacted chemicals toward, and reacted chemicals from the catalyst surface was provided. The amount of air which was blown into the reaction chamber amounted to

600 l/hr at a room temperature of 22°C and at a 350 mm chamber depth. The pH value of the resulting mixture of wastewater and catalyst was 5.5 and the COD amounted to 1,000 mg/l. The wastewater was aerated in the reaction chamber for 5 hours. At time intervals of 30 minutes, a test sample was withdrawn from the water in the reaction chamber and tested for its permanganate consumption (PV) and its COD. The test results are shown in Figure 7.10a. While the $KMnO_4$ consumption sank in the first 30 minutes only from 720 to 620 mg/l as a result of catalytic oxidation and then remained constant, the COD decreased steadily. After about 3 hours of aeration, the COD had decreased around 40% (referred to as a chemical breakdown of 40% in the contaminants).

FIGURE 7.10: WATER QUALITY PARAMETERS AS A FUNCTION OF TIME

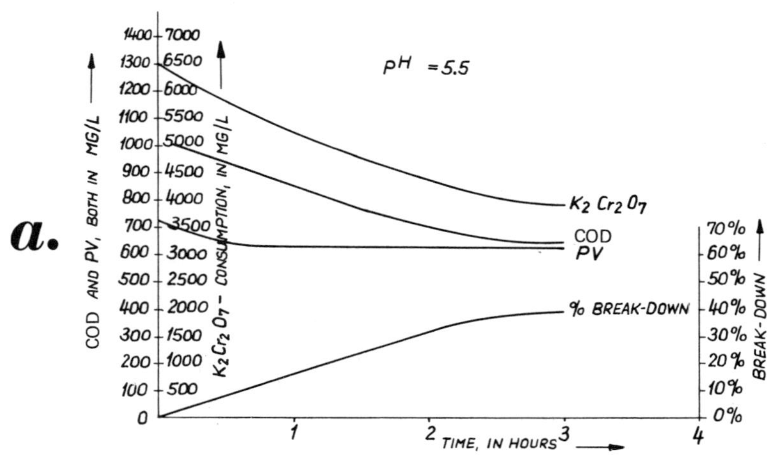

Example 1

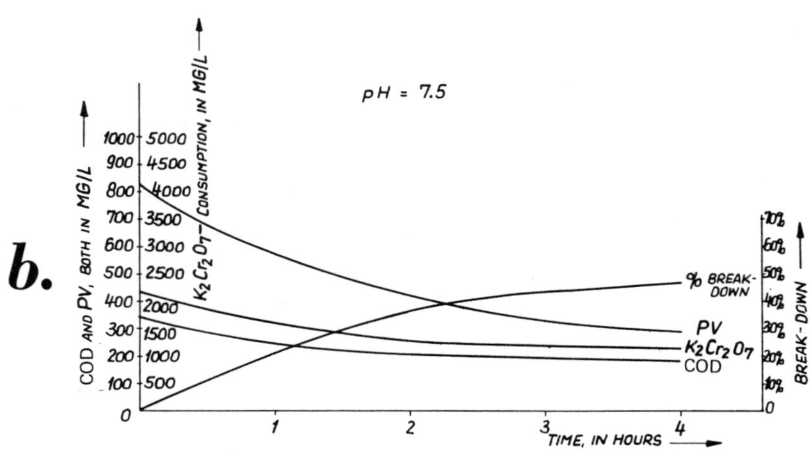

Example 2

(continued)

FIGURE 7.10: (continued)

c.

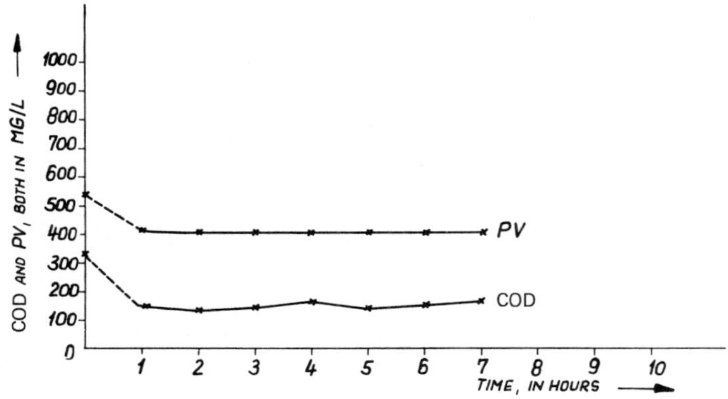

Example 3

Source: B. Hoke; U.S. Patent 3,817,862; June 18, 1974

After the filtration in the activated carbon filter, the water was only faintly colored and was practically free of solids. While the wastewater had tended to give enormously strong foaming when undergoing biological treatment, no foaming occurred in the reaction chamber of the process, even when the spraying device was shut off. It was found that the addition of sodium hypochlorite resulted in no further change in the COD and no destroying of the residual color in the water after the catalytic oxidation. Additionally presented in Figure 7.10a is the $K_2Cr_2O_7$ consumption for this example.

Example 2: It being discovered that the COD of the wastewater in Example 1 could be decreased, it was next desired to determine whether the water could be further improved by the process, after having first been subjected to biological treatment. To this end, the reaction chamber was filled with 30 liters of water from the outlet of the final clarification basin. The pH of the water was 7.5. The water had a COD of 340 mg/l. The potassium permanganate consumption lay at 830 mg/l. The test results are illustrated in Figure 7.10b. In contrast to the results of Example 1, here the catalytic oxidation of the process caused the $KMnO_4$ consumption to decrease steadily until after 4 hours a PV of 290 mg/l had been reached.

The COD decreased in the same time from 340 to 180 mg/l, corresponding to a breakdown of 160/340 or almost 50% after 4 hours. As determined by Example 1, and also in this example, the most favorable reaction time lies between 3 and 4 hours. After this time, the breakdown curves flatten out. This is illustrated, for example, in Figure 7.10a. The observation that a reaction time of 3 to 4 hours was most favorable for the apparatus and test conditions being used was confirmed by tests with other wastewaters. The water of this example was finally passed through the activated carbon filter to remove catalyst particles. The end product was clear and had a weakly colored appearance. Visibility lay at 45 cm. A subsequent treatment of the water with sodium hypochlorite gave, as in Example 1, no further reduction of the COD.

Example 3: After the batch tests in Examples 1 and 2 using the inlet and outlet waters of a biological clarification plant, it was desired to determine the effect of continuous treatment of the outlet water from the clarification plant using the techniques of the process. To this end, a larger amount of water, about 100 liters, was collected at the outlet of the plant, in order that ample wastewater of constant properties would be available for flowing through the test apparatus. The pH of the water amounted to 7.4, the $KMnO_4$ consumption was 540 mg/l, and the COD lay at 315 mg/l. By means of a small laboratory pump, this wastewater was pumped out of a storage vessel into the reaction chamber at the rate of 15 liters per hour.

For a flowthrough rate of 15 liters per hour and a chamber volume of 53 liters, a calculated residence time of water in the chamber equal to 3.5 hours is obtained. The specific amount of carbon amounted to 24 grams of carbon per liter of wastewater. The loading was consequently almost 50% higher than in the batch tests. Every hour, test samples were collected at the outlet of the reaction chamber and tested for COD and $KMnO_4$ consumption. The test results are shown in Figure 7.10c. The breakdown values obtained correspond to those already determined in the batch tests. In the case of COD, breakdown values of 45 to 53% were obtained. The samples were collected before the activated carbon filter, but the filtering changed the COD only slightly.

The content of permanganate-consuming substances, however, in contrast to the results obtained in Example 2, decreased only around 20%. It was further determined whether addition of sodium hypochlorite solution could provide an improvement of the water quality. With the addition of hypochlorite solution up to a concentration of 3 mg/l free chlorine, only a slight reduction of the COD was noted. In an available sludge treating plant using microorganisms, final parameters of COD and PV were reached after 5 days of treatment, in contrast to the reaching of the same final parameters after only 4 hours of treatment with the method of catalytic oxidation according to the process. This provided a direct comparability of the effectiveness of this process with another commonly used method.

Example 4: Another series of tests was carried out using the wastewaters from a sauerkraut factory. The test conditions were the same as in Example 1, except that no addition of sodium hypochlorite was used. In order to prevent fermentation during the test time, the wastewater was brought to a pH of about 10 using milk of lime. The COD in the raw water lay at 44,000 mg/l and the PV at 96,000 mg/l. In 12 hours, these had decreased to 7,400 milligrams per liter for the COD and 19,800 mg/l for the $KMnO_4$ consumption. The breakdown was thus about 83%.

Example 5: Here, the wastewater from a laundry for the ink-soiled cloths from a printing shop was treated. This water was contaminated with the dyes, solvents and wetting agents typical for printing shops. The test set-up as set forth in Example 1 was used. The wastewater, which had a dark-red to black color, had a pH of 8. The COD amounted to 6,250 milligrams per liter and the PV was 2,500 mg/l. In 10 hours of treatment, these quantities were as follows:

COD = 580 mg/l   PV = 240 mg/l

The breakdown was thus about 89.5%.

Example 6: Using the same experimental setup as described for Example 1, the wastewater from a chemical factory for the manufacture of textile impregnating agents was treated. Because of the high oil content, a precipitation (flocculation) with milk of lime, followed by a sedimentation, was carried out before the catalytic oxidation. The raw water had a COD of 7,400 mg/l. The clarified water present after the precipitation and sedimentation had a COD of 5,200 mg/l and a PV of 7,600 mg/l.

After catalytic oxidation with aeration over a period of 4 hours under the conditions as in Example 1, the COD had sunk to 1,200 mg/l, which corresponds to a 78% decrease. With reference to the raw, unclarified water, this represents a decrease of 84%. The water was then subjected again to catalytic oxidation, for another 3 hours, which brought the COD down another 300 mg/l, i.e., the COD sunk from 1,200 to 900 mg/l.

Example 7: In the same experimental set-up as described for Example 1, the wastewater from a leather factory was treated by the catalytic oxidation method of the process. The original COD was 1,700 mg/l and the original PV equaled 1,800 mg/l. After an oxidation period of 4 hours, the following values were measured:

COD = 220 mg/l PV = 383 mg/l

This corresponds to a COD decrease of 87%.

Example 8: The wastewater from a laundry for cloths used for cleaning machinery was first treated in an emulsion-breaking plant. In this way, the COD of the water was reduced from 9,900 to 1,200 mg/l. After an aeration period of 4 hours under the experimental conditions of Example 1, the COD decreased to 450 mg/l. If only 2 hours aeration time was used, the COD was 570 mg/l; with an aeration period of only 1 hour, the resulting COD was 660 mg/l.

In all of Examples 1 through 8, the same activated carbon was used, except that after every experiment a catalyst wear of 5 to 10% by weight was compensated for by the addition of an appropriate quantity of new carbon. In all experiments, the process yielded very good results. In a comparison with a biological treating operation run in parallel with the treatment of this process, the same breakdown effect was achieved with this process that took days to obtain with the biological treatment.

Along with small operational costs, only a fraction of the investment required for biological treating plants is required for the process. Moreover, because the process relies purely on chemical action and not on the action of living organisms, the troublesome stoppages met in even the best biological processes are avoided.

Example 9: The wastewater from a plant processing and coloring animal skins was treated according to the process in the manner described below. This water had a dark-red appearance when diluted. Undiluted it was black and could not be seen through. In its undiluted state, its COD was 1,200 mg/l. The undiluted water was mixed with powdered activated carbon in the ratio of 300 grams of carbon per cubic meter of water.

The resulting mixture was then allowed to fall through a cooling tower utilizing a countercurrent air draft. The fill of the tower was designed to prevent its collecting dirt. After passing through the tower, the water was taken from the collecting basin and pumped back to the top of the tower. A portion of this recirculated water equal to the amount of water coming into the mixing basin was directed off to a filter, with the filtrate being passed to a drainage ditch. The carbon powder sludge collecting in the filter was recirculated, with less than 5% loss, back into the mixing basin before the top of the tower.

A stirrer provided for good mixing of wastewater and carbon in the mixing basin. The wastewater, originally strongly loaded with wetting agents and dyes as well as other organic contaminants, left the filter clear and colorless, it did not foam, and its COD lay at 300 milligrams per liter. Due to the quantity of water recirculated by the pump from the collecting basin back to the top of the tower, a theoretical aeration time of 4 hours was calculated.

Example 10: In this example, a series of tests was carried out where Examples 1 through 9 were varied by replacing the activated carbon by ion exchanging materials as catalysts. Significant reductions in the COD and PV were obtained in each case; in most of the cases, the results corresponded to those obtained with the activated carbon catalyst. In decoloring, the ion exchanging materials had to be changed more often and regenerated. The COD values obtained were in each case within the values prescribed for the treatment of the particular wastewaters.

It will be understood that the above description of the process is susceptible to various modifications, changes and adaptations, and the same are intended to be comprehended within the meaning and range of equivalents of the appended claims. In the Examples 1 through 9 described, activated carbon was used in the form of carbon particles in sizes from 1 to 4 mm or in the form of carbon powder in particles sizes below 0.3 mm, preferably from 0.1 to 0.3 mm. In Example 9 activated carbon powder was used exclusively as catalyst. The different types can be used as mixtures. In Example 10, the ion exchange material used was the resinous adsorbent Lewatit MP 500 A.

The stirrers used in the examples mentioned were so-called ships propellers as well as simple bladed propellers, the speed of which was varied from 100 to 1,500 rpm. The diameter of these propellers amounted from 60 to 100 mm. Two-blade as well as six-blade propellers were used. It was found, however, that the stirring speed exerted no influence on the reaction rate of the floated catalyst.

COMPANY INDEX

The company names listed below are given exactly as they appear in the patents, despite name changes, mergers and acquisitions which have, at times, resulted in the revision of a company name.

INVENTOR INDEX

U.S. PATENT NUMBER INDEX

NOTICE

Nothing contained in this Review shall be construed to constitute a permission or recommendation to practice any invention covered by any patent without a license from the patent owners. Further, neither the author nor the publisher assumes any liability with respect to the use of, or for damages resulting from the use of, any information, apparatus, method or process described in this Review.

POLLUTION CONTROL IN METAL FINISHING 1973

by Michael R. Watson

Pollution Technology Review No. 5

Electroplating and other metal finishing waste streams are significant contributors to water pollution, either directly by their content of toxic and corrosive materials, such as cyanide, acids, and metals, or indirectly through the deleterious effect these effluents exert on sewage treatment systems.

Federal, state and municipal regulations fixing the allowable concentrations have been established and enforcement may be expected to become increasingly strict.

There is ample technology available for treating chromium and cyanide rinse waters to any required degree of detoxification. The problems facing the plater are:

1. Volume and composition of the rinse waters.
2. EPA requirements for an acceptable waste effluent.
3. Capital outlay for treatment facilities.
4. Operating costs for the treatment processes.
5. Disposal costs of unusable recovered materials.
6. Credits from the recovery of metals and decrease in sewage fees.

Solutions to these problems are afforded by this book which is based on various government surveys and reports with practical examples detailed from late U.S. patents. A partial and condensed table of contents follows. Chapter headings are given here, followed by examples of important subtitles.

ISBN 0-8155-0505-1 295 pages

OIL SPILL PREVENTION AND REMOVAL HANDBOOK 1974

by Marshall Sittig

Pollution Technology Review No. 11

Energy Technology Review No. 2

Ocean Technology Review No. 1

This handbook describes and expounds the most feasible and sophisticated technologies available at present for dealing with a menace of worldwide proportions, i.e. marine pollution caused by accidental contamination with crude or (less often) refined petroleum oils. To make matters worse the crude oil has a specific gravity very close to that of seawater.

Oil slicks, tar balls, and dispersed hydrocarbons are found on the surface as well as on the bottom of the ocean. Very little has been accomplished so far about relieving the latter calamity. Once the spilled oil has found something solid to cling to—it may be the sand of a beach, underwater rocks, wooden structures, the feathers of a gull, or the hair and skin of a bather—it does not readily let go.

Aside from most rigorous safety regulations and preventive measures now in force for tankers, offshore drilling, and seabed mining (still in its infancy), the worldwide effort towards combating oil spills is best illustrated by the myriad, albeit rudimentary, devices being designed by companies engaged in the development of containment booms and other mechanical containment apparatus for dealing with oil slicks on the surface of the sea.

Research continues to obtain improved performance, but, because of natural turbulences and gulfstream-type currents, the effectiveness of floating booms and other recovery systems is severely limited.

Once a spill has been contained and the sea is calm, cleanup measures can be instituted. The two major techniques used today are mechanical gathering and recovery with adsorbents. Mechanical recovery or skimming may be used without restriction in carrying out the contingency plan, while adsorbents may be used only if these materials do not by themselves or in combination with the oil increase the pollution hazard. Straw, manufactured fibers, and natural clays may be mixed with the oil slick, and collected. Straw is considered the most cost-effective agent, because it floats readily even when wet, holds five times its weight in oil and costs little.

Other methods include use of dispersing agents, sinking agents, biological agents and burning. Approval by the appropriate governments is required for each of these measures and is practically never granted for U.S. Coastal Waters. It has been suggested that relatively small beaches can be cleaned by feeding the contaminated sand through the municipal incinerator.

Even a relatively small oil spill may have effects that are catastrophic and long-lasting, but prompt application of the methods described in this handbook will at least help to prevent successive biological and degradative changes of the marine biota and the ecological entity.

In this handbook are condensed vital data from numerous government and other sources that are unusually difficult to find and to pull together. Important oil containment and removal processes are interpreted and explained by examples from recent U.S. patents.

A partial and condensed table of contents follows here. Numbers in parentheses indicate the number of processes or designs per topic. Chapter headings are given, followed by examples of important subtitles.

EFFECTS OF OIL SPILLS ON THE ENVIRONMENT
- Effects on Man
- Effects on Agriculture
- Effects on Industry
- Effects on Wildlife
  - Waterfowl
  - Shellfish
  - Fish
  - Marine Mammals
  - The Marine Food Chain
- Effects on Marshes

PREVENTION AND CONTROL OF SPILLS FROM OFFSHORE OIL PRODUCTION
- Mud Cleaning
- Leakage Control at Well Head (23 processes)
- Wastewater Purification (3)

PREVENTION AND CONTROL OF SPILLS FROM OIL TRANSPORT
- Oil Transfer Operations
  - Pre-Transfer Conference
  - Transfer Procedures
  - Vessel-to-Shore Transfer
  - Shore-to-Vessel Transfer
  - Miscellaneous Transfer Precautions
  - Terminal Facilities
  - MacLean Process
  - Wittgenstein Process
- Tank Cleaning (3)
- Ballast Disposal (3)
- Oil Removal from Tanker Accidents (4)
- Drydock Operations

PREVENTION AND CONTROL OF SPILLS FROM OIL REFINING
- Treatment of Oily Wastewaters (3)

PREVENTION AND CONTROL OF SPILLS OF INDUSTRIAL OILS
- Treatment of Oily Wastewaters (2)

OIL SOURCE DETECTION, IDENTIFICATION AND MONITORING
- Detection and Monitoring (2)
- Spill Quantity Estimation
- Sampling
- Tagging
- Identification

ORGANIZATION FOR OIL SPILL CONTROL
- The On-Scene Commander
- Industry Spill Cleanup Cooperatives
- Relations with Government Organizations
- Financing
- Use of Independent Contractors

OIL CONTAINMENT
- Air Barriers (7 designs)
- Chemical Barriers (4 processes)
- Floating Booms (103 designs)
- Combined Containment and Removal Devices (18 designs)

OIL REMOVAL FROM OPEN WATER SURFACES
- Weir Devices (16)
- Floating Suction Devices (14)
- Sorbent Surface Devices (32)
- Vortex Devices (6)
- Screw Device
- Jet Devices (2)
- Other Devices (4)
- Skimming Vehicles (49 designs)

OIL REMOVAL FROM STREAMS, LAKES, HARBORS
- Proprietary Processes (6)

OIL REMOVAL FROM UNDERGROUND WATERS
- Proprietary Processes (2)

OIL/WATER SEPARATOR DEVICES
- Proprietary Devices and Processes (18)

PHYSICAL AND CHEMICAL TREATMENTS OF OIL SPILLS
- Dispersion Treatments (3)
- Sorption Processes (33)
- Sinking Processes (4)
- Gelling/Coagulation Processes (8)
- Magnetic Separation (3)
- Combustion Processes (14)

OIL REMOVAL FROM BEACHES
- Spill Control Measures
- Chemical Control Agents
- Combustion Methods
- Culpepper Process
- Froth Flotation Treatment

DISPOSAL OF RECOVERED SPILL MATERIAL

ECONOMICS OF OIL SPILL PREVENTION AND CONTROL

FUTURE TRENDS

ISBN 0-8155-0543-4 **466 pages**

POLLUTION CONTROL IN THE PETROLEUM INDUSTRY 1973

by H. R. Jones

Pollution Technology Review No. 4

The petroleum industry can logically be divided into three major divisions: (1) Production, (2) refining, (3) subdividing and distributing. Production includes the operations involved in locating and drilling oil fields, removing oil from the ground, pretreatment at the well site, and transporting the crude to the refinery. The extent of pretreatment depends on the type of crude oil, and usually comprises removal of gas and brine. Refining includes the operations necessary to convert the crude into saleable products. Subdividing and distribution involves handling the finished petroleum products.

All of these activities are or can be sources of emissions into the air and water. This book shows you practical technology for avoiding or reducing these harmful emissions, and how to clean up after unavoidable damage has been done. There is also a separate chapter on the costs and economics of pollution control.

Chapter headings are given here, followed by examples of important subtitles.

ISBN 0-8155-0494-2 **322 pages**

POLLUTION CONTROL AND CHEMICAL RECOVERY IN THE PULP AND PAPER INDUSTRY 1973

by H. R. Jones

Pollution Technology Review No. 3

The pulp and paper industry has serious problems to solve, both in the air pollution and water pollution areas.

Many effluent wastes from paper mills are biodegradable, but treatment costs are increasing, effluent discharge requirements are becoming more stringent, and urbanization increasingly limits the availability of land. Economic control procedures applied by the industry, therefore, form an essential part of the pollution control procedures.

In this book are condensed vital data from government sources of information that are scattered and difficult to pull together. Important processes are interpreted and explained by examples from 54 U.S. patents. One should have to go no further than this condensed information to establish a sound background for action towards combating pollution in the pulp and paper industry.

A partial and condensed table of contents follows. Chapter headings are given, followed by examples of important subtitles.

ISBN 0-8155-0479-9 **337 pages**

POLLUTANT REMOVAL HANDBOOK 1973

by Marshall Sittig

The purpose of this handbook is to provide a one volume practical reference book showing specifically how to remove pollutants, particularly those emanating from industrial processes. This book contains substantial technical information.

This volume is designed to save the concerned reader time and money in the search for pertinent information relating to the control of specific pollutants. Through citations from numerous reports and other sources, hundreds of references to books and periodicals are given.

In this manner this book constitutes a ready reference manual to the entire spectrum of pollutant removal technology. While much of this material is presumably available and in the public domain, the locating thereof is a tedious, time-consuming, and expensive process.

The book is addressed to the industrialist, to local air and water pollution control officers, to legislators who are contemplating new and more stringent control measures, to naturalists and conservationists who are interested in exactly what can be done about the effluents of local factories, to concerned citizens, and also to those eager students who can foresee new and brilliant careers in the fields of antipollution engineering and pollution abatement.

During the past few years, the words "pollution", "environment" and "ecology" have come into more and more frequent usage and the cleanliness of the world we live in has become the concern of all people. Pollution, for example, is no longer just a local problem involving litter in the streets or the condition of a nearby beach. Areas of the oceans, far-reaching rivers and the largest lakes are now classified as polluted or subject to polluting conditions. In addition, very surprisingly, lakes and streams remote from industry and population centers have been found to be contaminated.

This handbook therefore gives pertinent and concise information on such widely divergent topics as the removal of oil slicks in oceans to the containing of odors and particulates from paper mills.

Aside from the practical considerations, including teaching you where to look further and what books and journals to consult for additional information, this book is also helpful in explaining the new lingo of pollution abatement, which is developing new concepts and a new terminology all of its own, for instance: "particulates, microns, polyelectrolytes, flocculation, recycling, activated sludge, gas incineration, catalytic conversion, industrial ecology, etc."

In order to have a safe and healthful environment we must all continue to learn and discover more about the new technology of pollution abatement. Every effort has been made in this manual to give specific instructions and to provide helpful information pointing in the right direction on the arduous and costly antipollution road that industry is now forced to take under ecologic and sociologic pressures. The world over, technological and manpower resources are being directed on an increasing scale toward the control and solution of contamination and pollution problems.

In the United States of America we are fortunate in receiving direct help from the numerous surveys together with active research and development programs that are being supported by the Federal Government to help industry and municipalities control their wastes and harmful emissions.

A partial and condensed table of contents is given here. The book contains a total of 128 subject entries arranged in an alphabetical and encyclopedic fashion. The subject name refers to the polluting substance and the text underneath each entry tells how to combat pollution by said substance:

CARBON MONOXIDE
CARBONYL SULFIDE
CEMENT KILN DUSTS
CHLORIDES
CHLORINATED HYDROCARBONS
CHLORINE
CHROMIUM
CLAY
COKE OVEN EFFLUENTS
COLOR PHOTOGRAPHY EFFLUENTS
COPPER
CRACKING CATALYSTS
CYANIDES
CYCLOHEXANE OXIDATION WASTES
DETERGENTS
DYESTUFFS
FATS
FERTILIZER PLANT EFFLUENTS
FLOUR
FLUORINE COMPOUNDS
FLY ASH
FORMALDEHYDE
FOUNDRY EFFLUENTS
FRUIT PROCESSING INDUSTRY EFFLUENTS
GLYCOLS
GREASE
HYDRAZINE
HYDROCARBONS
HYDROGEN CHLORIDE
HYDROGEN CYANIDE
HYDROGEN FLUORIDE
HYDROGEN SULFIDE
IODINE
IRON
IRON OXIDES
LAUNDRY WASTES
LEAD
LEAD TETRAALKYLS
MAGNESIUM CHLORIDE
MANGANESE
MEAT PACKING FUMES
MERCAPTANS
MERCURY
METAL CARBONYLS
MINE DRAINAGE WATERS
NAPHTHOQUINONES
NICKEL
NITRATES
NITRITES
NITROANILINES
NITROGEN OXIDES
OIL
OIL (INDUSTRIAL WASTE)
OIL (PETROCHEMICAL WASTE)
OIL (PRODUCTION WASTE)
OIL (REFINERY WASTE)
OIL (TRANSPORT SPILLS)
OIL (VEGETABLE)
ORGANIC VAPORS
OXYDEHYDROGENATION PROCESS EFFLUENTS
PAINT AND PAINTING EFFLUENTS
PAPER MILL EFFLUENTS
PARTICULATES
PESTICIDES
PHENOLS
PHOSGENE
PHOSPHATES
PHOSPHORIC ACID
PHOSPHORUS
PICKLING CHEMICALS
PLASTIC WASTES
PLATING CHEMICALS
PLATINUM
PROTEINS
RADIOACTIVE MATERIAL
RARE EARTH
ROLLING MILL DUST & FUMES
ROOFING FACTORY WASTES
RUBBER
SELENIUM
SILVER
SODA ASH
SODIUM MONOXIDE
SOLVENTS
STARCH
STEEL MILL CONVERTER EMISSIONS
STRONTIUM
SULFIDES
SULFUR
SULFUR DIOXIDE
SULFURIC ACID
TANTALUM
TELLURIUM HEXAFLUORIDE
TETRABROMOETHANE
TEXTILE INDUSTRY EFFLUENTS
THIOSULFATES
TIN
TITANIUM
TRIARYLPHOSPHATES
URANIUM
VANADIUM
VEGETABLE PROCESSING INDUSTRY EFFLUENTS
VIRUSES
ZINC

ISBN 0-8155-0489-6

528 pages

POLLUTION CONTROL IN THE ORGANIC CHEMICAL INDUSTRY 1974

by Marshall Sittig

Pollution Technology Review No. 9

Wastes from plants manufacturing identical compounds may be quite dissimilar, because of the difference in processes and raw materials. Since the organic chemical industry is now largely petrochemical, preference has been given to waste treatment from such operations.

Detailed treatability evaluation of each waste stream has become a prerequisite for any profitable product.

Physical methods for waste and by-product removal include gravity separation, air flotation, filtration, evaporation and adsorption techniques.

Chemical methods include precipitation, polyelectrolyte treatment, oxidation and other chemical conditioning such as neutralization.

Biological treatment methods comprise activated sludge and its modifications, trickling filters, aerated lagoons, and waste stabilization ponds.

This book intends to assist the chemical engineer in treating chemical waste products and effluents in conjunction with prudent raw material choices and thus bring about process economics. It presents condensed vital data from government and other sources of information that are scattered and difficult to pull together.

A partial and condensed table of contents follows here.

Dinitrotoluene from Nitrotoluene

Disulfoton from Ethanol

Dyes and Pigments from Aromatics

Epichlorohydrin from Allyl Chloride

Ethylbenzene from Benzene

Ethyl Chloride by
Hydrochlorination of Ethylene
Chlorination of Ethane
Hydrochlorination of Ethanol

Ethylene or Propylene from Paraffins

Ethylene Dichloride from Ethylene

Ethylene Glycol from Ethylene Oxide

Ethylene Oxide from Ethylene

Formaldehyde from Methanol

Long Chain Alcohols from
Ethylene Oligomers

Malathion® from Methanol

Methanol from Natural Gas

Methylamines from Methanol + Ammonia

Methyl Bromide from Methanol

Methyl Methacrylate from
Acetone Cyanohydrin

Nitrobenzene from Benzene

Nitrochlorobenzene from Chlorobenzene

Nitroparaffins from Paraffins

Oxo Products from Olefins

Parathion from p-Nitrophenol

Perchloroethylene from Propane

Phenol from Chlorobenzene

Phenol from Cumene

Phorate from Ethanol

Phosgene from CO + Cl_2

Phthalic Anhydride from
Naphthalene or Xylene

Propylene Glycol from Propylene Oxide

Propylene Oxide from Propylene

Styrene from Ethylbenzene

SNG from Crude Oil

2,4,5-T from Trichlorophenol

Terephthalic Acid from Xylene

Tetraethyl Lead from Ethyl Chloride

Trichloroethane from Vinyl Chloride

Trichloroethylene by
Chlorination, then by
Dehydrochlorination of Acetylene
or by Oxyhydrochlorination
of Dichloroethane

Trifluralin from Perfluoromethylchlorobenzene

Vinyl Acetate from
Ethylene and Acetic Acid

Vinyl Chloride from Acetylene

Vinyl Chloride from Ethylene Dichloride

AIR POLLUTION CONTROL
Halogen Acids
Halogenated Hydrocarbons
Halogens
Hydrocarbons
Particulates
Sulfur Compounds
Coking of Coal

WATER POLLUTION CONTROL
Primary
Oil Separation
Equalization
Neutralization
Sedimentation
Flotation
Flocculation
Nutrient Addition
Secondary
Activated Sludge
Extended Aeration
Trickling Filters
Aerated Lagoons
Waste Stabilization Ponds
Chemical Oxidation
Nitrification—Denitrification
Tertiary
Chemical Precipitation
Gas Stripping
Microstraining
Carbon Adsorption
Electrodialysis
Ion Exchange
Evaporation
Reverse Osmosis
Chlorination
Rapid Sand Filtration
Sludge Handling
Aerobic Digestion
Anaerobic Digestion
Wet Oxidation
Thickening
Lagooning
Sand Drying Beds
Vacuum Filtration
Filtration
Land Disposal
Incineration
Sea Disposal
Ultimate Disposal
Thermal Oxidation
Deep Well Disposal

OVERALL WATER POLLUTION CONTROL MODELS
BPCTCA: Best Practicable Control Technology Currently Available
BATEA: Best Available Technology Economically Available
BADCT: Best Available Demonstrated Control Technology

THE ECONOMICS OF POLLUTION CONTROL

FUTURE TRENDS

ISBN 0-8155-0536-1

305 pages